RESEARCH METHODS IN OUTDOOR STUDIES

Over the last two decades Outdoor Studies has emerged as an innovative and vibrant field of study. This is the first book to offer a comprehensive appraisal of established and cutting-edge research methods as applied to Outdoor Studies.

Covering qualitative, quantitative and mixed methods, the book examines key methodologies, themes and technologies such as digital research, mobile methodologies, ethnography, interviews, research design, research ethics and ways of disseminating research.

Featuring contributions from leading researchers from a variety of disciplinary backgrounds, this is an essential text for any Outdoor Studies course or for researchers looking for innovative and creative research techniques.

Barbara Humberstone is Professor of Sociology of Sport and Outdoor Education at Buckinghamshire New University, UK, and Visiting Professor at Plymouth Marjon University, UK. She is also Editor-in-chief of the *Journal of Adventure Education and Outdoor Learning*. Her research interests include: Embodiment, alternative/nature-based physical activities and life-long learning, wellbeing and outdoor pedagogies, and social and environmental justice.

Heather Prince is Professor at the University of Cumbria, UK. She is interested in pedagogic practice in outdoor and environmental education, including the design of higher education courses and support for research programmes, students and staff. Her research interests are in school-based outdoor learning, sustainability and adventure. She is Associate Editor of the *Journal of Adventure Education and Outdoor Learning* and Principal Fellow of the Higher Education Academy, UK.

ROUTLEDGE ADVANCES IN OUTDOOR STUDIES

Series Editors:
Barbara Humberstone, Buckinghamshire New University, UK.
Peter Higgins, University of Edinburgh, UK.

The *Routledge Advances in Outdoor Studies* series is a comprehensive and expanding book list that encompasses and integrates theoretical, practical and political aspects of outdoor studies, including education, leisure/recreation, adventure, therapy, nature-based sport, the environment (land & sea), sustainability, social justice and professional practice. In bringing together these dimensions in new and creative forms, this series will highlight new and innovative national and international research, enable greater accessibility to key critical developments, and strengthen interconnections between the underpinning disciplines.

Available in this series:

Routledge International Handbook of Outdoor Studies
Edited by Barbara Humberstone, Heather Prince, Karla A. Henderson

Research Methods in Outdoor Studies
Edited by Barbara Humberstone and Heather Prince

https://www.routledge.com/Routledge-Advances-in-Outdoor-Studies/book-series/RAOS

RESEARCH METHODS IN OUTDOOR STUDIES

Edited by Barbara Humberstone and Heather Prince

LONDON AND NEW YORK

First published 2020
by Routledge
2 Park Square, Milton Park, Abingdon, Oxon OX14 4RN

and by Routledge
52 Vanderbilt Avenue, New York, NY 10017

Routledge is an imprint of the Taylor & Francis Group, an informa business

British Library Cataloguing-in-Publication Data
A catalogue record for this book is available from the British Library

Library of Congress Cataloging-in-Publication Data
A catalog record has been requested for this book

ISBN: 978-0-367-18870-2 (hbk)
ISBN: 978-0-367-18883-2 (pbk)
ISBN: 978-0-429-19900-4 (ebk)

Typeset in Bembo
by Newgen Publishing UK

CONTENTS

PART II
Qualitative methodologies – choosing an appropriate approach 45

FIGURES

TABLES

CONTRIBUTORS

Linda Allin is Associate Professor and Director of Learning and Teaching in the Department of Sport, Exercise and Rehabilitation, Northumbria University, UK. Her main research focuses on women's experiences in sport and the outdoors, alongside pedagogical research in higher education. She is on the Editorial Board of the *Journal of Adventure Education and Outdoor Learning*.

Martha Bell is an independent sociologist conducting social and community research on contract in Dunedin, New Zealand. She met her co-authors through outdoor instructing, co-founding Women Outdoors NZ and researching feminist outdoor leadership.

Eric Brymer is Reader at Leeds Beckett University, UK. He specialises in researching the reciprocal nature of wellbeing outcomes from nature-based activities. Eric's expertise includes qualitative and mixed methods research design. He also holds research positions in health and outdoor studies at Queensland University of Technology, Australia and the University of Cumbria, UK.

David A.G. Clarke is Lecturer in Outdoor Studies at the University of the Highlands and Islands, UK. His research interests focus broadly on the blurring of environmental education research practice, philosophical theory and life experiences. More specifically he is interested in the philosophy of affect in relation to environmental degradation and climate change; immanent ethics and immanent ontology in all areas of education; and creative practices of inquiry for affective encounters, including writing, photography, songwriting and filmmaking.

Ben Clayton teaches socio-cultural issues in sport and research methods at Buckinghamshire New University, UK. He has published widely on the broad topic of gendered sport and has a particular interest in the use of fictional forms of representation to show experiences of sport participation.

Emily Coates earned her PhD at Buckinghamshire New University, UK, examining the experiences of traditional climbers with young children and has published papers about lifestyle sports and emerging methodologies in outdoor studies.

Marg Cosgriff is a Senior Lecturer in Te Huataki Waiora Faculty of Health, Sport and Human Performance at the University of Waikato, New Zealand. She was involved in the founding of Women Outdoors NZ and has a long-standing interest in feminist outdoor research and leadership.

Carol Cutler Riddick is Professor/Chair, Department of Physical Education and Recreation at Gallaudet University in Washington, DC, USA. She is also employed as a health education specialist, director of an intergenerational day camp, and hospital administrator. Carol's research agenda focuses on examining the impacts programmes/activities have on enhancing the health of older individuals.

Ulrich Dettweiler is Associate Professor in Pedagogy at Stavanger University, Norway. His research focus lies in learning psychological and health-related aspects of teaching outside the classroom, with both qualitative and quantitative approaches. He serves as Associate Editor for the *Journal of Experiential Education* and has been editing a special issue "Epistemological and Ethical Aspects of Research in the Social Sciences" for *Frontiers in Psychology*.

Janet Dyment is Senior Lecturer and Deputy Head of School in the School of Education at the University of Tasmania, Australia. Janet's teaching and research focus on issues related to quality teaching and learning (in outdoor education and other learning areas). She is committed to helping initial teacher education students develop deep understandings of how they can have the most impact on student learning.

Kass Gibson is Senior Lecturer in Sports Coaching and Physical Education at Plymouth Marjon University, UK. Kass' research uses a range of sociological theories and research methodologies to understand the relationships and effects between different ways of knowing, meanings, experiences and practices in physical activity, sport, physical education and public health.

Kirsti Pedersen Gurholt is Professor of Outdoor Studies and Physical Education at Norwegian School of Sport Sciences, Norway. Her research interests comprise historical, cultural, gender and narrative studies employing ethnography, interviews, documents and visual methods. She was lead for the Norwegian involvement in the Erasmus+ Joint Master's degree programme Transcultural European Outdoor Studies (2011–2017) and was chair of the European Institute for Outdoor Adventure Education and Experiential Learning (2008–2012).

Tracy Ann Hayes is Lecturer in Health, Psychology and Social Studies at the University of Cumbria, UK. She embraces transdisciplinary methodologies, which utilise creative and narrative approaches to research nature, outdoor learning and play. She is Fellow of the Royal Geographical Society (RGS), Conference Officer for the (RGS) Geographies of Children, Youth and Families Research Group (GCYFRG) and Fellow of the Higher Education Academy, UK.

Carrie Hedges is Outdoor Learning Research-Practitioner Regional Hubs Project Coordinator for the Institute for Outdoor Learning based at the University of Cumbria, UK. She is also an upland and forest ecologist and a graduate researcher in Environmental Sciences at the university.

Peter Higgins is Professor at the University of Edinburgh, UK, where he teaches outdoor, environmental and sustainability education through classwork, practical and online approaches. His research is primarily in the boundary area between these fields. He is Director of the University's Global Environment and Society Academy, the United Nations University Regional Centre of Expertise in Education for Sustainable Development (Scotland), and Scottish representative on a related UNESCO programme.

Allen Hill is Principal Lecturer in Sustainability and Outdoor Education at Ara Institute of Canterbury, Christchurch, Aotearoa New Zealand. Allen's professional career, in both secondary and higher education can be characterised by an enduring commitment to the development of people coupled with a strong concern for issues of justice, sustainability, transformation and place. How education can engage people with meaningful outdoor learning experiences and contribute to a sustainable future through connecting people with each other and with the places they inhabit is at the heart of his research and teaching interests.

Mark Leather is Senior Lecturer in the Institute of Education, University of St Mark and St John, Plymouth, UK, where he specialises in adventure education, outdoor learning and experiential pedagogies. Mark's research and publications are intertwined with his outdoor studies practice.

lisahunter researches physical culture, outdoor spaces, embodiment, sex/ualities and pedagogies with a more recent focus in activism, documentary-as-method, sociomateriality and sensory methodologies. Based in Australasia, lisahunter teaches in teacher education and research higher degree supervision, with research projects in community activism for the Institute of Women Surfers (Oceania), moving meditation, and queer family engagement in physical culture.

Chris Loynes is Reader in Outdoor Studies at the University of Cumbria, UK. He consults in the UK and internationally for universities and experiential education organisations. He was recently an Educational Adviser working with the

Paul Hamlyn Foundation Initiative Learning Away. He currently lectures on the Erasmus Mundus MA Transcultural European Outdoor Studies. Chris is a Fellow of the Royal Geographical Society and the founding editor of the *Journal of Adventure Education and Outdoor Leadership* from 1980–2000. He is also the chair of the EOE Network.

Alison Lugg is Senior Lecturer in initial teacher education at RMIT University, Melbourne, Australia. She has over 20 years' experience in the fields of teacher education and outdoor education at La Trobe University, Bendigo, the University of Edinburgh and the University of Melbourne. Alison's research and publications span outdoor education curriculum, women's experiences of outdoor education, sustainability in outdoor education, interdisciplinary work in higher education and pre-service teacher professional development.

Jonathan Lynch is Postgraduate Director with The Mind Lab in New Zealand. His research interests include place-responsive education and post-qualitative methodologies. He was previously a Senior Lecturer in Outdoor Studies at the University of Cumbria, UK.

Pip Lynch is Consulting Researcher, Tutor and Writer in New Zealand. In prior roles she specialised in outdoor education and outdoor recreation scholarship and teaching in New Zealand and Norwegian universities. She was a founding member of Women Outdoors NZ and a member of Women Climbing NZ.

Liz Mallabon is Principal Lecturer in Outdoor Studies in the Department of Science, Natural Resources and Outdoor Studies at the University of Cumbria, UK.

Lois Mansfield is Professor and Director of Campus at Ambleside for the University of Cumbria, UK, and a geographer. She is an expert in upland resource management, which includes the application of both quantitative and qualitative approaches in order to appreciate the complexities of real world challenges. She is currently involved in the management of the Lake District World Heritage Site and National Park in which the campus resides.

Jamie Mcphie is Course Leader for the MA Outdoor and Experiential Learning course at the University of Cumbria, UK. His research interests lie in environmental arts and post-humanities, therapeutic landscapes and psychogeography, (currently) thinking with philosophies of immanence, such as Contemporary Animism, Feminist New Materialisms and New Science of the Mind.

Marcus Morse is Senior Lecturer in Outdoor Environmental Education at La Trobe University, Australia. Marcus' research interests are in the areas of outdoor environmental education and philosophy, experiences within nature and wild pedagogies.

Philippa Morse is a Lecturer in Outdoor and Environmental Education, at La Trobe University, Bendigo, Australia and a PhD candidate at the University of Tasmania, Australia. Philippa's research focus is on imagination and posthuman pedagogical approaches in outdoor environmental education.

Philip M. Mullins is Associate Professor of Outdoor Recreation and Tourism Management at the University of Northern British Columbia, Canada. He studies relational and ecological approaches to outdoor activities and skill development from a position of belonging and participation. He has extensive experience with and a love for field schools, and uses travel on the land as a method of research and teaching that embraces movement and the dynamism of outdoor environments. He strives to bring critical and creative approaches to outdoor recreation and outdoor education that engage contemporary socio-environmental issues.

Robbie Nicol is Senior Lecturer in Outdoor and Environmental Education at the University of Edinburgh, UK. His life motivation comes from the realisation that human activities are fundamentally altering the planet's ability to sustain all species including the human race in the long term. As an educator he believes that the outdoors provides places where individuals can rediscover their direct dependence on the planet through embodied experiences. His teaching and research interests are directed towards the theoretical development and practical implementation of Place-Based Education, and epistemological diversity particularly in the outdoors.

Rebecca Olive is Lecturer in the School of Human Movement and Nutrition Sciences at The University of Queensland, Australia. Her research has focused on women and lifestyle and action sports, in particular surfing, as well as applying theory to better understand ethnographic research methods in the outdoors and on social media. More recently, she has begun to explore how participation in lifestyle sports enhances our knowledge of nature and ecologies.

Suzanne Peacock is Senior Lecturer in Outdoor Education at Leeds Beckett University, UK. She has a particular interest in mental health and wellbeing, and the ability for adventure and the outdoors to facilitate this. Suzanne has completed her PhD, which explored the role of adapted sport and adventure in the recovery of military personnel. She also holds a Master's in Sport and Exercise Psychology.

Kathleen Pleasants is Lecturer in the Department of Outdoor and Environmental Education, La Trobe University, Bendigo, Australia. Her teaching includes OEE epistemology, curricula and pedagogy. Kathleen's research interests lie in understanding what is produced by research in OEE and questioning taken-for-granted beliefs.

Roger Scrutton is Honorary Research Fellow in Outdoor Education at the University of Edinburgh, UK. Previously on the academic staff in the School of Geosciences, where the benefits of fieldwork were clear, and with a lifetime engagement in outdoor pursuits, he moved to the School of Education to conduct quantitative research that would convince stakeholders of the benefits of outdoor education for personal development and academic achievement.

Heidi Smith is an academic at the University of Edinburgh, UK. The research she conducts lies within the interpretive paradigm, utilises case study, narrative, autoethnography and autobiographical methodology, and consistently relies on methods such as interviews, intentional conversations, observations and document collection. Her teaching has spanned outdoor learning, learning for sustainability, outdoor leadership and teacher education. Heidi is passionate about good pedagogy, the outdoors as a space for learning, students as partners in learning communities and transculturality in higher education.

Ina Stan is Senior Lecturer at Buckinghamshire New University, UK, in the Learning Development Unit. She is a Fellow of the Higher Education Academy and has conducted doctoral research on group interactions of primary school children in the outdoor classroom, focusing on the process of learning in outdoor education. While working as a research assistant, she also undertook a research study on the wellbeing of children in the outdoors, focusing on the impact of body image and risk on children's wellbeing. Her most recent research explored the impact of group work on students' learning experience in higher education.

Alistair Stewart is Senior Lecturer and past Programme Head of the Department of Outdoor and Environmental Education, La Trobe University, Bendigo, Australia. His teaching and research interests include poststructuralist curriculum inquiry and place-responsive pedagogy, with particular reference to natural and cultural history.

John Telford is Programme Leader for the graduate diploma in Adventure Education at Camosun College on Vancouver Island, British Columbia, Canada. His research interests include investigating human/more-than-human relationality, applying social theory frameworks to better understand outdoor education experiences, and bringing sustainable ways of being into all aspects of his life.

Sue Waite is a visiting Specialist at the University of Plymouth, UK where she was formerly Reader in Outdoor Learning in Plymouth Institute of Education. She is also a visiting Associate Professor at Jonkoping University in Sweden, chair of the Red Kite Academy Trust and member of the International School Grounds Alliance Leadership Council and Natural England Research Strategy Group on outdoor learning, health and wellbeing, working to develop curriculum-based outdoor learning.

Phil Waters is Co-founder and Creative Director of I Love Nature, a social enterprise in Cornwall, UK, that provides training, outdoor environmental education, play consultancy and research. Phil's work brings together play, narrative and nature within a form of praxis called *Narrative Journey*. This has more recently culminated in a doctorate with the European Centre for Environment and Human Health, University of Exeter Medical School, where his major research interests include children's nature-based play: stories and storying, environmental education for young children, and playful praxis in research.

Robyn Zink works for Enviroschools, which supports schools to embed sustainability across the whole school community. She has a long-standing interest in research and playing in the outdoors.

FOREWORD

The coming of age of research in outdoor studies

John Quay

The field of outdoor studies is maturing. Some might say it is coming of age. Evidence of this development is in the published record. A potted history of journals in the field reveals this growth. Perhaps the first journal bearing the name *Outdoor Education* was published between 1916 and 1918 by the School Garden Association of America. This signalled the broad emergence of an educational response to certain societal issues of the time, mainly to do with health and the increasing prevalence of city living. The message was for people to take advantage of the healthy living accessible outdoors, away from the more polluted and crowded indoor environments of the city, prone as they were to the transmission of disease. School gardens played a significant role in this endeavour, along with camps.

Fast forward to the 1960s and the world has changed significantly. There is growing awareness of issues concerning the health of the environment, which overlap with a love of the outdoors felt by those who recreate in natural places. The *Journal of Outdoor Education* begins publication in 1966, continuing until 1994, facilitated by staff at the Lorado Taft Field Campus of Northern Illinois University. Following this is the *Journal of Experiential Education*, first published in 1978 by the Association of Experiential Education. Next came the *Journal of Adventure Education*, first published in 1981 by the National Association for Outdoor Education in Great Britain. This journal has been through a range of title changes, including as the *Journal of Adventure Education and Outdoor Leadership*, now being known as the *Journal of Adventure Education and Outdoor Learning*. Drawing this potted history to a close is the *Australian Journal of Outdoor Education* which began in 1996 and is now called the *Journal of Outdoor and Environmental Education*, a publication of Outdoor Education Australia.

The development of these journals showcases the continuing maturation of the field of outdoor studies. The journal titles are windows into the conceptual positioning of the field, highlighting terms which play into various discourses: outdoor, adventure, environment, leadership, learning, education. Both people and place are important in conceptions of the outdoors, orientated around particular practices. Hence "the 'outdoors' may be perceived, in one sense, as an ideological space where people alone or together engage actively or passively with their 'environment'" (Humberstone, Brown & Richards, 2003, p. 7). This statement preferences the language of perceptions and ideas, informing engagements between people and environment that occur in connection with the outdoors. These engagements, these practices, are understood in certain ways, they have specific meanings attached to their enactment.

Various concepts have been used in attempting to comprehend these meanings. Many of these concepts are polysemous, with the meaning intended only accessible if one recognises the broader cognate positioning within a theoretical framework (Quay, Bleazby, Stolz, Toscano & Webster, 2018). In outdoor studies, this complexity can be seen in concepts such as place, space, nature, environment, land, wilderness, culture, experience, human, self, others, community, people; a list which hides within it the challenge of comprehending the relation between people and environment (to use but two of these words). More recently the phrase "more-than-human" (Abram, 1996) has come to the fore in an attempt to bridge some of these conceptual distinctions. Yet even this is mired in confusion. One interpretation is that, by coining this phrase, Abram is highlighting the importance of recognising that the "larger community includes, along with the humans, the multiple non-human entities that constitute the local landscape" (p. 6); whereas another interpretation positions more-than-human as referring to non-human entities.

This is one of the conundrums in meaning which the field of outdoor studies faces. However, the field has grown markedly in its comprehension of such meanings. Acknowledgment of these conundrums is evidence of a field which is searching, researching, for an understanding of the deeper issues that influence and inform practical conduct. The editors of this book assert in the introduction that their "aspiration for this book is to provide a stimulating and critical contribution to outdoor studies research". The many chapters offer a plethora of ideas that bring this aspiration to life in concrete terms; in other words, the authors share numerous examples from their own research experience which illuminate how to contribute to this research endeavour, through application of methods and methodologies they have used.

Methodologies broadly fall into one or other of the categories of qualitative and quantitative research. These are very different; and the sections of the book on design, conduct and communication of research support understanding of this difference. Methodology raises questions about research practices; however such questions cannot be separated from those pertaining to ontology and epistemology. Ontology highlights questions of being, methodology questions of doing, epistemology questions of knowing. All three are always in play in the conduct of any research project, and it is a significant strength of the research being undertaken in outdoor studies that these questions are acknowledged and engaged with, as is made plain in this book.

The depth of engagement in this book with the philosophical ideas supporting the design, conduct and communication of research is a sign of the coming of age of outdoor studies. The quality and quantity of outdoor studies research has grown over decades and across the world. This book is a testament to that fact. And while outdoor studies may seem to be a very small field in the scope of other research being conducted today, its importance will continue to increase as the understandings being generated inform the education of current and future generations, with the aim of positively influencing the wellbeing of all life on this planet.

Dr John Quay
Associate Professor
Melbourne Graduate School of Education
University of Melbourne | Vic 3010 Australia

References

Abram, D. (1996). *The spell of the sensuous: Perception and language in a more-than-human world.* New York, NY: Pantheon Books.

Humberstone, B., Brown, H. & Richards, K. (2003). *Whose journey? The outdoors and adventure as social and cultural phenomena.* Barrow-in-Furness: Fingerprints.

Quay, J., Bleazby, J., Stolz, S.A., Toscano, M. & Webster, R.S. (2018). *Theory and philosophy in education research: Methodological dialogues.* New York, NY: Routledge.

ACKNOWLEDGEMENTS

We would like to thank all contributors for making this book possible, and also our families, friends and colleagues for their support. Heather would like to thank her family – Ivan, Angus and Hal Walsh. Barbara and Heather would like to acknowledge their reciprocal thanks to each other in the editing of this research book and for the professional dialogue throughout the course of the project.

We are indebted to Taylor & Francis for accepting our submission to compile this book as part of the series *Advances in Outdoor Studies* and appreciative of the editing and production staff for their careful and timely support.

INTRODUCTION

Heather Prince and Barbara Humberstone

Research in outdoor studies has grown significantly, particularly since 2000 as the number of researchers and outputs have increased (Humberstone, Prince & Henderson, 2016; Prince Christie, Humberstone & Pedersen Gurholt, 2018) and this increase in research is arguably having influences on policy (see Chapter 29). Drawing together international researchers, this volume brings a variety of perspectives and methodologies to outdoor studies research. Each chapter provides diverse approaches to enable the exploration of questions, issues and hypotheses. These perspectives and approaches come from a variety of disciplinary and epistemological backgrounds. The chapters draw upon a wealth of theory and practice from research in social sciences some of which utilise applied research methodologies that are embedded in other disciplines whilst other chapters show the ways in which combinations of approaches provide for creativity in, and innovative approaches to, research. These chapters offer the reader valuable knowledge and understandings to explore research perspectives and to address realities emergent in the key threads of education, leisure, physical culture, sport, the outdoor environment and practice. We continue to maintain that, "The 'outdoors' may be perceived, in one sense, as an ideological space where people alone or together engage actively or passively with their 'environment'" (Humberstone, Brown & Richards, 2003, p. 7).

In 2016, Humberstone et al. (2016, p. 2), conceptualising 'outdoor studies', emphasised that, "terminology in any sphere … is governed by culture, policy drivers and history, with political, temporal, institutional, chronological and marketing determinants". This fluid and responsive stance has proved critical in examining the manifestation of research methods in the field and its derivatives three years later where emerging research includes drawing on mobilities and place-based approaches, embodiment and sensorial methodologies, narrative and stories. Research methods in outdoor studies span the cultural, political and social contexts through which diverse outdoor traditions have emerged.

Research methods are techniques or procedures used to gather and analyse data related to research questions or hypotheses (Crotty, 1998; Wahyuni, 2012); it has been suggested that these are a-theoretical (Sarantakos, 2005). However, like Sparkes (2015 p. 50) we argue that a "researcher's ontological, epistemological and methodological commitments will constrain which methods can be used". These philosophical positions involve epistemological beliefs (on the nature and construction of knowledge) and ontological assumptions (assumptions about the nature of reality). These give rise to methodological considerations (the approach to, and process of, gaining knowledge – a framework in which to conduct research within a particular paradigm, where a paradigm is a "basic belief system(s) or world view that guides the investigation" (Guba & Lincoln, 1994, p. 105). Further as Coates, Hockley, Humberstone and Stan (2016, p. 69) suggest, "Approaches to understanding and making sense of material and social phenomena are changing continuously through critical reflection and practice", not least in outdoor studies research.

As research methods are practical tools for carrying out research, they can be underpinned by different methodologies. Methods that are replicated in this way are illustrated in this compilation to facilitate deeper understanding and to demonstrate a range of applications in outdoor studies research in international contexts to "... fruitfully encompass(es) a broad range of approaches, foci and methods ..." (Humberstone, Prince & Henderson, 2016, p. 3). However, research design is frequently driven by context and potential outcomes, and in some cases will encourage particular methodological perspectives and so incorporate an appropriate range of research methods to explore the research questions, issues or hypotheses (Wahyuni, 2012).

Cutting-edge research in outdoor studies continues to be interdisciplinary and it is also transdisciplinary. "Transdisciplinarity has emerged in order to meet the promise of transcending disciplinary knowledge production in order to more effectively address real world issues and problems" (Leavy, 2016, p. 24; Humberstone, 2016). This encompasses important emerging research methods/methodologies in socio-cultural and socio-environmental areas of outdoor recreation, leisure and sporting activities Thus, research may be intrinsic within disciplines but should have wider meaning, significance, reach and impact. Knowledge-building and dissemination may be seen as holistic processes, which require innovation and flexibility (Wickson, Carew & Russell, 2006). Thus, *Research Methods in Outdoor Studies* includes chapters on publishing and dissemination, the research-practice nexus and concludes with a narrative around research influence and impact in respect of policy.

Our aspiration for this book is to provide a stimulating and critical contribution to outdoor studies research. The contributors here challenge and/or develop traditional approaches to research and in so doing highlight a diversity of research methods, underpinning methodologies and philosophical perspectives. The book is appropriate for established researchers wishing to refresh their knowledge and those contemplating undertaking research that may challenge conventional methodologies; It is also aimed at final year undergraduates embarking on an extended research project for the first time, taught postgraduate students, postgraduate and early career researchers. The volume comprises a range of methodologies and methods with contextual application across the globe balanced by more 'process' (how to) chapters.

Research Methods in Outdoor Studies is organised into five sections. The balance of the text is weighted in respect of qualitative approaches, a reflection, we feel, of the research output in the field, supported by a smaller section on quantitative and mixed methods, couched by contributions on the research process and disseminating research.

The first section, *Conceptualising and initiating the research process*, comprises research design, ethical issues and practicalities and debate on appropriate philosophical and methodological dimensions. Normative process is deliberately upended and challenged in Chapter 1, which examines contested theorisations and dualisms and seeks new and alternative ways of onto-ethico-epistemological thinking in outdoor studies. Ethical issues are central to all human and more-than-human interactions, not least in research. Chapter 2 considers the complexity of ethical practice and provides four scenarios that highlight dilemmas surrounding particular outdoor research. Codes of ethical practice are referred to and the workings of ethical panels are considered. Chapter 3 provides guidance on how to go about undertaking research in outdoor studies, reflecting upon research design and underpinning conceptual, philosophical and theoretical frameworks.

The next section is concerned with choosing an appropriate approach using qualitative methodologies. Phenomenology is explored as a philosophical tradition and a research methodology in Chapter 4. The challenges of data collection from individuals and groups in remote places and spaces for extended periods of time where participants are on the move are explored in Chapters 5 and 6. The pragmatics of authentic and naturalistic data collection necessitate a rethinking of the ways in which methods are applied towards complex new materialisms. Chapters 7 and 8 are exemplar case studies concerned with capturing the complexities of human interaction and the environment in two very different contexts. Chapter 7 focuses upon outdoor play and learning that involve children and young people actively moving through space and time, whilst Chapter 8 uses an ecosocial framework for researching outdoor sustainability education practice with teachers, which allows for uncertainty, emergence and collaborative interactions.

The next chapters (9–12) have their roots in interpretative, phenomenological paradigms (as does much qualitative research), which emerged in the latter half of the 20th century as alternative narrative approaches to positivistic dogma then prevalent in much social science research (Humberstone, 1997). Chapter 9 takes a traditional ethnographic approach highlighting the processes involved in exploring teaching and learning with 9–11 year old children in a small outdoor education centre. Chapter 10 focuses upon autoethnography as methodology, highlighting theoretical underpinnings and the significance of the senses and memory in outdoor research. The authors use different forms of narrative to engage the reader in place-based research. The early challenge to positivism largely spawned the reflexive researcher, whose research developed in different countries using diverse terms and terminology as Chapters 11 and 12 show. Chapter 11, based in Australian sociology, draws on feminist theory and practice and demonstrates through research in surfing cultures "that we (might usefully) think the social through ourselves, and explore the productive critiques of what it means to centralise our own subjectivity in our research". Chapter 12 is concerned with autobiography as research methodology, shown to have its roots in UK education and feminist theory and practice. Critical self-reflection is one way of knowing self and provides for reflexivity in research, which is crucial in interpretative, narrative research.

Section three, *Contemporary creative qualitative methods*, comprises chapters that examine relatively unexplored methods and methodologies in outdoor studies research. Chapter 13 exemplars creative nonfiction as a way of (re)-presenting research. The authors use their own story about the juxtaposition of parenting and 'serious' climbing to elucidate the what, why, and how of creative nonfiction. Chapter 14 explores shared-story approaches as ways in which research experiences can be analysed critically, understood and conceptualised to give 'testimony' to otherwise complex and unvoiced situations/narratives. In Chapter 15, digital narrative methodology is considered to explore human/nature interaction and the aesthetics of nature. Narrative theory is drawn upon and multisensory ethnography enables transformation of (becoming) researchers' sensorial, material and social engagement in outdoor field research. Feminist reflexivity underpins the collaborative letter writing and thematic analyses explored in Chapter 16. In a similar approach, through dialogue, but from a different theoretical perspective, Chapter 17 provides narrative between the authors through which they encourage the reader to write and think creative scholarship. Chapters 18–20 draw to varying degrees on mobilities methodologies conceptualised initially in social geographical research (Ingold, 2004) and perhaps less understood in outdoor studies. Located within this paradigm, Chapter 18 explores praxis as an approach to engagement in outdoor research through the outdoor journey. The walking interview as a mobilities methodology is presented in Chapter 19. The outdoors provides immense, varied and complex sensoria pungent for understanding and exploration. In Chapter 20 creative multimodal and mobile technologies with sensory-based methods provide for alternative methods of exploring human relationships with the more-than human. Creative artistic methods are used in Chapter 21 to de-privilege language and draw upon more aesthetic forms of understanding experience.

The fourth section provides readers with quantitative and mixed method approaches to case examples in the field. Chapter 22 responds to the call by many stakeholders and policy makers for evidence of the effectiveness of activities and interventions in the outdoors through the provision of metrics and measures. Chapter 23 focuses on scientific investigations through fieldwork in outdoor environments. Mixed method approaches are considered in the next two chapters. A call to reflect critically when considering a mixed method approach to research is made in Chapter 24. It cautions against "failing to adequately consider how different research methodologies present different ways of knowing" and argues that for credible mixed method research "accompanying evaluation, quality, and evidentiary criteria" should be "integrated in careful and respectful ways". Chapter 25 provides a framework for researchers interested in combining qualitative and quantitative techniques to explore phenomena and the practical implications of mixed method approaches in outdoor studies. Although this volume is short on quantitative research examples, Chapter 26 presents 'pure' quantitative analyses. This is provided through illustrated application to small samples where a hierarchy of analysis can represent the data in a better way and result in evidence being understood in constructivist terms. Section five focuses on disseminating, communicating and sharing research in outdoor studies. Dissemination through publishing, including a scaffolded approach for practitioner research and early career researchers, is examined in Chapter 27 together with detail in respect of metrics and evidence for the quality of research in the expanding outdoor field. Chapter 28 provides a model of research and practice based in the UK that involves practitioners, researchers, stakeholders and policy makers in sharing and communicating research for maximum influence and impact. The closing Chapter 29 is a professional narrative case around bringing research knowledge to the understanding of policy makers, particularly those in Scotland and provides an exemplar of how researchers in outdoor studies might go about influencing and shaping future policies in the various dimensions of outdoor studies.

Here, we present a collection of critical perspectives from a range of disciplines and geographical areas. Chapters in this book highlight the broad scope of contexts, understandings and approaches that comprise outdoor studies and cutting-edge research seeking to progress understanding and make contributions to knowledge. The compilation allows readers to identify ontologies and epistemologies relevant to their particular contexts and disciplines and to develop methods and methodologies that are relevant for their research questions. This research methods book hopes to encourage further research that enhances understanding of human transformation within and with the environment/more-than-human world. It points to co-constructed research in outdoor learning, outdoor education and outdoor experiences that identifies best practice, and participatory research in nature-based physical cultures that may reveal practices that lead to increasing ecological sensibilities and praxis.

Research in outdoor studies presents challenges and opportunities together with tensions, debates, contested viewpoints and a range of axiologies and, in this respect, is no different from other disciplines. It is a rich and diverse area that is fast expanding as it seeks to respond to public health and wellbeing, educational, cultural, social and environmental agendas. This book establishes the foundations of exciting research in outdoor studies, which is necessarily creative, critical, responsive and apposite.

References

Coates, E., Hockley, A., Humberstone, B. & Stan, I. (2016). Shifting perspectives on research in the outdoors. In B. Humberstone, H. Prince & K.A. Henderson (Eds.) *International handbook of outdoor studies* (pp. 69–78). London, New York, NY: Routledge.

Crotty, M. (1998). *The foundations of social research. Meaning and perspective in the research process.* London: SAGE.

Guba, E. & Lincoln, Y.S. (1994). Paradigmatic controversies, contradictions and emerging confluences. In N.K. Denzin & Y.S. Lincoln (Eds.) *The SAGE handbook of qualitative research* (3rd ed., pp.191–216). London: SAGE.

Humberstone, B. (1997). Challenging dominant ideologies in the research process. In G. Clarke & B. Humberstone (Eds.) *Researching women and sport* (pp. 199–211). London: Macmillan.

Humberstone, B. (2016). Section 6. Introduction – Transdisciplinary and interdisciplinary approaches to understanding and exploring outdoor studies. In B. Humberstone, H. Prince & K.A. Henderson (Eds.) *International handbook of outdoor studies* (pp. 421–424). London, New York, NY: Routledge.

Humberstone, B., Prince, H. & Henderson, K.A. (Eds.) (2016). *International handbook of outdoor studies.* London, New York, NY: Routledge.

Humberstone, B., Brown, H. & Richards, K. (2003). *Whose journey? The outdoors and adventure as social and cultural phenomena.* Barrow-in-Furness: Fingerprints.

Ingold, T. (2004). Culture on the ground. *Journal of Material Culture, 9*(3), 315–340.

Leavy, P. (2016). *Essentials of transdisciplinary research: Using problem-solving methodologies.* Oxford: Routledge.

Prince, H., Christie, B., Humberstone, B. & Pedersen Gurholt, K. (2018). Adventure education and outdoor learning: Examining journal trends since 2000. In P. Becker, C. Loynes, B. Humberstone & J. Schirp (Eds.) *The changing world of the outdoors* (pp. 144–159). Oxford: Routledge.

Sarantakos, S. (2005). *Social research* (3rd ed.). New York, NY: Palgrave Macmillan.

Sparkes, A. (2015). Developing mixed methods research in sport and exercise psychology: Critical reflections on five points of controversy. *Psychology of Sport and Exercise, 16*(3), 49–59.

UCSM (University College of St Martin). (1998). *Development plan* (unpublished). Lancaster.

Wahyuni, D. (2012). The research design maze: Understanding paradigms, cases, methods and methodologies. *Journal of Applied Management Accounting Research, 10*(1), 69–80.

Wickson, F., Carew, A.L. & Russell, A.W. (2006). Transdisciplinary research: Characteristics, quandaries and quality. *Futures, 38*(9), 1046–1059.

PART I

Conceptualising and initiating the research process

1

ENTANGLED PHILOSOPHICAL AND METHODOLOGICAL DIMENSIONS OF RESEARCH IN OUTDOOR STUDIES?

Living with(in) messy theorisation

Kathleen Pleasants and Alistair Stewart

Introduction

It is more or less foundational in outdoor studies[1] that human knowledge of the outdoors, outdoor activity and the natural environment lays, primarily, in and through sensed experience. However, while experience is important "within the fields of education and educational theory as well as in the larger context of social theory and philosophy", Roberts (2012) laments that "we have an entire field – that of experiential education – that uses the term almost automatically and without sufficient interrogation" (pp. 113 & 114). We would go further and suggest that in outdoor studies the terms 'experiential learning' and 'experiential education' are often used coterminously, despite not being synonymous.[2] Among other issues, this might reflect a lack of engagement with not only the practices of outdoor studies (including research practices) but also with the ontological, epistemological and axiological assumptions that inform it. In mapping experiential education, Roberts acknowledges that "it is simply not possible, nor wise ..., to aim for a 'complete' theoretically [sic] mapping of experience in education" and highlights "the importance of theoretical and philosophical explorations moving forward" (p. 114). In many ways it seems, as Braidotti (2013) writes, that "'theory' has lost status and is often dismissed as a form of fantasy or narcissistic self-indulgence" (p. 4).

St Pierre (2016) observes that "social science researchers often rush to application, to empirical method and methodology, before studying the history, philosophy, and politics of various empiricisms" (p. 111), and that "privileging practice over thought has a long history and is dominant in applied fields like education" (p. 111). The premise of this chapter is that we hold this concern for research within outdoor environmental education (OEE) and outdoor studies more broadly. As researchers, we share an interest in the normalisation that takes place with/in/through the rush to method and methodology.

Prompted by O'Toole & Beckett's (2013) discussion of the possibilities for research studies to establish "better theorization, or reasons for practices, than is taken for granted, or presently justified" (p. 57), in this chapter we are guided by the following questions. *How might outdoor studies attend to the entangled philosophical and methodological dimensions of research? How might we as researchers undertake research cognisant of the philosophical and political landscapes to which methods and methodologies belong? What new or alternative ways of thinking about ontology and epistemology of outdoor studies might be produced when guided by the "ethical imperative to rethink the nature of being to refuse the devastating dividing practices of the dogmatic Cartesian image of thought"* (St Pierre, Jackson & Mazzei, 2016, p. 99)?

Theorypractice framework – research assemblage

St Pierre (2016) draws on the work of Deleuze and Foucault to argue that theory and practice are inseparable, that one might write them together as *theorypractice*. However, St Pierre notes that

> the press to practice and methods-driven research in the social sciences distracts us from first attending to the onto-epistemological formations in which empirical practices are possible, and I think the rush to application is tripping us up now as we try to do this 'new' work. The structure of our humanist research methodologies simply cannot accommodate this new work. One can't carelessly use a concept from one 'grid of intelligibility' or 'system of thought' in another because the concept brings with it the entire structure in which it is imbricated with all that structure's assumptions about the nature of the world.
>
> *p. 112*

Similarly Jackson (2017) argues that a focus on method provides a normative form *to* our thinking in research within social sciences, and particularly qualitative research; "method supposedly, somehow, saves us from criticisms of credibility and reliability" (p. 666). Jackson, and others (see for example Gough, 2016; Lather, 1991, 2012, 2013; St Pierre et al., 2016), map how the dogmatic, orthodox Cartesian image of thought that drives much social science research is at work in normalising rules, conventions and predetermined outcomes. Drawing on Deleuze and Parnet, Jackson notes that this image of thought is machinic in the way it installs an apparatus of power, training thought to operate according to hegemonic norms, that it "is ready-made: already at work when we start to think, most of the time without our even knowing it" (p. 668). Jackson highlights that the Cartesian lineage has resulted in assumptions about what it is to think and know, and presumes that "thinking is natural, voluntary, and we all do it the same way" (p. 669). She draws on Deleuze (1994) who claims that this image of thought distorts what happens when we think through the postulates of recognition and representation. For Deleuze, recognition and representation are not thinking, and they function to deny difference.

A key point of much of Deleuze's work (and that of Deleuze and Guattari) was to encourage thought without image in order to promote thinking as an act of creation, not recognition.

Informed by the ideas above, in this chapter we employ Deleuze and Guattari's (1987) concept *assemblage* to frame our research *theorypractice*. As with many Deleuzo-Guattarian concepts. assemblage has been used in educational research as a means to bring elements together, to do something, to produce something (see for example Fox & Alldred, 2015; Gough, 2016; Jackson & Mazzei, 2012; Stewart, 2018). Deleuze and Guattari developed the concept of assemblage based on the French concept *agencement*, which entails the processes of organising, arranging and fitting together (Livesey, 2005). As Jackson and Mazzei (2013) note "an assemblage isn't a thing—it is the *process* of making and unmaking the thing" (p. 262, emphasis in original). For Deleuze and Guattari assemblages are "complex constellations of objects, bodies, expressions, qualities, and territories that come together for varying periods of time to ideally create new ways of functioning" (Livesey, 2005, p. 18).

We employ assemblage to highlight the enmeshed dynamics of ontology, epistemology, ethics, matter and agency as they relate to outdoor studies in an effort to prompt researchers to engage with philosophical and methodological shifts underway in related disciplines, and in doing so encourage creativity in reconceptualizing research, pedagogy and curricula within the field.

Background – methodological (and philosophical) developments

Pure empiricism takes it as read that sense experience is a better mirror of the world than pure thought, thus devaluing intuition. We see this reflected in what we might call the neoliberal obsession with qualification and measurement, along with descriptions of what constitutes scientific enquiry. One of the many impacts of quantitative imperialism within educational studies is expressed through government educational policies driven by testing, data and standards (Colebrook, 2017). Over the last 20+ years a growing body of research has highlighted how the social sciences, education, and qualitative research have been impacted by and caught up in the shifting projects of neoliberalism and positivism. Lather (2013) observes:

> the contest over the science that can provide the evidence for practice and policy pits the recharged positivism of neoliberalism against a qualitative 'community' at risk of assimilation and the reduction of qualitative to an instrumentalism that meets the demands of audit culture. To refuse this settlement is to push back in the name of an insistence on the importance of both epistemological and ontological wrestling in governmentality and calling out the unthought in how research-based knowledge is conceptualized and produced.
>
> *p. 636*

An important consideration for researchers, as they seek to serve and sustain the community and democracy, is that research method is political according to Lather (2013).

Lather (2006) describes being against the kind of 'methodolatry' in which "the tail of methodology wags the dog of inquiry" (p. 47). She discusses methodological fundamentalism with its links to pure empiricism and positivist research. Questions of truth, evidence, validity, reliability and criteria draw our attention to how assumptions that methods judged to be valid necessarily equate to valid findings needs questioning. Research assessment exercises that privilege randomised controlled trials and empirical research reflect a conception of education as the child of enlightenment requiring constant reformation, while references to evidence-based policy offer false hope for improvement based on rationality.

Research in outdoor studies is not immune to the effects of these developments. The material technologies of outdoor studies research may include visual images and written artefacts used to promote the field, activities, pedagogical and curricula choices. It is these technologies of virtual witnessing that are at the forefront in research centred on human experience. The use of data, journal entries, survey responses, or interview transcripts, for example, offer the illusion of logical progression and the voice of authority. Research in OEE tends to be dichotomised – either by the journals in which it is published, or by its authors – as either qualitative or quantitative. Mertens' (2005) description of educational research paradigms as being either post-positivist, constructivist or transformational has previously been used by editors (Thomas, Potter & Allison, 2009) of the three pre-eminent journals[3] in outdoor studies to classify research conducted therein (see Gough, 2016). This analysis did not include a discussion of methodological approaches or stances however. With some exceptions, few published research studies have explicitly acknowledged, or engaged with the socioeconomic, historical and political context in which they are conducted.

One of the contradictions of qualitative research in outdoor studies is that approaches that lay claim to a basis in interpretivist traditions often trust so completely in quasi-statistical analyses such as coding, so deeply embedded in positivism. St Pierre and Jackson (2014) describe how these practices have "been proliferated and formalized in too many introductory textbooks and university research courses" (p. 715). This form of pseudo-scientism presumably lends authenticity and validity to research for those deeply committed to traditional notions of pure empiricism. The separation of subject and object is reinscribed. St Pierre and Jackson (2014) identified two problems.

Firstly, the words of participants contained within interviews and field notes are given primacy over theory because of the privileging of face-to-face encounters. Here the Cartesian image of thought arises: the subject/object dualism that views the object (participant) existing on some separate plane – their natural setting perhaps – to the subject (researcher). Secondly, and related, as words are counted, reduced to themes, formalised through analysis, and written down, they become

> brute data that can be broken apart and decontextualized by coding – even using existing coding schemes from others' research projects. Once coded, words can be sorted into categories and then organized into 'themes' that somehow naturally and miraculously 'emerge' as if anyone could see them.
>
> *p. 716*

Of course, as MacLure describes (cited in St Pierre and Jackson, 2014), the act of transcribing data, and thereby transforming spoken word to text may be a way to get to know large quantities of material in depth by annotating and describing, hearing nuance, drawing links with and between ideas and thinking with theory.

Thinking with/in and through theory

Our concern is that while we may make our worlds, we do not make them alone, because we move within and through multiple worlds. Through the work of Haraway (1988) and others we have come to understand that knowledge is situated, interconnected, and indeed intra-connected. The possibilities for knowing are limited by our conceptions of what it means to be human and our capacities to recognise the blank spots and blind spots (Wagner, 1993) that exist in taken-for-granted discourses around knowledge and being. Braidotti (2013) reminds us that

> the question of method deserves serious consideration: after the official end of ideologies and in view of the advances in neural, evolutionary and bio-genetic sciences, can we still hold the powers of theoretical interpretation in the same esteem they have enjoyed since the end of the Second World War?
>
> *p. 4*

St Pierre (2016) describes how the ontological commitments of the new materialisms provide concepts for "understanding the agency, significance, and ongoing transformative power of the world – ways that account for myriad 'intra-actions' between phenomena that are material, discursive, human, more-than-human, corporeal and technological" (Alaimo & Hekman, in St Pierre et al., 2016, p. 101). Resisting the rush to application (St Pierre, 2016), we wish to foreground with some discussion – limited though it might be – about aspects of this thing called new materialism and what it might produce.

Ulmer (2017) prompts us to be mindful that attending "to materiality, vitalism, ecologies, flora, fauna, climate, elements, things, and interconnections has created openings across academic fields regarding who and what has the capacity to know" (p. 1). The work of scholars such as Braidotti, Haraway, Colebrook and Alaimo can help us understand that human-centred approaches to research, and reflected in the material technologies described above, are insufficient, that "material, ecological, geographical, geological, geopolitical and geophilsosophical" (Ulmer, 2017, p. 2) considerations are required for any deliberation of how we wish to be in and with the world. New materialisms raise questions for thinking through methodology. According to Ulmer (2017), they

> provide a rich, if imperfect, playground of theoretical concepts that extended across feminist, poststructural, post/critical, and post-qualitative approaches to research. With Karen Barad, matter mattered. It could diffract, intra-act, cut. It could spacetimematter. Catherine Malabou's plasticity described how phenomena could simultaneously give form, take form, and destroy form. Jane Bennett helped me think 'data' through vibrant political ecologies. Giles Deleuze and Félix Guattari suggested a lifetime of concepts with which to play, such as events, assemblages, folds, rhizomes, and lines of flight.
>
> *p. 3*

New materialisms are generative tools through which we might rethink our basic assumptions about the role and place of humans in a world made insecure by human existence and recognised in the affects of the Anthropocene. St Pierre et al. (2016) make the point that

> if humans have no separate existence, if we are completely entangled with the world, if we are no longer masters of the universe, then we are completely responsible to and for the world and all our relations of becoming with it. We cannot ignore matter (e.g., our planet) as if it is inert, passive, and dead. It is completely alive, becoming with us, whether we destroy or protect it.
>
> *p. 100*

Research challenges us to avoid the trap of (un)knowingly reiterating redemptive narratives through qualitative research by asking the kinds of onto-ethico-epistemological questions that reveal how normative truths about bodies, subjectivities and privilege are constituted and contested (Barad, 2003). This twists the focus of much qualitative research away from the emphasis on a hermeneutics of lived experience that has privileged the interpretive subject and towards the material-discursive relations that make the life of the body and its movement (im) possible (Butler, in Fullagar, 2017).

New materialist commitments invite us to vacate the ethico-onto-epistemological constraints of constructivist and interpretivist research and negotiate the trope of theory/practice divides. In the remainder of this chapter we traverse ideas and literature that we hope will encourage an undoing of the normative assumptions about research, philosophies and methods.

Living with(in) messy theorisation – signposts for rethinking outdoor studies research

Gough (2016) has written that shifts in methodological research of the social sciences, particularly addressing new materialism and new empiricism, have largely been overlooked in outdoor environmental education. Gough's appraisal of these approaches in relation to OEE research provides a significant foundation for this chapter. Gough, and many of the researchers cited here, call for thought experiment(s) driven by curiosity about what it means to think thinking differently. The various imperatives become entwined in ethico-onto-epistomology in which the empirical and the material are imbricated (St Pierre et al., 2016).

Opportunities for research reside in thinking with/in/through postparadigmatic and new materialisms that will require attention to a priori assumptions and procedures. By directing attention to the in-between spaces and assemblages of diverse agential elements (Bennett, 2010) possibilities emerge. We offer the following suggestions, none of which are mutually exclusive, nor conclusive.

Engaging with postparadigmatic materialisms

Barad's (2007) concept of agential realism disrupts the belief in independently existing individuals by "framing theories of subjectivity and agency within intra-active, relational entanglements" (Lather, 2016, p. 126). In new materialist thinking the taken-for-granted ascendency of humanism in qualitative research is turned around: there is a "co-implication of humans with non-human matter" (Davies, 2018, p. 114) and we are challenged to reconsider the status of life in terms of bioethical and biopolitical issues, including our understanding of thought and human existence. An ethico-onto-epistemological shift might find outdoor studies researchers rethinking agency from the position of human exceptionalism, towards an orientation that insists "on the radical distribution of agency as the effect of collaborations, as opposed to being set forth from human intentions: everything is active in cultural-natural-technological collectives, and anything present is therefore potentially agentic" (Weedon, 2015, p. 15). For example, Rautio (2013) demonstrates how "autotelic practices of humans such as picking up and carrying stones are potentially relevant in further understanding and conceptualising the ways in which humans are nature in relation to and constituted by all other animate or non-animate co-existing entities" (p. 394).

Lather (2016) calls for both more and other than reflexivity and points to Somerville's (2013) collaborative study of drought that "generates a radical alternative methodology across worlds that cannot know one another" (p. 127). In such an approach matter is more than things, and the complexity that surfaces when we think of the world in new materialist terms demands renewed attention to the material conditions that are geopolitically and socioeconomically implicated and excluded (see Coole & Frost, 2010; Fenwick & Edwards, 2011; Hultman & Lenz Taguchi, 2010).

Think and do theorypractice

Some thinkers challenge the *theorypractice* divide evident in the teaching of method/methodology and prevalent in outdoor studies discourse. In Deleuzean terms, theory is analogous to a toolbox, in that "it has to be used, it has to work" (Deleuze, 2004, p. 208). Re-imagine "what method might *do*, rather than what it *is* or *how to do* it" (St Pierre et al., 2016, p. 105, emphasis in original). Jackson and Mazzei (2012) illustrate how qualitative researchers can "use theory to think *with* their data (or use data to think *with* theory) in order to accomplish a reading of data that is both *within and against interpretivism*" (p. vii)[4]. Jackson and Mazzei (2013) note that "data is partial, incomplete, and is always in a process of retelling and remembering" (p. 262). The transcription of interviews, for example, can become vignettes, sense events forming part of a non-hierarchical rhizomatic research assemblage (Masny, 2015).

When "theories of the subject shift from an epistemology of human consciousness to a relational ontology" (Lather, 2016, p. 125) multiplicities are produced. Gough (2016) refers to Malone's (2016) work with children and dogs in Bolivia as an effective deployment of theories of intra-action and to Mcphie and Clarke's (2015) use of immanence in environmental education. Elsewhere, Clarke and Mcphie (2014) have critiqued relational understandings of the world promulgated by literature that invokes machinic Earth systems and prioritises phenomenological human-centred representations. They draw from Deleuze and Guattari (1987) to

> contend that relational approaches which highlight our 'relationship', or 'connection', or even 'disconnection' to 'nature', and, indeed, the concept of 'nature' itself, ultimately depict falsely boundaried entities … [and] offer animism as a way of seeing founded on an ontology of immanent materiality as one direction through this problem.
>
> *p. 199*

Mikaels and Asfeldt (2017) deploy a relational materialist approach (after Hultman & Lenz Taguchi, 2010) together with the concept of entanglement in their study with university students journeying in the Canadian Rockies. They describe how

> a decentring of humans in favour of mutual and relational engagements with matter and the more-than-human, in combination with place-stories and outdoor skill development that involves reading the land from embodied learning with/in its natural~cultural history, opens up new possibilities for embodied relations to place(s).
>
> *p. 2*

Concept as method

In the introduction to a Special Issue of *Qualitative Inquiry* dedicated to using concept as method, Lenz Taguchi and St Pierre (2017) comment:

> With Colebrook's encouragement, we wondered whether it would be possible to tilt educational inquiry, deeply imbricated for some time with the

> Cartesian image of thought that grounds the social sciences, their concepts, and their methodologies, toward philosophy, its concepts, and its conceptual-based practices.
>
> *p. 643*

Lenz Taguchi and St Pierre (2017) observe that "what a concept *is* and what it can *do* changes from discipline to discipline" (p. 644, emphasis in original). Colebrook (2017) draws on Deleuze and Guattari's (1994) notion of concept to encourage researchers to rethink method(s). Deleuzo-Guattarian concepts do not operate independently, but cross-pollinate during a process of becoming other. Concepts are not definitive, but they provide new ways of thinking, lines of flight. Colebrook (2002) reminds us that concepts offer "this power to move beyond what we know and experience to think how experience might be extended" (p. 17). Masny (2015) provides an example of how we might disrupt orthodox notions of qualitative research by engaging with "concepts emerging from Multiple Literacies Theory (reading, text, sense, toolbox, theory and praxis), and rhizoanalysis (de- and re territorialisation, assemblage, lines of social formation: molar (rigid), molecular, lines of flight)" (p. 2).

Read, read, read and look to literature in other fields

Gough and Whitehouse (2018) call attention to how, in environmental education research making reference to new materialist feminism, the work of both Barad and Haraway is sometimes overlooked. They argue that this matters because, among other reasons,

> It is important to draw on all the available ideas more conclusively, comprehensively, and coherently to advance thinking in environmental education research. We suggest environmental education researchers should draw more inclusively on international sociological and cultural thinking across ecofeminism and material discursive analyses.
>
> *p. 344*

Feminist orientations in outdoor studies might seek to critique and reframe universalist assumptions underpinning normative thinking that are "premised on an unacknowledged white, male, heterosexual, able-bodied subject who exists apart from non-human nature" (Fullagar, 2017, p. 250).

This is not a matter of (re)creating knowledge about women or lauding them for their contributions in the outdoors and the broader field of outdoor studies (see for example, much of Gray & Mitten, 2018). Instead as demonstrated by Humberstone, Caniglia, Riley and Ward in that volume, it is a move "beyond dualistic categories and with reference to gendered practices of othering difference, as well as those that diffract and trouble the normative" (Grosz, 1994; Barad, 2007) by exploring the generative possibilities and "politics of possibility" for creating "other ways of knowing and being gendered subjects" (Fullagar, 2017, p. 250). Braidotti (2013) argues that such an approach goes beyond critiques that unceasingly reassert overgeneralised explanations for social problems and reify processes through concepts such as neoliberalism and patriarchy.

Closing comments

Our discussion has deliberately sought to highlight both the challenges and generative capacity of undertaking research with, against and across traditions. We agree with Lather (2006) who argues that "facing the problems of doing research in this historical time, between the no longer and the not yet, the task is to produce different knowledge and produce knowledge differently" (p. 52).

Notes

1 Although our preferred term is outdoor environmental education (OEE), we use 'outdoor studies' in this chapter to refer to both it and other variations of terminology that may abound in lieu of foreclosing on *a* definition, and to be consistent with other chapters in this volume.
2 Experiential education is concerned with a philosophy of education – sometimes occurring in the form of outdoor learning – whereas experiential learning refers to the individual learning process, and the learning context.
3 These are the: *Journal of Outdoor and Environmental Education* (Australia), previously the *Australian Journal of Outdoor Education*; *Journal of Adventure Education and Outdoor Learning* (UK); and, the *Journal of Experiential Education* (USA).
4 There is an extensive literature on why, what and how data might be thought differently. See for example the Special Issue of *Qualitative Inquiry*, 2014, *20*(6).

References

Barad, K. (2003). Posthumanist performativity: Toward an understanding of how matter comes to matter. *Signs: Journal of Women in Culture and Society, 28*(3), 801–831.

Barad, K. (2007). *Meeting the universe halfway: Quantum physics and the entanglement of matter and meaning*. Durham, NC: Duke University Press.

Bennett, J. (2010). *Vibrant matter: A political ecology of things*. Durham, NC: Duke University Press.

Braidotti, R. (2013). *The posthuman*. Hoboken, NJ: Wiley.

Clarke, D. & Mcphie, J. (2014). Becoming animate in education: Immanent materiality and outdoor learning for sustainability. *Journal of Adventure Education and Outdoor Learning, 14*(3), 198–216.

Colebrook, C. (2002). *Gilles Deleuze*. London: Routledge.

Colebrook, C. (2017). What is this thing called education? *Qualitative Inquiry, 23*(9), 649–655.

Coole, D. & Frost, S. (Eds.). (2010). *New materialisms: Ontology, agency, and politics*. Durham, NC: Duke University Press.

Davies, B. (2018). Ethics and the new materialism: A brief genealogy of the "post" philosophies in the social sciences. *Discourse: Studies in the Cultural Politics of Education, 39*(1), 113–127. doi:10.1080/01596306.2016.1234682

Deleuze, G. (1994). *Difference and repetition*. New York, NY: Columbia University Press.

Deleuze, G. (2004). Intellectuals and power [an interview with Michel Foucault] (M. Taormina, Trans.). In D. Lapoujade (Ed.) *Desert islands and other texts: 1953–1974* (pp. 206–213). Los Angeles, CA, New York, NY: Semiotext(e).

Deleuze, G. & Guattari, F. (1987). *A thousand plateaus: Capitalism and schizophrenia* (B. Massumi, Trans.). Minneapolis, MN: University of Minnesota Press.

Deleuze, G. & Guattari, F. (1994). *What is philosophy?* (H. Tomlinson & G. Burchell, Trans.). New York, NY: Columbia University Press.

Fenwick, T. & Edwards, R. (2011). Considering materiality in educational policy: Messy objects and multiple reals. *Educational Theory, 61*(6), 709–726.

Fox, N.J. & Alldred, P. (2015). New materialist social inquiry: Designs, methods and the research-assemblage. *International Journal of Social Research Methodology, 18*(4), 399–414.

Fullagar, S. (2017). Post-qualitative inquiry and the new materialist turn: Implications for sport, health and physical culture research. *Qualitative Research in Sport, Exercise and Health, 9*(2), 247–257. doi:10.1080/2159676X.2016.1273896

Gough, N. (2016). Postparadigmatic materialisms: A "new movement of thought" for outdoor environmental education research? *Journal of Outdoor and Environmental Education, 19*(2), 51–65.

Gough, A. & Whitehouse, H. (2018) New vintages and new bottles: The "Nature" of environmental education from new material feminist and ecofeminist viewpoints. *Journal of Environmental Education, 49*(4), 336–349. doi:10.1080/00958964.2017.1409186

Gray, T. & Mitten, D. (Eds.). (2018). *The Palgrave international handbook of women and outdoor learning*. Cham: Palgrave Macmillan.

Grosz, E.A. (1994). *Volatile bodies: Toward a corporeal feminism*. Bloomington, IN: Indiana University Press.

Haraway, D. (1988). Situated knowledges: The science question in feminism and the privilege of partial perspective. *Feminist Studies, 14*, 575–599.

Hultman, K. & Lenz Taguchi, H. (2010). Challenging anthropocentric analysis of visual data: A relational materialist methodological approach to educational research. *International Journal of Qualitative Studies in Education, 23*(5), 525–542. doi:10.1080/09518398.2010.500628

Jackson, A.Y. (2017). Thinking without method. *Qualitative Inquiry, 23*(9), 666–674.

Jackson, A.Y. & Mazzei, L.A. (2012). *Thinking with theory in qualitative research: Viewing data across multiple perspectives*. London: Taylor & Francis Group.

Jackson, A.Y. & Mazzei, L.A. (2013). Plugging one text into another: Thinking with theory in qualitative research. *Qualitative Inquiry, 19*(4), 261–271.

Lather, P. (1991). *Getting smart: Feminist research and pedagogy with/in the postmodern*. New York, NY: Routledge.

Lather, P. (2006). Paradigm proliferation as a good thing to think with: Teaching research in education as a wild profusion. *International Journal of Qualitative Studies in Education, 19*(1), 35–57.

Lather, P. (2012). The ruins of neo-liberalism and the construction of a new (scientific) subjectivity. *Cultural Studies of Science Education,* 7(4), 1021–1025. doi:10.1007/s11422-012-9465-4

Lather, P. (2013). Methodology-21: what do we do in the afterward? *International Journal of Qualitative Studies in Education, 26*(6), 634–645.

Lather, P. (2016). Top ten+ list: (re)thinking ontology in (post)qualitative research. *Cultural Studies ⇔ Critical Methodologies, 16*(2), 125–131.

Lenz Taguchi, H. & St Pierre, E.A. (2017). Using concept as method in educational and social science inquiry. *Qualitative Inquiry, 23*(9), 643–648.

Livesey, G. (2005). Assemblage. In A. Parr (Ed.) *The Deleuze dictionary: Revised edition* (2nd ed., pp. 18–19). Edinburgh: Edinburgh University Press.

Malone, K. (2016). Theorizing a child–dog encounter in the slums of La Paz using post-humanistic approaches in order to disrupt universalisms in current 'child in nature' debates. *Children's Geographies, 14*(4), 390–407. doi:10.1080/14733285.2015.1077369

Masny, D. (2015). Problematizing qualitative educational research: Reading observations and interviews through rhizoanalysis and multiple literacies. *Reconceptualizing Educational Research Methodology, 6*(1), 1–14. doi:10.1007/BF03401009

Mcphie, J. & Clarke, D. (2015). A walk in the park: Considering practice for outdoor environmental education through an immanent take on the material turn. *Journal of Environmental Education, 46*(4), 230–250. doi:10.1080/00958964.2015.1069250

Mertens, D.M. (2005). *Research and evaluation in education and psychology: Integrating diversity with quantitative, qualitative, and mixed methods* (2nd ed.). Thousand Oaks, CA: Sage.

Mikaels, J. & Asfeldt, M. (2017). Becoming-crocus, becoming-river, becoming-bear: A relational materialist exploration of place(s). *Journal of Outdoor and Environmental Education, 20*(2), 2–13. doi:10.1007/BF03401009

O'Toole, J. & Beckett, D. (2013). *Educational research: Creative thinking and doing* (2nd ed.). Melbourne: Oxford University Press.

Rautio, P. (2013). Children who carry stones in their pockets: On autotelic material practices in everyday life. *Children Geographies, 11*(4), 394–408. doi:10.1080/14733285.2013.812278

Roberts, J.W. (2012). *Beyond learning by doing: Theoretical currents in experiential education.* New York, NY: Routledge.

Somerville, M.J. (2013). *Water in a dry land: Place learning through art and story.* New York, NY: Routledge.

St Pierre, E.A. (2016). The empirical and the new empiricisms. *Cultural Studies ⇔ Critical Methodologies, 16*(2), 111–124.

St Pierre, E.A. & Jackson, A.Y. (2014). Qualitative data analysis after coding. *Qualitative Inquiry, 20*(6), 715–719.

St Pierre, E.A., Jackson, A.Y. & Mazzei, L.A. (2016). New empiricisms and new materialisms: Conditions for new inquiry. *Cultural Studies ⇔ Critical Methodologies, 16*(2), 99–110.

Stewart, A. (2018). A Murray Cod assemblage: Re/considering riverScape pedagogy. *The Journal of Environmental Education, 49*(2), 130–141. doi:10.1080/00958964.2017.1417224

Thomas, G., Potter, T. & Allison, P. (2009). A tale of three journals: A study of papers published in AJOE, JAEOL and JEE between 1998 and 2007. *Australian Journal of Outdoor Education, 13*(1), 16–29.

Ulmer, J. (2017). Posthumanism as research methodology: Inquiry in the Anthropocene. *International Journal of Qualitative Studies in Education, 30*(9), 832–848. doi:10.1080/09518398.2017.1336806

Wagner, J. (1993). Ignorance in educational research: Or, how can you *not* know that? *Educational Researcher, 22*(5), 15–23.

Weedon, G. (2015). Camaraderie reincorporated: Tough Mudder and the extended distribution of the social. *Journal of Sport & Social Issues, 39*(6), 431–454.

2

ETHICAL ISSUES AND PRACTICALITIES IN OUTDOOR STUDIES RESEARCH

Barbara Humberstone and Carol Cutler Riddick

The concept of ethics is interpreted and debated in a variety of ways, yet the nexus among the varying definitions and common understandings is that of the "justification of human action" (Schwandt, 2007, p. 71). Research ethics are thus concerned with justification of researchers' actions. As Lahman, Geist, Rodriguez, Graglia and DeRoche (2011, p. 1398) point out, researchers need to consider and critically reflect upon, "how those actions affect participants, participants' families, the researcher, the research community, and the public research consumers. This justification is thought to be one that is right, proper, or moral" and we address this also to the non-human. How research is conducted is a matter of informed choice and decision-making; from the topic that is selected to how the study is designed and ultimately reported. Nevertheless, some possible choices and actions are considered unethical and so unacceptable, including causing physical and/or psychological harm to participants; acts of dishonesty (including data fabrication, plagiarism); and unjust discrimination in the selection of research participants. Riddick and Russell (2015, pp. 199–214) provide clear guidelines on practically addressing ethical responsibilities. Here we provide a brief overview of the background and current thinking on research ethics, principles of ethical research (with scenarios that illustrate some dilemmas in outdoor studies) and draw attention to research ethical review bodies.

Perspectives on research ethics

Ethical codes were developed as a consequence of the outcry regarding global events and medical interventions that could not be justified, and which undermined human dignity and value. These initially were largely to do with the misuse of medicine or medical interventions such as the Nazi atrocities in World War II. The response led to the Nuremburg Code of 1947 outlining ethical principles for human experiments. In the 1960s and 1970s came a number of notorious studies in bio-medical, psychological and social science research that fell well below acceptable ethical and moral standards, causing harm to participants (see Lahman et al., 2011). This resulted in the publication in the United States of the Belmont Report, 1979, which advocated three primary ethical principles: beneficence, respect and justice for human research, which form the basis of current ethical standards (Israel & Hay, 2006). From philosophical debates about ethics emerged a number of theoretical positions including rights-based theory, virtue-based theory and feminist ethics, principled theories and composite principled ethics (McNamee, Olivier & Wainwright, 2007).

As we show, ethical decision-making in research is an ongoing process, informed by the researcher's theoretical background, experience and professional codes of practice, as well as specific research context and setting. In addition to basic ethical principles, Lahman et al. (2011) propose applying "culturally responsive relational reflexive ethics" which draw attention to dilemmas that a researcher encounters throughout a project and are informed by social justice values (e.g. issues around "ethnicity, age, sexual orientation, and power", p. 1398).

Like McNamee et al. (2007, p. 47), we emphasise and illustrate here the complexity and fluidity of ethics:

> In the application of ethics, it seems that there is an expectation that the final and transcendent resolution of ethical disputes is possible. However, ethical systems do not exist in order to eliminate ethical discourse. Rather they provide working frameworks for each discourse to be considered in particular situations that pose ethical problems.

Ethical principles that guide research

Ashworth, Maynard and Stuart (2016), drawing upon key ethical considerations, provide an informative critical framework for 'doing' research in outdoor studies which we draw upon here, illustrated with scenarios. These scenarios emphasise the point made by Ashworth et al., (2016, p.199) that "Although these (ethical) considerations can act as guidelines, ensuring these are critical rather than taken for granted is necessary; the context of outdoor studies is both diverse and complex, and applications can unearth dilemmas in practice". According to McCrone (2002), ethical behaviour in research is underpinned by considerations of six fundamental principles that are the foundation to human service codes of ethics: respect, honesty, beneficence, justice, non-maleficence, and competence. These principles can be recast into four key ethical deliberations: informed consent; honesty, gain and justice; risk of harm; and confidentiality and anonymity.

Informed consent

Respect is an underpinning principle in all behaviour with both humans and the non-human. The right of individuals to decide about whether, and to what extent, they want to participate in a study is central. For much research, the ethical expectation is that a researcher clearly informs potential participants (as well as all stakeholders) about the purpose(s) of the study, the possible risks and benefits to participation, and their right to refuse or cease participation in the study at any time. Nevertheless, in certain situations seeking informed consent may be considered impractical or inappropriate (Spicker, 2011). For instance, research panels may waive the necessity of using informed consent when researching covertly 'underground' cultures or if such consent is likely to influence the outcome of psychological research but not anticipated as being detrimental to participants (Rosenthal & Rosnow, 2007).

As an acknowledgement of a potential study participants' autonomy, informed consent has now become common practice and normally requested by research ethics panels/committees (REPs) around the world. Informed consent is the ethical commitment to ensuring that a potential participant has enough information about the study to make a sound decision about participating and is aware they can freely withdraw at any time up to a given date. Consent forms should be written or conveyed in the language that matches the age and abilities of the specified group to be studied. Particularly, considerations need to be made for 'specific communities', 'vulnerable populations' and children in terms of informed consent and throughout the research process. Wright and O'Flynn (2012, p. 70) draw attention to how research can be undertaken that is respectful of indigenous communities by consulting and involving them from inception to the reporting.

Ashworth et al. (2016, p. 199) propose the following should be considered when planning an outdoor study:

- Purpose – including research questions and the intended use of the research;
- Confidentiality and anonymity – including the use of pseudonyms (or code names);
- Safeguarding and disclosure (of risk or criminality);
- Security of the data – including who will have access to it, where it will be stored and for how long, and how it will be disposed of;
- Participants' access to their own data at any time;
- Participant opportunity to review their data (participant validation);
- Participants' right to withdraw their information at any point;
- How the research will be written up and who will see it.

When one author (Barbara) undertook her doctoral research over 30 years ago in the UK there was no formal educational ethical committee at the university, but ethical issues were central expectations throughout when undertaking ethnographic projects (Fetterman, 1989; Hammersley & Atkinson, 1983). Verbal explanations were given to participants and they were asked for their permission. The only written agreement sought was from the head of the outdoor centre where the research took place. REPs in education did not then exist. About ten years later she was asked to sign a form to agree to be interviewed for a PhD project. She chose not to sign, but agreed to be audio interviewed, thus providing implied consent. Around this time, more formalised ethical procedures in educational and sport-related social research developed and provoked our and others' concern that reliance on 'tick box' procedures for ethical approval should not compromise ongoing ethical considerations (Sparkes & Smith, 2014; Wiles, Crow, Charles & Heath, 2007).

As Ashworth et al. (2016) point out, outdoor studies is often about spontaneity. Decisions need to be made as the research unfolds. Scenario 1 highlights the sensitivity of the researcher and how consent may need to be elicited as the research evolves.

SCENARIO 1: RESPECTING THE INTEGRITY OF THE CHILD, AN ETHICAL DILEMMA DOING RESEARCH WITH CHILDREN OUTDOORS

Karen Barfod, senior lecturer, VIA University College, Denmark

During my doctoral work, I was studying teaching in the outdoors (Barfod & Daugbjerg, 2018), with the teachers as my study object. On that occasion, I was planning observations on five schools, working with learning outside the classroom, and I wanted to study how the teaching was performed by the teacher in a natural setting (Hammersley, 2013, p. 13). To support my memory with more than my field notes, I asked the parents during a parents evening if I could take photographs during the lessons, and all parents gave me their written permission. I decided to use my small hand held camera, to disturb the teaching as little as possible. So far so good. But in one class, at my very first visit, just after I had presented myself and my research object ("I am studying your teacher and how she teaches, not you"), an 11 year old boy told me he hated pictures and wouldn't be photographed. He was hiding his face in the hood of his jacket, and interfering with the educational situation.

Should I just carry on, knowing I had the legal right from his parents to take the pictures, ignoring him and getting my data, or should I bias my research by only taking pictures without him, this changing my focus of observation?

I believe children should be respected too. So I asked him if it was all right that I just took pictures during the day as I planned. And then, at the end of the day, showed him the pictures, and gave him the opportunity to choose if the pictures should be deleted or not. He agreed, and after the first session where he was actually asked which pictures I should delete, he relaxed. He never asked me to delete any, and at the next session he came to me and said it was all right to take pictures.

Honesty

Honesty, or fidelity, deals with establishing a relationship of trust between the researcher and the participant. Deception in the form of misrepresentation is to be avoided (Monette, Sullivan & DeJong, 2008), although there can be situations where misrepresentation may be prudent, if there is no harm to the participants. An illustration of misrepresentation is an outdoor recreation/tourism piece of reported research which raised, in the outdoor community, considerable discussion. Moeller, Mescher, More and Shafer (1980a) compared campers' attitudes towards pricing using both formal and informal interviews (the latter were carried out using incognito interviewers who posed as campers). Different answers were provided, depending on whether the person was interviewed by a self-identified interviewer versus a supposed "camper". The deception used in this study was subsequently questioned (Christensen, 1980) and an ensuing controversy erupted (Moeller, Mescher, More & Shafer, 1980b).

Along with the ethical question of whether the Moeller et al. (1980a) study violated the honesty principle, we might also consider a methodological issue. If an overt direct question is asked, is the response based on what the respondent feels the questioner wants to hear? Or does the informal or natural situation provide more authentic responses (i.e. evoke what the respondent thinks/feels at the time)? These sorts of questions, raised in social science more widely, contributed to the development of interpretative oriented research where the researcher explores and tries to make sense of the everyday experiences of the participants in natural environments/situational contexts and soon became drawn upon in outdoor studies research (see Chapter 9).

Gain and justice

The ethical principle of beneficence refers to what individuals will gain from participating in the study. Riddick and Russell (2015) suggest that potential participants be informed beforehand what sort of benefit they might receive if agreeing to be in the study. Some (market) research provides incentives (e.g. money or objects) for participation; we discourage this, as do most REPs.

Further, Ashworth et al. (2016, p. 202) address the need that justice is used when conducting outdoor studies. Cozby and Bates (2011) point out justice addresses the issues of fairness in receiving the benefits of research as well as bearing the burdens of accepting risks. Justice has been defined as the "fair, equitable, and appropriate treatment in light of what is due or owed to persons" (Sylvester, Voelkl & Ellis, 2001). Ashworth et al. (2016) recommend that the purpose and who benefits from the research need to be reflected upon at the start of the research. They ask is it ethical to "take knowledge and experience from one stakeholder for the gain of others" (p. 200)? They give an example of using participant data without consent in order to gain funding for an organisation.

Risk of harm

'Do no harm' is the ethical principle of non-maleficence. Research in outdoor studies should be executed in a way that causes no harm to participants. Anticipation of any potential risks as perceived from the perspective of each stakeholder is paramount. Potential risks include physical, psychological/emotional harm, and loss of privacy. This is very real particularly when undertaking extended fieldwork-research as in expeditioning and in outdoor residential settings. Scenario 2 illustrates how risk of harm, may gradually emerge during fieldwork.

SCENARIO 2: DEPARTURES AND CONFLICTING VALUES DURING FIELDWORK

Phil Mullins, Associate Professor, Outdoor Recreation and Tourism Management, UNBC, Canada

I work with small groups of participants during fieldwork, sometimes in remote locations for long durations. I take an anthropological and phenomenological approach to praxis: working alongside participants to understand their world and our field, and then trying to communicate that to a wider audience. Research, therefore, becomes quite reliant on each participant. This has raised two ethical issues. First, the important right of participants to withdraw from a trip and/or the research, second, working alongside participants who say or do things that challenge my own values and ethics.

A participant's choice to withdraw affects the research, use of resources, and other participants. Those remaining may want the person to stay, or resent their departure. These consequences make such a decision, potentially, very difficult for participants to make. Obviously, however, a group must be open to accommodating someone's desire to leave. Socially, the group needs to be open to receiving such a decision. Pragmatically, plans and means need to be in place to adapt routes, access, and data collection to accommodate a departure. Before agreeing to join a study, however, participants must have a clear and accurate understanding of the fieldwork, its challenges, and the difficulties of withdrawing. This ethical and contractual obligation enables their free and informed consent, and likely mitigates withdrawals while establishing commitment.

On occasion, participants use derogatory language, or uphold stereotypes, different social and gender norms, or challenging ethics. When this happens, I try to consider the severity of the offence given potential impacts on the research, whether the behaviour is relevant to the research and change we are seeking, and if an opening exists to address it constructively. Other participants might resist, or join in. I have not had to ask for a withdrawal, though that is a possibility. As data, such behaviour might be powerful and telling. Therefore, I might consider how best to approach change: should I confront the issue with participants, or/and through analysis and discussion in a report? I also have obligations to protect all research participants while finding appropriate and accurate ways to represent them, and the phenomenon investigated, in my reports. Doing so can require careful work both in the field, and when writing.

As this scenario shows, decisions made in the field are not easy or as simple as REPs might suppose. It also highlights the tensions for the researcher between making immediate decisions and their potential impact on the overall research project.

Confidentiality and anonymity

Psychological and physical harm also may unfold due to breaches in confidentiality or anonymity. Confidentiality is about promising not to identify a response or observation as belonging to a particular person. Researching small special groups such as say high altitude climbers who are disabled, or teaching and learning within an outdoor centre or school, may pose difficulty as they may be more readily identified. This needs careful consideration as to how the findings are represented. Anonymous means there is no way any response can be connected with a specific respondent. That is, no information/identifiers (such as name, phone number, address, date of birth) is recorded that could be used to identify an individual participant.

Confidentiality is considered an extension of the right to privacy. Confidentiality refers to information, privacy refers to persons. Scenario 3 draws attention to ethical decision-making around a young child's need for privacy set against the researcher's need to adhere to safeguarding protocol (see also Chapter 7). Ethical issues when working with children are complex and will involve thorough consideration by the REP before approval.[1]

SCENARIO 3: ETHICS IN THE MOMENT

Philip Waters, Creative Director, I Love Nature CIC, Cornwall, UK.

I'm sat in the school office with a child; she's 7 years old. We're facing a computer watching films recorded on a chest-worn camera depicting her first-person viewpoint. I'm interested in what extended meaning she can add to her play. She selects a sequence where she is having a dialogue with another child about a plan they must have previously concocted. They *plan* to kidnap another child at the setting by persuading him to come over to her house, where they will trap him in a box, put him on a boat, float him down a river and set the boat on fire. She's not, of course, intent on committing such a crime. It's merely an idea in which they are both dialogically playing out. Nevertheless, she is so concerned that teaching staff might overhear the dialogue that she gets up from her chair and closes the office door.

In moments like these, safeguarding protocols flash-up like emergency warnings in your head. INSIST THE DOOR BE OPEN! And for sure, within some contexts as a practitioner who has worked with children for 23 years, I might insist the door be open. But not in this instance.

First, I'm protected as a researcher by the fact this whole episode is being filmed by a camera mounted on top of the computer, providing evidence of probity. However, that doesn't drive my resistance to the alarms ringing in my head. My intuition, feelings and empathic senses as a practitioner in this moment, with this child, reading her, sensing her, appreciating her dilemma, are what guide my response. I recognise that I have also promised to protect her views, interests and opinions from others in line with her informed consent. She has put her trust in me, in that moment, to respect her privacy. In the language of the United Nations Convention on the Rights of the Child (1989), it is a conflict between protection rights and participation rights (Waters & Waite, 2016). Protection rights would employ safeguarding principles as guaranteed within ethics codes of conduct. Participation rights afford her the privacy she requests as a fellow human being. One acts as a procedural framework, the other as a philosophy for the relational aspects of conducting research with others. Both are necessary and must be balanced in ethical research conduct.

Once information has been collected, there is an ethical responsibility to protect it to assure confidentiality. Identifiers should be removed from the interviews, questionnaires, or observation forms in such a way that data cannot be linked to an individual (Fowler, 2009). One way to address this is to record the name or other identifying information at the end of the form. That way, the identifiers can be readily separated from actual survey responses or observational data. In narrative research this de-identification may be observed through the presentation of creative nonfiction stories (see Chapter 13). Generally, not even the evaluator or researcher should be able to link the data to a specific individual. Study participants should likewise be assured that any information collected from or about them will be held in strict confidence. In some biographical interviews, however, it may be appropriate for the interviewed person to be identified with their agreement. Cultural dimensions should be considered as Western ethical norms may not be universal and what is acceptable may be questioned in indigenous groups.

Ashworth et al. (2016) point to their experience of young people requesting to be identified in the write up as they are proud to have taken part, and their dilemma of whether to maintain anonymity for participant protection or allow them to control the decision. They suggest aligning with stakeholders' and organisational policies on safeguarding and disclosure to help with researcher decision-making.

Purpose and practice of REPs

Confidence in, and support of, research requires trusting to the integrity of researchers. Globally governmental regulations require universities and other organisations to establish review boards for research involving humans and animals.[2] The basic function of such entities is to review and approve research submissions to determine if the proposal is aligned with a multitude of ethical principles, including if the researcher is competent enough to carry out the proposed study. Investigator(s) are required to clarify, defend or amend their research proposal in order to meet any ethical concerns. Lahman et al. (2011, p. 1397) draw attention to the way in which these boards are often not received well by many researchers finding them "rule-bound", "tedious" and "from hell".

Effective boards require members not only to be experienced researchers but also to have undertaken some form of ethics training. In many countries panel members are advised to take online training modules such as, 'Ethics. Good Research Practice'.[3] These provide opportunities for members to practise or refresh their decision-making in relation to ethical codes of practice and to familiarise themselves with different epistemological perspectives in research and the ethical codes associated with them. Professional codes of ethics encompass the norms, values and principles that govern professional conduct and relationships with clients, colleagues, and society (Kornblau & Burkhardt, 2012)[4].

Whilst some ethics board members can be dogmatic, many will provide support to researchers. Ethics board members are generally quite knowledgeable in one or more types of research paradigms and methods. Bingham (2018) draws attention to the complexities involved when submitting his proposal to the REP in Scenario 4.

SCENARIO 4: THE SUBMISSION

Kevin Bingham, lecturer, Sheffield Hallam University, UK

I submitted my ethics application three times. It was concerned with investigating a small group of urban explorers. Initially my research was considered dangerous (to myself and the people I intended to observe), outside the rules of the law and morally wrong.

Twice my ethics application, which was supported with relevant literature, was rejected. I was left feeling increasingly frustrated because it felt as though my arguments were being overlooked – or even ignored. I argued that the worlds urban explorers create for themselves exist regardless of whether we think they're moral or not, and that any prejudices need to be set aside when exploring 'deviant' forms of leisure. I also made it clear that trespass is not a criminal offence according to UK law, although it currently remains a civil offence. I also emphasised that I was aware of what constituted illegal behaviour and that I would not engage in or encourage such conduct (breaking and entering for example). However, I argued that even if my participants might choose to deviate from the law, it's not my role to influence their decisions. I would simply observe them and experience their world first-hand. As an urban explorer before deciding to conduct a study, I was already knowledgeable about the relevant risks, dangers and laws.

By attempt three, I was briefly tempted to go ahead with the study because I felt I was going around in circles. However, I was successful in gaining ethical approval. A major argument was that the research environment is unpredictable and people do not act out consistent behaviours or identities. I successfully emphasised that it was neither wrong nor unethical to witness things which may appear potentially deviant so long as all the participants in the study are aware you may use the experience for research purposes. It was a long hard slog to reach the point where my application was approved and I was then able to go out into the field.

Criteria used by REPs in reviewing research proposals is essentially to determine that: risks to research participants are minimal; benefits of participating in the study are greater than the risks; vulnerable populations (such as children, institutionalised individuals) are sensitively selected and dealt with; and participants are extended the right to informed consent.[5]

As Bingham (2018) and others (Lincoln & Cannella, 2016) attest, research ethics are complex and involve critical thinking, managing dilemmas, and ongoing decision-making on the part of both the researcher and the ethics board.

Concluding reflections

Researching outdoor studies involves continuous, dynamic ethical decision-making which is influenced by theoretical frameworks of the researcher, their knowledge and experience and chosen ethical codes of practice. As our scenarios show, ethical concerns stem from respect for participants and recognition of the complexity and fluidity of human and environmental interactions. It is almost inevitable that problems, issues and dilemmas will emerge, and informed ethical decisions based upon researchers' particular onto-epistemological standpoints will be made (McNamee et al., 2006). Challenges when embarking on outdoor research involve applying ethical judgements over the course of the research endeavour. It is a mindful and ongoing process, from the beginning phases of conceptualisation to the reporting end stage (Ramcharan & Cutcliffe, 2001). Outdoor studies is fluid and dynamic with risk and safety as frequent concerns. We, like Lahman et al. (2011, p. 1398) advocate, for outdoor studies research, mirroring "culturally responsive reflexive ethics". This means providing strong, well-argued submissions to REPs, some of which may not be particularly flexible, or familiar with the issues/contexts/norms, and sometimes there may be a tendency to apply a generalised/universal approach to approving. We promote ethical decision-making that, as Ashworth et al. (2016, p. 205) argue, is "based on best judgement rather than hard-and fast rules".

Notes

1 See https://learning.nspcc.org.uk/media/1504/research-ethics-committee-guidance-applicants.pdf
2 Office for Human Research Protections at www.hhs.gov/ohrp/regulations-and-policy/decision-charts/index.html#.VH8iemEo7-g.email
3 See www.epigeum.com/ Websites that provide helpful information: Resources for Research Ethics Education (http://research-ethics.net/).

4 Some relevant professional code of ethics can be found at: Canadian Evaluation Society (https://evaluationcanada.ca/), the Australasian Evaluation Society (https://aes.asn.au/resources/ethical-guidelines-2.html), the British Sociological Association (www.britsoc.co.uk/media/24310/bsa_statement_of_ethical_practice.pdf), and the British Educational Research Association (www.bera.ac.uk/wp-content/uploads/2018/06/BERA-Ethical-Guidelines-for-Educational-Research_4thEdn_2018.pdf?noredirect=1).
5 Guidelines on human subject protection is available at www.hhs.gov/ohrp/international/compilation-human-research-standards/index.html. An international listing of 27 social-behavioural standards exists at www.hhs.gov/ohrp/international/social-behavioral-research-standards/index.html.

Acknowledgement

We wish to thank the writers of the scenarios for their generosity in sharing them and their helpful comments on the chapter; and Dr Ben Clayton, Chair of Bucks New University Research Ethics Panel, for his insightful feedback.

References

Ashworth, L., Maynard, L. & Stuart, K. (2016). Ethical considerations in outdoor studies research. In B. Humberstone, H. Prince & K.A. Henderson (Eds.) *International handbook of outdoor studies* (pp. 198–206). Oxford, New York, NY: Routledge.

Barfod, K.S. & Daugbjerg, P. (2018). Potentials in Udeskole: Inquiry-based teaching outside the classroom. *Frontiers in Education, 3*(10). https://doi.org/10.3389/feduc.2018.00034

Bingham, K. (2018). Trespassing in (un)familiar territory: Knowing "the other" in ethnographic research. *International Journal of Sociology Research 1*(2), 122–137.

Christensen, J. (1980). A second look at the informal interview. *Journal of Leisure Research 12*(2), 183–186.

Cozby, P. & Bates, S. (2011). *Methods in behavioral research* (11th ed.). Boston. MA: McGraw-Hill Higher Education.

Fetterman, D.M. (1989). *Ethnography: Step-by-step*. London: SAGE Publications.

Fowler, F. (2009). *Survey research methods* (4th ed.). Thousand Oaks, CA: SAGE Publications.

Hammersley, M. (2013). *What is qualitative research?* London: Bloomsbury.

Hammersley, M. & Atkinson, P. (1983). *Ethnography: Principles in practice*. London: Routledge.

Israel, M. & Hay, I. (2006). *Research ethics for social scientists*. London: SAGE Publications.

Kornblau, B. & Burkhardt, N. (2012). *Ethics in rehabilitation: A clinical perspective* (2nd ed.). Thorofare, NJ: Slack Incorporated.

Lahman, M.K.E., Geist, M.R., Rodriguez, K.L., Graglia, P. & DeRoche K.K. (2011). Culturally responsive relational reflexive ethics in research: The three Rs. *Qualitative & Quantitative: International Journal of Methodology, 45*(6), 1397–1414.

Lincoln, Y.S. & Cannella, G.S. (2016). Ethics and the broader rethinking/reconceptualization of research as construct. In N.K. Denzin & M.D. Giardina (Eds.) *Ethical futures in qualitative research: Decolonizing the politics of knowledge* (pp. 67–84). Oxford: Routledge.

McNamee, M., Olivier S., and Wainwright, P. (2006). *Research ethics in exercise, health and sport sciences*. London: Routledge.

McCrone, W. (2002). Law and ethics in metal health and deafness. In V. Guttman (Ed.) *Ethics in mental health and deafness* (pp. 38–51). Washington, DC: Gallaudet University Press.

Moeller, G., Mescher, M., More, T. & Shafer, E. (1980a). The informal interview as a technique for recreation research. *Journal of Leisure Research, 12*(2), 174–182.

Moeller, G., Mescher, M., More, T. & Shafer, E. (1980b). A response to "A second look at the informal interview". *Journal of Leisure Research, 12*(2), 187–188.

Monette, D., Sullivan, T. & DeJong, C. (2008). *Applied social research: Tool for human services* (7th ed.). Belmont, CA: Thompson Wadsworth.

Ramcharan, P. & Cutcliffe, J. R. (2001). Judging the ethics of qualitative research: Considering the 'ethics as process' model. *Health and Social Care in the Community, 9*(6), 358–366.

Riddick, C. & Russell, R. (2015). *Research in recreation, parks, sport, and tourism* (3rd ed.). Champaign, IL: Sagamore Publishing.

Rosenthal, R. & Rosnow, R. (2007). *Essentials of behavioral research: Methods and data analysis* (3rd ed.). New York, NY: McGraw-Hill.

Schwandt, T. (2007). *The SAGE dictionary of qualitative research*. Los Angeles, CA: SAGE Publications.

Sparkes A.C. & Smith, B. (2014) *Qualitative research methods in sport, exercise and health.* London: Routledge.

Spicker, P. (2011). Ethical covert research. *Sociology, 45*(1), 118–133.

Sylvester, C., Voelkl, J. & Ellis, G. (2001). *Therapeutic recreation programming: Theory and practice.* State College, PA: Venture Publishing.

United Nations. (1989). *The United Nations convention on the rights of the child.* London: Unicef.

Waters, P. & Waite, S.J. (2016). Towards an ecological approach to ethics in visual research methods with children. In D. Warr, M. Guillemin, S. Cox, and J. Waycott (Eds.) *Ethics and visual research methods: Theory, methodology and practice* (pp. 117–129). London: Palgrave Macmillan.

Wiles, R., Crow, G., Charles, V. & Heath, S. (2007). Informed consent and the research process: Following the rules or striking balances. *Sociological Research Online, 12*(2), 83–95.

Wright, R. & O'Flynn, G. (2012). Conducting ethical research. In K. Armour & D. Macdonald (Eds.) *Research methods in physical education and youth sport* (pp. 66–78). Oxford: Routledge.

3

DESIGNING EFFECTIVE RESEARCH PROJECTS IN OUTDOOR STUDIES

Heather Prince and Liz Mallabon

Research needs to be systematic and rigorous but it also provides an opportunity to explore in depth a new or existing area of interest. As such the process of designing a research project should be exciting, with the potential to create new knowledge, read widely about a chosen area, evaluate critically sources of information and previous research, and subject your final output to scrutiny. Although there are many resources available in support of navigating the research process (for example, Bell, 2014; Gilbert, Camiré & Culver, 2014; Locke, Spirduso & Silverman, 2014), these are relatively generic in nature.

This chapter will examine design principles for successful and effective research projects in outdoor studies through the lenses of academic researchers, including as supervisors and managers of dissertations, doctoral research and commissioned research projects, and students and their experiences of carrying out research projects. It will identify enablers, pitfalls, challenges and positive experiences of design, underlying philosophical assumptions and our reflections on the student journey. We acknowledge an Anglocentric and UK perspective and recognise that degrees involving research may be structured and often defended differently across the globe. However, we aim to support current and future researchers in planning for success and efficacy in research projects in outdoor studies through a pragmatic and philosophical approach to design.

Taught degree research projects

In many cases, higher education students of taught Bachelor's and Master's degrees have tended to concentrate their planning at the outset on the content, area of research interest and possibly the research questions. Whilst this can result in successful projects, our experience is that students are not always able to translate these ideas quickly into secure research design. In some higher education institutions

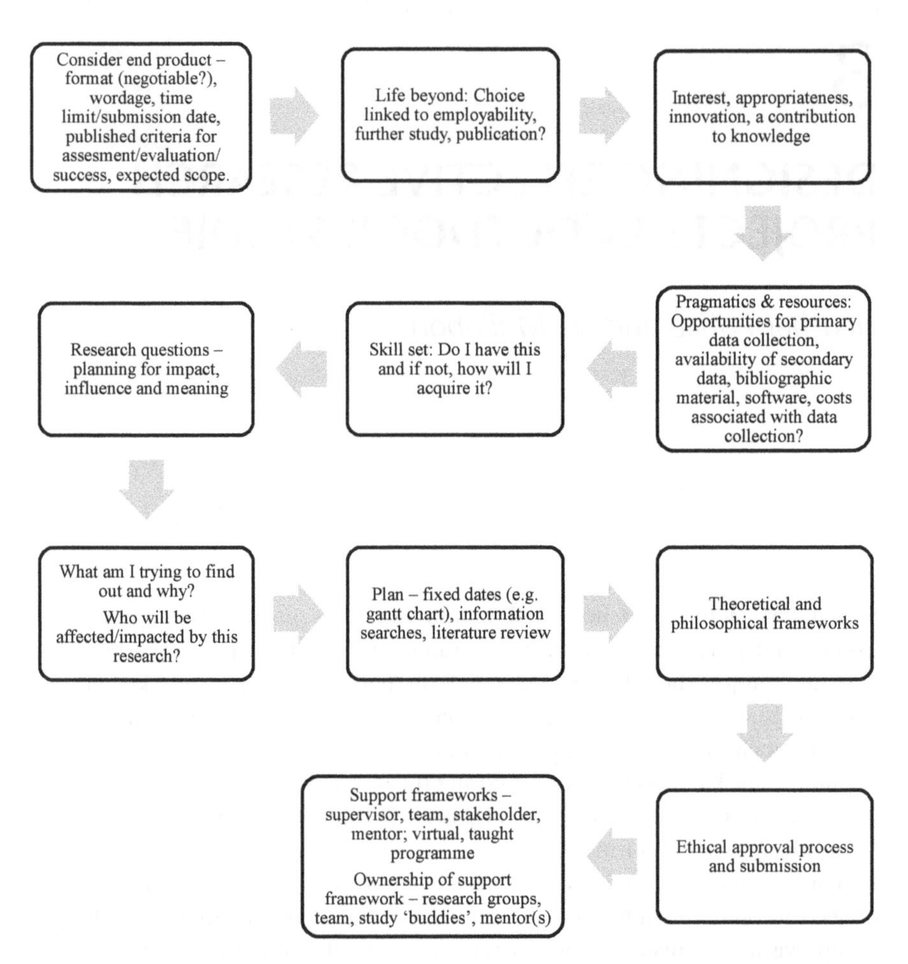

FIGURE 3.1 Pathway analysis: Undergraduate and taught postgraduate dissertations and theses (open and closed choices of project)

and research institutes, students may choose a project from a range of suggested titles/research areas, usually linked to a supervisor and reflecting his or her interests. These types of project may benefit from a clear research design although may compromise student autonomy or creativity in direction at a later stage.

Figure 3.1 identifies a pathway analysis for the design of taught degree research projects. It highlights the process and the areas of questioning and self-analysis that might be considered at each stage.

Enablers

A key enabler is the researcher or student themselves. Most taught degree courses encompass a dissertation module towards the end of the programme to provide students with the opportunity to design and conduct a substantial piece of independent, supervised research and to reflect critically on their findings with respect to current practice in the field, although there may be scaffolded skill development in the form of smaller scale projects and/or a research methods module prior to this. Almost all undergraduates and postgraduates report that they consider research to be the most challenging and rewarding aspect of their higher education programme (Haigh, 2013). Many taught students felt they might not be able to complete their dissertation but on finishing felt, "a sense of pride and achievement – here it is, my own bound research project" ('Daisy', undergraduate student).

As such, the concept of the intellectually independent (or semi-independent) learner is assumed and means in practice that it is the student that drives and takes ownership of the project. It is important for the researcher to ascertain the parameters and expectations for the project and if these are not clear, to seek guidance and support in this area. Reflexivity (Steier, 1991) is a key skill and accurate self-review in terms of self-analysis of skills and abilities at the outset and at various stages of the project is important, so that relevant gaps might be identified and supported.

Others play a key role. A supervisor, mentor or tutor might be appointed or act in that role – use him, her or them! Learning communities might be established by a tutor within a modular structure but a group established by students themselves often constitutes a valuable framework of support through social media, online discussion groups, real time meetings, work groups in a shared space or just social interactions in shared student residences. Friends and family can be a tremendous support, encouraging a student to persevere and maintain momentum, provide respite if necessary, ground them in real life and support through words and actions, for example in proofreading, through what is often a long and sustained process.

Project area. "If your research focuses on a topic that you are passionately interested in, you are more likely to be motivated from the outset and your motivation is more likely to be sustained" (Haigh, 2013, p.62). Choosing a research topic or area if this is an option, can be daunting for some and we will return to this later. This, of course, needs to be balanced with the expected outcomes, parameters or criteria, availability of data and originality.

Proposal/ethical approval. These may be considered challenges but they comprise key elements of the design process. A written or verbal proposal that needs approval by a more experienced researcher will give the impetus for articulating an overview of the project and the opportunity to gain feedback. A non-specialist perspective is often much valued by researchers as it provides the opportunity to articulate your research to a 'lay' person in comprehensible language, and the potential for objective questions to be raised. For example, in respect of a project involving analysis of photographs that participants posted on social networking sites following an outdoor experience through content analysis and photo-elicitation, the ethics panel suggested recruiting participants after they had taken part in the experience and then inviting them to take part in the research if they had taken photographs. This more naturalistic approach also met with approval from the research subjects as one participant commented, "We took photographs anyway, so the research did not influence or impinge on our experience". Ethical approval is usually required in some form for research projects involving humans, animals and sometimes for environmental projects as well, and you will need to state that your research has been approved by a panel or committee in the final output. It is worth exploring the form of this and the timeline, so that it can be embedded in the research design.

Pitfalls

Too narrow a research question or too broad a focus for the nature of the project (more common) are two of the main pitfalls that mitigate against successful projects. For example, the perspective of a single instructor on the value of challenge courses might be too limited, but the place of outdoor learning in schools might be too broad for most research projects. However, this might be contingent on the nature of the data collection (for example, the single instructor might have experience of a number of contexts and the place of outdoor learning in schools might be illustrated by a case study – albeit ill-defined research questions). It is important to discuss your ideas with a more experienced researcher who has knowledge of the expected parameters to ascertain its feasibility and whether or not it is likely to meet the guidelines for the project.

Duplication of research might not be an issue as more data in time or space might give more credence to a previously defined question, issue or hypothesis. However, there are examples when a piece of research might have little relevance in this respect: for example, if a national governing body (NGB) is already reviewing their awards structure and has new proposals in place, there is little point in investigating people's ideas on the current awards and suggesting changes at that time. If this is an area of interest, it might be worth considering contacting the NGB to see if there is any further research that would be valuable to them (but see challenges below), thus increasing impact.

A research project will usually require collection or collation of primary and/or secondary data. It is important through a design process to ensure that data collection is feasible or accessible in the time available, or to have an alternative plan if it is not or if it is delayed for any reason. This might link to a fixed event (e.g. an adventure race, a residential experience, a volunteering opportunity). Ensure that you know what the deadlines are and, if data gathering is scheduled to take place close to or after the event, a change of plan in the design of the project is required!

Time management is important but it might be not in your control, so planning a timeline for research may realise the full potential of a project. A gantt chart (www.gantt.com) or similar is useful in outlining the various stages of a project in the design process.

Challenges

Stakeholder timescales/access and/or environmental variables and how these match to the project are all to be considered. Many projects in conjunction with a third party or stakeholder that also are to be submitted as dissertations with a fixed timescale have been problematic or even unsuccessful due to a misalignment of deadlines. For example, teachers may not be able to facilitate data collection in a busy school environment, or answer emails or electronic surveys; stakeholders, or the administrative functions within them, might not have time or devote time to retrieving systems data. If your research involves a specific population (for example, disabled climbers), will you be able to access them and, if you can, will you be able to ensure that the ethics of your research are acceptable? If you want to collect data on snowsports in an area with unreliable snow cover, what will happen if there is no snow in the year of your research? If it is on river paddling, what will happen if there is too much water or not enough?

The output of research is usually judged in terms of metrics – a figure or set of figures (see Chapters 22, 27), criteria and/or descriptors. In the design process, it is worth building in time and planning to accommodate the compilation of the output, be it a dissertation, thesis, presentation, publication or other format. Too often the metric measuring the output does not measure the research undertaken because it concentrates on the final output, which is usually either text or visual in nature. Trevor, a postgraduate student, wrote "I wish I'd started writing earlier. I can't believe the amount of time it's taken to write". A good verbal, written or visual output needs careful checking, proofreading and organisation. Let the end product justify the quality of the research.

Whilst self-reflection is important, questioning self-belief can be problematic. Research projects are usually extended pieces of work, and sustaining interest and quality throughout is sometimes difficult, particularly when there are challenges in execution; this is often attributable to wavering self-belief. Having a mentor or another individual or group that you can discuss this with is critical to maintaining momentum. Consider also presenting your work within a comfortable environment, for example, a department seminar or local conference, where you will get constructive feedback on your research. This is worth building into an ongoing design framework.

Inexperienced researchers are often bewildered by the terminology of the research process and confused by concepts such as paradigms, methodologies and methods and how and where they are positioned in varied theoretical frameworks. It is well worth spending time reading around your subject in books and journals to get a feel for different conceptualisations.

Postgraduate researchers

Many of the themes in the design pathway analysis (Figure 3.1) are also applicable to doctoral (PhD or professional doctorate) and MPhil or MRes researchers, although normally they will already have some experience of research design, albeit at a different level. However, doctoral programmes in particular tend to be more extensive and are subject to a longer timescale than taught programmes. They therefore incorporate the need for sustained research, which can challenge motivation, self-belief and work-life balance to a greater extent, raising the need for possible changes of direction as a result of data, agendas, stakeholder requests and other factors.

For example, it could be that your research requires use of a certain instrument to collect the data and you cannot find a suitable one, or the one available is inappropriate for your context (perhaps aimed at adults rather than children, for example) and therefore, you need to spend time designing an instrument and testing it before data can be collected. Your agenda might have been changed by a political agenda outwith your control, for example, the UK's exit from the European Union. You might find that a wider contact with an extended sample questions your approach, definitions or terminology. For example, in a research project exploring young people's 'connection' with nature, even though the literature supports such an expression, through primary data collection it is evident that most of the sample population understand 'connectivity' to be via smart phones, and therefore a change of emphasis to 'relationships' with nature is a more accurate reflection of the research (see Hayes, 2017).

There may be expectation of ongoing outputs: presentations to the research team, reports, teaching, publications and monitoring of progress. Thus, the need for doctoral researchers to have support in writing and development of confidence in writing, linked to academic identity, has been recognised (Fergie, Beeke, McKenna & Creme, 2011). The expectations that the research constitutes critical academic discourse and is a substantial contribution to knowledge illustrate a higher level of research than for Bachelor's or Master's students.

Commissioned research projects

Most of the research design components will have been presented at the proposal or bidding stage of funded research and the project management may be a greater challenge to some than the research itself. The timescale and the integration of data collection, analysis and reporting usually is a priority for more experienced researchers who will be familiar with the research methodologies and methods, although it is important to check that the nature of the outputs expected by the commissioner/funder are specified. Livingston (2017) introduces the notion of a 'supercomplexity' paradigm to reflect the position of researchers undertaking commissioned research in shifting and challenging situations with multiple influences. It may be pertinent to ascertain the scope and flexibility of the research and level of consultation needed if there are challenges in implementing the accepted proposal or bid, and any flexibility in timelines.

Developing the design

As mentioned above, terminology for some neophyte researchers can be challenging. Beyond the initial focus of the research and framework for the particular process, there needs to be a definition of a question, hypothesis or issue. The means by which the research project develops is guided fundamentally by philosophical assumptions, or epistemological beliefs on the part of the researcher. Epistemology is the nature and construction of knowledge (what is there to know in the world and who can know). All researchers are to some extent guided by these principles and it is the adoption of their key assumptions that is central to becoming a competent researcher (Denzin & Lincoln, 2017). Thus, for the outsider to fully understand the research, acknowledgement of the philosophical assumptions of the research is essential (Sparkes & Smith, 2014). Indeed, the philosophical framework of the research is inseparable from the research strategy undertaken, and it may be useful to state explicitly the underpinning paradigm of the research in order to assess critically the value of its contribution (Guba & Lincoln, 1994). The chosen theoretical paradigm of any research thus forms the foundation for the work and is normative. Guba and Lincoln (1994, p.105) define a paradigm as the "basic belief systems or world view that guides the investigation". Thus, we conduct a research project using a particular paradigm because it encapsulates assumptions that we believe in about the world and that are supportive of our values (Sparkes & Smith, 2014).

Maxwell (2011) argues for the complexity of research and the notion that there are many paradigms:

> philosophical stances and assumptions, like theories, are lenses through which we view the world. These lenses are essential for our understanding, but the views they provide are fallible and incomplete, and we need multiple lenses to attain more valid, adequate, in-depth knowledge of the phenomena we study.
>
> *p. 29*

Researchers making different paradigmatic assumptions result in the world being seen differently, because the underpinning epistemology is intimately linked to the ontology (belief about the nature of reality; the filters through which we see and consider the world (Allison, 2000)), methodology (the approach to, and process of, gaining knowledge – a framework that helps you decide how you are going to engage with the world and the methods you will use). Consequently research may be conducted differently depending upon onto-epistemological underpinnings (Sparkes & Smith, 2014). Central to this process is the adoption of certain assumptions in relation to ontological, epistemological and methodological questions: "essentially, ontological assumptions give rise to epistemological assumptions which have methodological implications for the choices made regarding particular techniques of data collection" (Sparkes, 1992, p.14). Ontological assumptions are individuals' beliefs about the nature of the social world around them and the nature of their existence, whilst epistemological assumptions refer to the nature of knowledge and how it is gained (Sparkes, 1992). In other words, the researcher needs to be aware of which paradigmatic assumptions or world view underpins their research, although this may change as they progress through their research.

The positivist paradigm emerged as the dominant lens with its epistemological assumptions of objectivity from Descartes in 1637. Quantitative researchers explain, control and predict phenomena and follow a realist ontology, assuming a single, uniform and objective reality exists, independent of a person. They adopt a dualist and objectivist epistemology, which means the researcher and the researched are independent and the researcher can study something without being influenced by it or it influencing them. They tend to favour an experimental or manipulative approach. A quantitative researcher may state a hypothesis or question in propositional form, which is then subject to testing empirically to accept or reject, verify or falsify, under controlled conditions. Data are normally numerical in nature and subject to statistical testing and the researcher is assumed to be value-free and unbiased (see Chapter 26).

In contrast, Kant in 1781 introduced the concept of the world being understood in ways other than the direct objective methods of positivism, leading to the early developments in the qualitative school of thought. During the 1860s and 1870s Dilthey further developed the qualitative school of thought and, in particular, the interpretive school, through his emphasis on 'understanding'. Interpretive researchers acknowledge the role that positivistic approaches play within the study of the physical world, but contest that they are too simplistic when addressing the social world (Sparkes, 1992). Through this interpretive paradigm, a qualitative researcher with a relativist ontology assumes that social reality is constructed by humans and can become multifaceted. Multiple, subjective realities exist in the mind that shape and construct social reality. Qualitative thinkers adopt a subjectivist, transactional and constructionist epistemology, whereby the researcher and the researched cannot be separated because values influence the interpretation (axiology). They see and try to understand the world from the participant's perspective and may favour a hermeneutical and dialectical approach, acknowledging their assumptions and values through reflective accounting.

As Sparkes & Smith (2014) summarise,

> The basic philosophical differences … lead to quantitative and qualitative researchers developing different research designs, using different techniques to collect different kinds of data, performing different types of analyses, representing their findings in different ways, and judging the 'quality' of their studies using different criteria.
>
> *p. 14*

Within the paradigm debates of research, purists have emerged. The qualitative purist rejects positivism, arguing that context-free generalisations are neither desirable nor possible, and that the subjective knower is the only reality. In contrast the quantitative purists maintain the need for social science inquiry to remain objective (Onwuegbuzie, Johnson & Collins, 2009). It was Weber (1864–1920) who tried to bridge the gap between positivism and interpretivism, by stressing the importance of interpretation as well as observation in understanding the social world. Weber's initial intuitions have evolved to an anti-purist approach, modifying the idea of a qualitative versus quantitative debate in favour of combining the approaches, a mixed methods approach: "A false dichotomy exists between qualitative and quantitative approaches and that researchers should make the most efficient use of both in understanding social phenomena" (Creswell & Poth, 2017, p.176). In a mixed methods approach (see Chapters 24 and 25) both inductive and hypothetico-deductive (or supported) reasoning can complement each other (McAbee, Landis & Burke, 2017). In inductive reasoning, theories are formulated by drawing general inferences from particulars or data, while hypothetico-deductive reasoning involves testing a priori hypotheses by collecting data and determining the degree to which the hypotheses are rejected (as in a null hypothesis).

Reflections on the student journey in designing an effective research project

We, as academics, encourage students on taught degree programmes for whom the dissertation is a mandatory component, to reflect on areas of personal interest in outdoor studies. From this and through discussion with peers and tutors, the objective is to be able to identify an area of interest suitable for a dissertation and to write a proposal, or to develop their thoughts further. For some this is easy and focused in the first instance but for others, "like so many (other) students we meet at this stage in the research process, (her) focus could at best be described as 'all over the place'" (Gilbert et al., 2014, p. 62). All ideas can be developed and focused with support, but we like to enable the student to own the idea if possible, not the tutor.

Some students struggle to articulate ideas and tutorial time can draw out interests, motivations, future career pathways and pragmatics. One way forward is to suggest reading, for example, previous dissertations (particularly the Suggestions for Future Research sections), a recent major monograph in outdoor studies covering a range of areas (e.g. Becker, Humberstone, Loynes & Schirp, 2018; Humberstone, Prince & Henderson, 2016; Jeffs & Ord, 2017; Pike & Beames, 2013) or recent past issues of a relevant journal, such as the *Journal of Adventure Education and Outdoor Learning* (see also Chapter 27). Another, linked to the literature, is to identify current key research topics of research groups and academics' interests or recent or future conference themes.

Other students have identified a research area but it is far too extensive for a dissertation or to generate a research question. Recent examples of these have been: "the primary school outdoor curriculum", "mountain-biking, climbing or open boating as those are my main outdoor interests", "environmental activism" and "anxiety". In this case, directing students to the literature and discussing ideas and their feasibility in terms of generating a research question and collecting data will help. Often, a student will come with two or more valid proposals, unable to decide between them and again, feasibility of the project in the time available, impact and originality are major considerations.

Even after identifying a research area, question(s) and philosophical assumptions, some students are overwhelmed by the task ahead. One way to break this down is to write or draw a simple flow diagram on one sheet of A4 paper (or equivalent) to develop the design further. Students may also be encouraged to keep a research diary as they progress and to write down their motivations and rationale for choosing the topic – this can become part of the dissertation as appropriate.

Conclusion

Outdoor studies is a wide-ranging area of pedagogy, research and practice. It is interdisciplinary and transdisciplinary and the breadth of research areas it encompasses is substantial. It is often a challenge for an experienced researcher to narrow the focus and identify the appropriate philosophical assumptions, methodologies and methods and an appropriate design, let alone a neophyte student seeking to write a dissertation. It is hoped that this chapter provides support in the early part of the research process through pragmatic steps, tips and a definition of terminology for researchers, which will result in the design of effective research projects. The research process is an exciting journey and it is hoped that this chapter has provided sound knowledge for developing your own research adventures.

References

Allison, P. (2000). *Research from the ground up: Post-expedition adjustment.* Occasional papers 1. Ambleside: Brathay Trust.

Becker, P., Humberstone, B., Loynes, C & Schirp, J. (Eds.) (2018). *The changing world of outdoor learning in Europe.* Oxford, Los Angeles, CA: Routledge.

Bell, J. (2014). *Doing your research project: A guide for first-time researchers* (6th ed.). Maidenhead: McGraw-Hill.

Creswell, J.W. & Poth, C.N. (2017). *Qualitative inquiry and research design: Choosing among five approaches* (4th ed.). London: SAGE.

Denzin, N.K. & Lincoln, Y.S. (2017). *The SAGE handbook of qualitative research* (5th ed.). Los Angeles, CA: SAGE.

Fergie, G., Beeke, S., McKenna, C. & Creme, P. (2011). "It's a lonely walk": Supporting postgraduate researchers through writing. *International Journal of Teaching and Learning in Higher Education 23*(2), 236–245.

Gilbert, W., Camiré, M. & Culver, D. (2014). Navigating the research process. In L. Nelson, R. Groom & P. Potrac (Eds.) *Research methods in sports coaching* (pp. 55–65). Oxford, New York, NY: Routledge.

Guba, E. & Lincoln, Y.S. (1994). Paradigmatic controversies, contradictions and emerging confluences. In N.K. Denzin & Y.S Lincoln (Eds.) *The SAGE handbook of qualitative research* (3rd ed., pp. 191–216). London: SAGE.

Haigh, N. (2013). A framework and process for conceptualising and designing a new research project. In R. Donnelly, J. Dallat & M. Fitzgerald (Eds.) *Supervising and writing good undergraduate dissertations* (pp. 19–42) Ebook: Bentham Science Publishers.

Hayes, T.A. (2017). *Making sense of nature: A creative exploration of young people's relationship with the natural environment* (Unpublished doctoral dissertation). University of Cumbria, UK.

Humberstone, B., Prince, H. & Henderson, K.A. (Eds.) (2016). *International handbook of outdoor studies.* Oxford: Routledge.

Jeffs, T. & Ord, J. (Eds.) (2017). *Rethinking outdoor, experiential and informal education.* Oxford: Routledge.

Livingston, K. (2017). Undertaking commissioned research in education: Do research paradigms matter? In L. Ling & P. Ling (Eds.) *Methods and paradigms in education research* (pp. 206–219). Hershey, PA: IGI Global.

Locke, I.F., Spirduso, W.W. & Silverman, S.J. (2014). *Proposals that work: A guide for planning dissertations and grant proposals* (6th ed.). Thousand Oaks: SAGE.

Maxwell, J.A. (2011). Paradigms or toolkits? Philosophical and methodological positions as heuristics for mixed methods research. *Mid-Western Educational Researcher, 24*(2), 27–30.

McAbee, S.T., Landis, R.S. & Burke, M.I. (2017). Inductive reasoning: The promise of big data. *Human Resource Management Review, 27*(2), 277–290.

Onwuegbuzie, A.J., Johnson, R.B. & Collins, K.M. (2009). Call for mixed analysis: A philosophical framework for combing qualitative and quantitative approaches. *Journal of Multiple Research Approaches, 3*(2), 114–139.

Pike, E.C.J. & Beames, S. (2013). *Outdoor adventure and social theory.* Oxford: Routledge.

Sparkes, A. (1992). The paradigms debate: An extended review and a celebration of difference. In A. Sparkes (Ed.) *Research in physical education and sport: Exploring alternative visions* (pp. 9–60). London: Falmer Press.

Sparkes, A.C. & Smith, B. (2014). *Qualitative research methods in sport, exercise and health.* Oxford, New York, NY: Routledge.

Steier, F. (1991). *Research and reflexivity.* London: SAGE.

PART II

Qualitative methodologies – choosing an appropriate approach

4

PHENOMENOLOGICAL APPROACHES TO RESEARCH IN OUTDOOR STUDIES

John Telford

For researchers in the field of outdoor studies phenomenology has obvious appeal given the claims made about its potential for coming to some sort of understanding about what it is that participants experience and the meanings that they construct or interpret from their experiences. However, there are significant differences within the broad field of phenomenology that are important to be aware of in order to carry out research that is methodologically coherent and therefore trustworthy. Phenomenology is a term that has been variously employed to describe a philosophical tradition, a research paradigm, a social science standpoint, a framework for selecting research methods, and an organising theme within qualitative research (Brown, Jeanes & Cutter-Mackenzie 2014, p. 28; Creswell, 1994). As a result, there can be confusion surrounding the way the word is being used when talking about research. In this chapter I will consider two major uses of the term: a philosophical tradition and, following from that, a research methodology. As we shall see as the chapter unfolds there is great diversity within both of these areas. There is no one phenomenology. In order to offer guidance on using phenomenological approaches in research I outline some key aspects of the philosophical roots of phenomenology. This informs a discussion on the varieties of phenomenological approaches to research which in turn leads to a consideration of associated research methods.

Why phenomenology?

Before entering into a discussion of the philosophical roots of phenomenology I would first like to elucidate why it might be considered relevant to research in the field of outdoor studies. The concerns of research in outdoor studies often pertain to experiences that participants have had that are difficult to describe. Often what we are looking at, drawn to, wondering about, or trying to explain to others is an experience that we have had ourselves or have had reported to us by others that was in some way deeply meaningful. They are often the sorts of experience that are incredibly difficult to give expression to. Broadly speaking, phenomenology as philosophy gives primacy to subjective consciousness as a means of understanding experience and existence. In turn, phenomenological approaches to research seek to "explore, describe, and analyze the meaning of individual lived experience" (Marshall & Rossman, 2011, p. 19) and inquire into "the very nature of a phenomenon, for that which makes a some-'thing' what it is and without which it could not be what it is" (van Manen, 1997, p. 10). If we are looking for an approach to research that offers insights into particular experiences and the meanings that participants draw from them as they engage with the world then phenomenology seems to have the potential to do this. In the following section I will lay out the philosophical foundations that might justify such claims.

Philosophical foundations of phenomenology

Laying out the philosophical roots of phenomenology is no easy task. Moran (2000, p. 2) states that philosophers who are related in some way to phenomenology display significant diversity in their interests, in how they understand the key issues, in how they carry out the practice of phenomenology, and how they understand the relationship between phenomenology and philosophy more broadly. In this section I shall focus primarily on the work of Husserl (e.g. 1960; 1970a, 1970b) as he is commonly accepted as the founder of phenomenology as a philosophy, and in understanding his work we are able to better comprehend the directions in which others took phenomenology.

Husserl's original motivation to develop a philosophy of phenomenology was his deep concern about the vulnerability of the scientific method to produce valid and objective results (Romdenh-Romluc, 2011, pp. 5–7). The perceived vulnerability was based on Husserl's belief that a person's experiences are 'given' in such a way that she/he has immediate knowledge of them without need to rely on any presuppositions. I immediately 'know' pain and joy, I immediately know that I am seeing something or smelling something. However, the actual existence of that which is causing me pain or joy, or which I see or smell is not immediately given to me. I have to assume that they exist in reality and assume that the contents of my experience are a result of my interactions with those objectively, externally existing things. Therefore, because scientific investigation begins with a process of observing and measuring a world external to immediately given experience it is predicated upon an *assumption* that that which we experience does actually exist and therefore can be investigated. Husserl's response to this vulnerability as he saw it was to develop a philosophy that would provide science with a firm foundation – a justification for the existence of the world – from which it could operate to generate valid and objective knowledge.

The starting point for Husserl's project to provide science with a solid foundation had to be one that did not rely on presuppositions. Therefore, it had to begin with the content of human experience because only experience can be considered as being given immediately. Hence the use of the term phenomenology – a Greek root of *phenomenon* (meaning 'what appears') and *logy* (meaning 'study') – which leads us to understand phenomenology as the study of that which appears or, in philosophical terms, the study of that which appears to consciousness (Velasquez, 1997, p. 265). Furthermore, investigating the content of experience could only be through a process of description; not explanation. Any attempts at explanation would require hypothesising on the basis of presuppositions and conjecture.

Husserl's insistence on description may appear counter-intuitive. How could any attempt to produce objective knowledge use a method grounded in individual, subjective perspective? However, the sort of description that Husserl envisaged was not one of superficial activity in the world based on workaday thinking. Phenomenology would dig deeper by employing what Husserl referred to as the *Transcendental-Phenomenological Reduction.* This entails bracketing what Husserl referred to as the *natural attitude*: our everyday assumption that the world exists and functions in the ways that the presumptions of the science of Euro-centric culture tells us. The term natural attitude is coherent with that of *naturalistic inquiry*, which is applied to the natural and human sciences. These are processes of inquiry that work within a paradigm of Galilean science that deems only that which is quantifiable as real and assumes the world adheres to universal and measurable laws of nature.

So what does this mean exactly? How does this transcendental process work? The Transcendental-Phenomenological Reduction means that we bracket (also referred to as a process of epoche), or suspend, what we have developed as our natural attitude as regards the existence of the world. We do not adopt a complete disbelief, just a temporary suspension of belief. In doing so we also abrogate any attachment to descriptions of the world that adhere to the rationality of Western scientific logic or theory. What remains after doing this, according to Husserl, is pure experience. In this state, consciousness can be understood as the *Transcendental Ego*. Romdenh-Romluc (2011) describes this as "subjectivity that is not part of the causal order or dependent on any aspect of the world" (p. 8). This is the *Lebenswelt*; the world as we experience it prior to having our understanding of it narrowed by a restrictive Galilean understanding of rationality, where only the measurable is real and measurability is the defining characteristic of rational inquiry. Husserl's intention to set aside this form of rationality – due to it being incomplete as opposed to entirely inaccurate – offered a much more expansive opportunity to understand the world. Doing so allowed for the investigation of all the other aspects of existence that make up its holistic entirety: matters of value, issues of morality, questions of emotion, instances of creativity and spontaneous inspiration, and so on. Husserl's phenomenological project would also allow for the investigation of the essential structures of these aspects of the *Lebenswelt* that Galilean science condemned as either non-existent or unimportant.

However, carrying out Transcendental-Phenomenological Reduction is not sufficient to achieve the aims of phenomenology because it only reveals the particular structures of individual consciousness. By contrast, Husserl's intention was to provide a universal explanation of consciousness in general. To do this the phenomenologist must move towards discovering the *essential* qualities of consciousness through a subsequent process of *Eidetic Reduction*. This is another process of bracketing but this time it is a bracketing of the specific details that are particular to an experience. The process requires that one imagine the experience that is the subject of inquiry, identify a particular characteristic of that experience, vary that characteristic in some way, and then check to see if that has altered the nature of that experience beyond which it is no longer identifiable as an experience of that type. The researcher needs to continually vary each characteristic until the essence is arrived at and do the same for every imaginable feature of the experience under investigation. When this process is satisfactorily completed the researcher will have arrived at the essential structure of consciousness revealed through that experience.

In contrast to Husserl, Heidegger's hermeneutic phenomenology, also referred to as philosophical hermeneutics or existential phenomenology (Brown et al, 2014, p. 29), took phenomenology in a direction that placed greater emphasis on the interpretive nature of understanding lived experience. Heidegger asserted the continuity of consciousness with the world and saw the main task of phenomenology as investigating the meaning of 'being-in-the-world' (Cohen & Omery, 1994). His hermeneutic phenomenology moved away from an emphasis on pure description and embraced the agential involvement of humans in the world (Overgaard & Zahavi, 2009, p. 95). Personal experience is inexplicably implicated in the process of phenomenological philosophy; it is not possible to separate conscious thinking from the world in order to be an external spectator. Husserlian bracketing is replaced by a hermeneutic process of reciprocity between pre-understanding and understanding (Dowling, 2007, p. 134) which, when translated into research methodology, allows for the acknowledgement of influential pre-judgements as well as the benefit of contextual expertise in the interpretive act.

Merleau-Ponty (e.g. 1962; 1969) considered his work on phenomenology as extending Husserl's developments. Of particular note is his attention to continuing what he saw as Husserl's gradual distancing of phenomenology from its quasi-Cartesian roots (Romdenh-Romluc, 2011, p. 10). Husserl's *Lebenswelt* of his later work is conceptually much more subjective and inter-subjective in nature in comparison to the ideas of his early phenomenology that sought to produce universal truths in support of an objective science. The *Lebenswelt* is necessarily particular to place, time, context and culture. Nevertheless, he does not fully relinquish the transcendental aspect of his thinking (Overgaard & Zahavi, 2009, p. 93). Merleau-Ponty (1962) diverges from this Husserlian transcendental position with his realisation that, "The most important lesson which the reduction teaches us is the impossibility of a complete reduction" (p. xiv). Complete phenomenological reduction is impossible because human beings are not transcendental subjects. We are always enmeshed in the world. Our body/mind is always interacting with it, always in continuous and concurrent activity with it. In this we can see the influence of Heidegger's thinking on Merleau-Ponty. Merleau-Ponty does, however, continue with a revised concept of Transcendental-Phenomenological Reduction that brackets the Galilean scientistic perspective of considering only that which is objectively measurable to be real. Such an act reveals what Merleau-Ponty describes as the *phenomenal field* which, through a further process, leads to the *Lebenswelt.* Merleau-Ponty (1962) admits that this is no easy task: "Nothing is more difficult than to know precisely *what we see*" (p. 58). This challenge aside, it is important to recognise that Merleau-Ponty's phenomenology is not aiming at revealing the universal essences of consciousness through a Transcendental Ego as seen in Husserl's early work. Merleau-Ponty moves towards a position of understanding the meaning of the lived experience as experienced by the human subject.

In this section I have outlined the early foundations of phenomenology as seen in Husserl's work and some of the developments made by Heidegger and Merleau-Ponty. As a philosophy, phenomenology is concerned with providing descriptions of the general ontological and epistemological characteristics of experience and consciousness. In the next section I will discuss some of the ways in which phenomenology as philosophy translates into the sphere of social science research.

Phenomenology as research methodology

So where does this philosophical understanding of phenomenology direct us in terms of research? Well, first and foremost it puts individuals' experience of living at the front and centre of scientific inquiry. Scientific explanations, in Merleau-Ponty's (1962) opinion are second-order expressions. Each person's own experiences of phenomena, screened of distracting theoretical presuppositions, are first-order expressions.

> To return to things themselves is to return to that world which precedes knowledge … and in relation to which every scientific schematization is an abstract and derivative sign-language, as is geography in relation to the countryside in which we have learnt beforehand what a forest, a prairie, or a river is.
>
> *Merleau-Ponty, 1962, pp. viii–ix*

Moran (2000, pp. 2–4) states that a defining characteristic of phenomenology is its diversity; there is no consensus among philosophers as to what phenomenology is and correspondingly to how it is done. Unsurprisingly, this complexity carries over into phenomenology's adoption as a research methodology. Broadly speaking we can identify two streams of thought around and between which a variety of perspectives are taken. On the one hand, there is a Husserlian perspective that positions phenomenological inquiry as a purely descriptive process. On the other hand, there is a hermeneutic perspective that posits that living and being are always and unavoidably interpretive (Dowling, 2007, p. 134). This basic difference becomes manifest in the choice of methods of phenomenological inquiry employed and the degree to which they are employed. For example, the former perspective would emphasise the use of phenomenological reduction and free imaginative variation (similar to eidetic variation) to arrive at an understanding of the essence of an experience (Polkinghorne, 1989; Spiegelberg, 1982; van Manen, 1997). The latter perspective, by contrast, would engage in a more hermeneutic process of interpretation that draws on the knowledge and understanding that individuals bring to an experience. The question of emphasis reflects the foundational philosophical understanding. Can detached description of pure conscious experience of a phenomenon be arrived at in order to reveal the objective structures of a phenomenon? Or, is that epistemologically impossible and so the focus must be on an individual's subjective understanding of their experience of a phenomenon?

Methods of phenomenological research

For all that is written about phenomenology in the social sciences there is proportionally very little that gives clear guidance on how to carry out a phenomenological research inquiry (Eberle, 2012; Errasti-Ibarrondo, Diez-Del-Corral & Arantzamendi, 2018; Groenewald, 2004). One might well level that very accusation at this chapter. However, what I hope I have done in this chapter is provide sufficient clarification of key foundational principles that the reader may then use as a guide towards appropriate methods. Some guidance on methods is given below but they will need to be investigated in much greater detail in order to apply them appropriately to any given research inquiry. Other chapters in this book lay claim to phenomenological underpinnings and these, along with wider literature more generally, need to be carefully consulted. It is worth noting, however, that at least part of the reason for the lack of clear guidance regarding methods is intentional. Some authors (e.g. Hycner, 1999; van Manen, 1997, 2017a) express great concern about the dangers of undermining the integrity of phenomenological inquiry by focusing too much on specific methods. Their fear is that this will undermine a methodological responsiveness that is a characteristic of the philosophical foundation of phenomenology. Van Manen (2017b) states that, "the serious student of phenomenology should be cautious and shy away from simplistic schemes, superficial programs, step-by-step procedures and cookery book recipes that certainly will not result in meaningful insights" (p. 5). The intent is that researchers will ground themselves in the philosophical foundations and through this understanding develop their own methods that are appropriate to the inquiry. The philosophy is the method.

Where there is general agreement among phenomenologists is that language is at the heart of phenomenological inquiry – particularly written descriptions. Seidman (2006, p. 8) asserts that the use of language as a means of symbolically representing experience is a fundamentally human act, and throughout history it has been the primary method of making sense of lived experience. Very often phenomenological inquiry will involve interviewing as a prime means of generating data. Interviews are then transcribed. The interviewer aims to assist the participant in eliciting as detailed a description of the phenomenon as possible. Interviews would usually be open-ended and unstructured, giving both researcher and participant time to allow for in-depth exploration. What this means in terms of time will depend on the subject of the inquiry and the ability of the research participants to self-reflect and communicate their thoughts. This in turn has implications on who might be a suitable research participant. Van Kaam (1966) proposes that research participants be able to express themselves verbally without difficulty, recognise and verbalise emotions, and make connections between emotions and related experiences. Similarly, Colaizzi (1978) suggests that, "Experience with the investigated topic and articulateness suffice as criteria for selecting subjects" (p. 58). Given the difficulties that some participants may have with the requirements of Colaizzi (1978) and van Kaam (1966), it may well be that the researcher should expand their repertoire of possible methods. There are many experiences that are difficult to adequately translate into literary form even for articulate adults. When working with participants (adults or children) who are less adept with spoken or written language as a means of expression, it may be that arts-based approaches such as photography, sketching, painting, dance, sculpture, poetry (though literary it offers the opportunity to be less literal) should be explored as possible methods that provide a bridge towards the gradual, facilitated development of written accounts that can then be used for analysis. Interviews may also take place over two or three occasions with each session building on the previous one. Face-to-face interviews are one way of doing this, but it could equally happen via telephone or an online video link, or through an exchange of emails, or perhaps a combination of any and all. Colaizzi (1978) points to the use of several possible data sources: written descriptive protocols, dialogical interviews, observation combined with perceptual description, and imaginative presence leading to phenomenological description. Phenomenology is an approach that is open to any number of methods of data generation.

It will come as no surprise at this point to say that the role of bracketing or epoche is contested in phenomenological research. Who does it? Researchers? Participants? Both? Is it used before data generation begins or only before the data analysis? What defines bracketing? Is it simply a process of removing the researcher's preconceptions in an attempt to prioritise the participants' voices? Or is it more philosophically fundamental, implicating concepts of the phenomenal field and the *Lebenswelt* as intended by Husserl or Merleau-Ponty? From either perspective, one of the keys to revealing the full richness of an experience is through the adoption of a non-theoretical and unprejudiced standpoint leading to the creation of a relationship of deep understanding, empathy and indwelling with research participants. In so doing, the researcher is able to gain insights into the experience (and possibly the meanings they create depending on the degree to which a hermeneutic perspective is adopted). Crotty's (1996) observation that researchers who are nurses view bracketing as the defining feature of phenomenological research is one example of how different aspects of the broad phenomenological tradition are elevated or reduced in a given field. This leads to strong debates on issues of legitimacy and rigour (e.g. van Manen, 2017b). For example, one can see how, in the absence of other grounding, philosophically anchored principles, bracketing can become synonymous with the sort of researcher reflexivity (Dowling, 2007) that is a commonly expected element of almost any form of qualitative research. Differences of opinion regarding the role and centrality of bracketing are grounded in a fundamental difference of opinion on the primary focus of phenomenological research that again is rooted in philosophical principles. Is the aim to discover the essence of a phenomenon as experienced by human beings or is it to investigate the interpretive meanings that individuals ascribe to an experienced phenomenon?

When it comes to data analysis, whatever the methodological tradition being followed, van Manen's (2017c) following advice should be kept in mind: "It is not sufficient for an author to list some dubious themes that are primarily rephrased texts from interview transcripts as research 'findings' as is only too often done" (p. 776). Van Manen (2017b) also warns against the expectation that from a description of lived experience meanings and insight will simply pour out. This is not the case. Whatever the approach taken, phenomenological inquiry will require deep and close familiarity with the data. The collection of data sources will need to be read and reread until a clear sense of the experience is felt and then some form of bracketing and reduction or hermeneutic engagement will need to take place. Hycner (1999) outlines a process that includes bracketing the researcher's preconceptions and carrying out phenomenological reduction, delineating units of meaning, clustering units of meaning to form themes, and extracting general and unique themes which will contribute to the creation of a composite summary of the phenomenon. The phenomenological work that has emerged from the field of psychology, particularly from the Duquesne school at Pittsburgh University (e.g. van Kaam, 1966; Colaizzi, 1978; Giorgi, 1994) holds to a very Husserlian approach. The emphasis here is strongly on description of phenomena achieved through a process of intuiting essences as shown in consciousness (Dowling, 2007, p. 135). Polkinghorne (1989) summarises the approaches of van Kaam, Colaizzi, and Giorgi as consisting of three steps: dividing phenomenological descriptions into units; transforming units into psychological and phenomenological concepts; and bringing the transformed units together into a coherent description of the experience. However, this is a very summarised version. Colaizzi (1978), for example, details at least seven steps that relate specifically to working with participants' written protocols alone.

Conclusion

There is much more to discuss than space allows for in this chapter. For example, how does one manage the vast reams of data that could quite easily be generated in a phenomenological study? How exactly does one recognise an 'essence'? What exactly is a 'meaning unit' (Colaizzi, 1978)? What form should the final discussion or representation of the results of the study take? These are all fundamental questions and need to be thoughtfully responded to by examining the available literature, some of which is referred to here. It is axiomatic to say that the methods of inquiry or research must be appropriate to the matter or subject of our inquiry. However, the warning is perhaps particularly worth bearing in mind when attempting to carry out a phenomenological research inquiry due to the diversity of philosophical positions that inform an equally wide array of methodological practices. Ensuring the compatibility of philosophical principles with methodological perspective will allow for the selection and implementation of appropriate methods. My hope is that this chapter has gone some way to assisting in that understanding and will encourage anyone interested in phenomenological forms of research to engage deeply with as much of the available literature as possible in order to develop their expertise in what is a highly nuanced sphere.

References

Brown T., Jeanes R., Cutter-Mackenzie A. (2014). Social ecology as education. In B. Wattchow, R. Jeanes, L. Alfrey, T. Brown, A. Cutter-Mackenzie, J. O'Connor. (Eds.). *The Socioecological Educator* (pp. 23–45). Dordrecht: Springer. doi:10.1007/978-94-007-7167-3_2

Cohen, M.Z. & Omery, A. (1994). Schools of phenomenology: implications for research. In J.M. Morse (Ed.) *Critical issues in qualitative research methods* (pp. 136–156). Thousand Oaks, CA: Sage Publications.

Colaizzi, P.E. (1978). Psychological research as the phenomenologist views it. In R.S. Valle & M. King (Eds.) *Existential-phenomenological alternatives for psychology* (pp. 48–71). New York: Oxford University Press.

Creswell, J. (1994). *Research design: Qualitative and quantitative approaches*. Thousand Oaks, CA: SAGE.

Crotty, M. (1996). *Phenomenology and nursing research*. Melbourne: Churchill Livingstone.

Dowling, M. (2007). From Husserl to van Manen. A review of different phenomenological approaches. *International Journal of Nursing Studies 44*, 131–142.

Eberle, T. (2012). Phenomenology and sociology: Divergent interpretations of a complex relationship. In H. Nasu & F.C. Waksler (Eds.) *Interaction and everyday life. Phenomenological and ethnomethodological essays in honor of George Psathas* (pp. 135–152). New York, NY: Lexington Books.

Errasti-Ibarrondo, B., Jordan, J.A., Diez-Del-Corral, M.P. & Arantzamendi, M. (2018). Conducting phenomenological research: Rationalizing the methods and rigour of the phenomenology of practice. *Journal of Advanced Nursing, 74*, 1723–1734. doi:10.1111/jan.13569

Giorgi, A. (1994). A phenomenological perspective on certain qualitative research methods. *Journal of Phenomenological Psychology, 25*(2), 190–220.

Groenewald, T. (2004). A phenomenological research design illustrated. *International Journal of Qualitative Methods, 3*(1). Article 4.

Husserl, E. (1960). *Cartesian meditations: An introduction to phenomenology* (D. Cairns, Trans.). The Hague: Martinus Nijhoff.

Husserl, E. (1970a). *The idea of phenomenology*. The Hague: Martinus Nijhoff.

Husserl, E. (1970b). *The crisis of the European sciences and and transcendental phenomenology: An introduction to phenomenological philosophy*. (D. Carr, Trans.). Evanston, IL: Northwestern University Press (Original work published 1936).

Hycner, R.H. (1999). Some guidelines for the phenomenological analysis of interview data. In A. Bryman & R.G. Burgess (Eds.) *Qualitative research* (Vol. 3, pp. 143–164). London: Sage.

Marshall, C. & Rossman, G.B. (2011). *Designing qualitative research* (5th ed.). Thousand Oaks, CA: SAGE Publications.

Merleau-Ponty, M. (1962). *Phenomenology of perception*. (C. Smith, Trans.). London: Routledge.

Merleau-Ponty, M. (1969). *The visible and the invisible*. (A. Lingis, Trans.). Evanston, IL: Northwestern University Press.

Moran, D. (2000). *Introduction to phenomenology*. London: Routledge.

Overgaard, S. & Zahavi, D. (2009). Phenomenological sociology: The subjectivity of everyday life. In M.H. Jacobsen (Ed.) *Encountering the everyday: An introduction to the sociologies of the unnoticed* (pp. 93–115). Basingstoke: Palgrave Macmillan.

Polkinghorne, D. (1989). Phenomenological research methods. In R.S. Valle & Halling, S. (Eds.) *Existential-phenomenological perspectives in psychology. Exploring the breadth of human experience* (pp. 41–60). New York, NY: Plenum Press.

Romdenh-Romluc, K. (2011). *Routledge philosophy guidebook to Merleau-Ponty and phenomenology of perception*. London: Routledge.

Seidman, I. (2006). *Interviewing as qualitative research: A guide for researchers in education and the social sciences*. New York: Teachers College Press.

Spiegelberg, H. (1982). *The phenomenological movement*. Dordrecht: Martinus Nijhoff.

van Kaam, A. (1966). *Existential foundations of psychology*. Pittsburgh, PA: Duquesne University Press.

van Manen, M. (1997). *Researching lived experience: Human science for an action sensitive pedagogy*. London, ON: Althouse Press.

van Manen, M. (2017a). Phenomenology and meaning attribution. *Indo-Pacific Journal of Phenomenology, 17*(1), 1–12. doi:10.1080/20792222.2017.1368523

van Manen, M. (2017b). Phenomenology in its original sense. *Qualitative Health Research, 27*(6), 810–825.

van Manen, M. (2017c). But is it phenomenology? *Qualitative Health Research, 27*(6), 775–779.

Velasquez, M. (1997). *Philosophy: A text with readings* (6th ed.). Belmont, CA: Wadsworth Publishing.

5

A CRITICAL EXAMINATION OF THE PLACE OF INTERVIEWS IN OUTDOOR STUDIES RESEARCH

Allen Hill, Philippa Morse and Janet Dyment

Introduction

Interviews have long been a central feature of qualitative research. Armed with interview schedules ranging from a single open-ended inquiry through to highly structured questions, qualitative researchers have used interviews to gain insight into the social world. Indeed, Peraklya and Ruusuvuori (2011) suggest that "most qualitative research is probably based on interviews" (p.529) because interviews have an element of convenience, they allow researchers to access peoples' perspective and experiences, and help to overcome distance in time and space. The 'interview' is employed in so many different qualitative methodologies yet is less often singled out for critical examination on its own. This chapter takes up this challenge and encourages researchers in outdoor studies to carefully consider the role of interviews in their research.

The history of qualitative interviewing is one that can be characterised by contestation, change and now exploration (Fontana and Frey, 2005). Epistemological and ontological shifts have brought into question the legitimacy of positivist interview accounts that seek to portray a mirror reflection of the social world (Miller & Glassner, 2016) in favour of seeing interviews as situated, symbolic, discursive and representative narrative accounts. The way interviews are constructed, conducted and analysed has changed to better account for the contexts and subjectivities of interviewers and participants, including the more-than-human world. Interviews as personal narratives are "negotiated texts, sites where power, gender, race and class intersect" (Denzin & Lincoln, 2011, p. 416). The scope of interviews has expanded through researchers exploring creative ways of interviewing in order to include multiple diverse voices.

The implications of this history for outdoor studies research means that it is incumbent on researchers to not only understand the techniques of interviewing, but to also articulate a strong rationale for *why* interviews are appropriate and *how* the gathering, interpretation and representation of interview data articulates with the epistemological and ontological positioning of the researcher. Through this chapter we seek to stimulate researchers in the field of outdoor studies to critically examine the place of interviews in their research through three key questions:

1. What is the subject in/of the interview?
2. What are the possibilities for dialogue in the conception, conduct, and analysis of interviews?
3. How can interviews in outdoor studies take into account the agency of place and the more-than-human world?

We are mindful that there is a plethora of research methods texts which discuss interviews in depth. To minimise duplication, we have avoided description of different types of interviews, or the mechanics of conducting interviews. Rather, we challenge researchers with the questions above through looking critically at practical examples from our own research where we have employed interviews to explore the lived experience of educators, students, and the places they inhabit.

What is the subject in/of the interview?

An important consideration for researchers is to have an in-depth understanding of the role of the subjects in and of the interview process. In a humanistic sense, this forces us to consider the subjectivities of both the researcher and the participant(s) in an interview. In a posthuman sense, as we will discuss later in the chapter, the events and relationships between researcher, participants, *and* the more-than-human world are also crucial to consider.

The way knowledge production and research are perceived, conceived and received has changed greatly from the paradigm wars of the late 20th century. Historically, advocates of qualitative research had to fight for the legitimacy of data collection methods such as interviews, and to reconceptualise the very notion of data amongst critiques from positivist researchers committed to upholding the privilege of the scientific method. Amongst these 'battles' were debates about the validity and reliability of people's accounts of the social world creating tensions between objectivity and subjectivity within the research process. In the one corner, positivists couldn't reconcile how a qualitative researcher could come to any objective conclusion or 'truth' about the social world when it relied solely on the subjective viewpoints of people. In another corner, constructivists and post-structuralists contested that the very notion of objective truth obtained through interviews was not only out of reach, it was nonsensical; that all knowing was inextricably bound historically, politically, socially and culturally (Fontana & Frey, 2005).

Such tensions tended to dominate the historical development of qualitative research and therefore interviewing as a method is well documented across journals such as *Qualitative Inquiry* and the *International Journal of Qualitative Studies in*

Education. More than 20 years ago Scheurich (1995) provided a postmodern critique of interviews as a realist account and warned against decontextualised and sanitised representations of the participant subject through interview text. The wrestle with interviews as a key method in qualitative knowledge has also been present through multiple editions of the *SAGE Handbook of Qualitative Research* (Denzin & Lincoln, 2011). One chapter by Fontana and Frey (2005), for example, deals specifically with the historical development of the interview, charting its trajectory away from a supposedly neutral tool used to gather objective data for scientific purposes to one that is inherently contextual, complex, subjective, and politically laden. Fontana and Frey go as far as labelling the goal of scientific neutrality in interviewing as "largely mythical" (p. 116).

Miller and Glassner (2016) wrestle with the notion of finding realities in and through interviews in a way that moves beyond objectivist-constructivist dualisms. In doing so they acknowledge that no research can provide a "mirror reflection of the social world" (p. 53) yet they remain firm in their belief that interviews are still a valuable tool to gather evidence of "what happens" (p.52) within social worlds. In taking this position, Miller and Glassner argue that interviews can result in knowledge of the social world that goes beyond the subjective interactions between the interviewer and the interviewee. That is, the stories that interviewees share about their perspectives, experiences and beliefs have value in helping to better understand social worlds.

The ability of interviews to capture an external reality, authentic account, or naturally occurring data must be carefully scrutinised as researchers approach interviews. In a note of caution, Miller and Glassner (2016) argue that qualitative researchers should abandon the dualistic tendency to classify interviews as either entirely subjective or objective. Such ambiguity may not be a bad thing. It may help researchers to recognise that any 'reality', or any individual's experience or construction of reality, has to move beyond just a singular description of that reality. This recognition forces researchers to carefully consider the *subject*, as historically, socially, culturally and linguistically constituted.

Here the work of St Pierre (2011, 2014), and recent developments related to post-qualitative research, are useful to help us better understand the subject in qualitative interviewing. In her critique of conventional humanist qualitative inquiry, St Pierre (2011) is highly critical of a conception of human beings as subjects who 'know' primarily through rational deduction. Rather, St Pierre positions the human being subject as *entangled*.

> The implications of entanglement are staggering. If one no longer thinks of oneself as "I" but as entangled with everyone, everything else – as haecceity, as assemblage – what happens to concepts in social science research based on that "I" – the *researcher*, the *participant, identity, presence, voice, lens, experience, positionality, subjectivity, objectivity, bias, rationality, consciousness, alienation, reflexivity, freedom, transformation, dialogue?* In space-time how does one think of *research design, research process, timeline, narrative, cause and effect*

> … In entanglement, how does one think about face-to-face methods like *interviewing* and *observation,* methods that privilege *presence.*
>
> *p. 619*

Considering the interviewee as entangled may help us deal with those earlier critiques of Scheurich (1995) warning against decontextualising the interview subject in a futile search for realist accounts of the social world. Researchers themselves are also entangled subjects as St Pierre points out, which has implications for many of the concepts that are taken for granted in qualitative social science research. The implication for researchers employing interviews therefore, is to recognise that, not only are individuals shaped by their social worlds, but that they are entangled with everyone and everything, including the more-than-human world.

In the case studies following we are mindful that we have not fully engaged the implications of St Pierre's (2011, 2014) theoretical provocations and highlight that post-qualitative ideas are further explored in, for example, Chapter 1 and Chapter 17 of this handbook. However, a useful starting point for researchers to employ interviews in outdoor studies is to carefully consider the role of the subject in the interview design, conduct and analysis. This can involve critically reflecting on questions such as:

- Who/what are the subjects in interviews?
- How is the entanglement among researcher, participant and the more-than-human world unpacked, acknowledged and represented?

In an effort to consider these questions in practice, the following two sections explore ways that subjectivity and entanglement might be considered in the interview process. The first considers the possibilities of dialogue in interview-focused data collection and analysis. The second unpacks how interviews might take account of the more-than-human world.

What are the possibilities for dialogue in the conception, conduct and analysis of interviews?

One possibility that might help us to unpack issues of subjectivity in research is to consider how dialogue might feature in the conception, conduct, and analysis of interviews. Most people would consider a conversation between two people dialogue. But do researchers consider interviews to be dialogues? This may seem a simple question but it has multiple layers. For example, what would constitute a dialogical interview? Why might dialogue be important in an interview anyway? When does the dialogue start or finish? Should analysis also be dialogical? Who are the 'subjects' that are part of the dialogue? And perhaps most importantly, how does power influence dialogue in interviews?

The notion of dialogue is not new. As interviews came to be accepted as a data collection method in the 1980s, they were often conceived as a more progressive, personalised, authentic, relational and dialogical way to explore a subject's world than positivist questionnaires or experiments (Kvale, 2006). Those familiar with the critical qualitative inquiry work of Carspecken (1996) will be aware of his advocacy for collection of dialogical data, where the researcher and participants work collaboratively to create authentic narratives. Likewise, authors such as Kincheloe and McLaren (2005) whose thinking around critical pedagogy and qualitative research urged researchers to adopt an evolving criticality which reflexively unpacked the ways in which power circulates in the research process. More recently, Sullivan (2012) drew on Michael Bakhtin's ideas to propose a dialogic approach to qualitative data analysis. A common theme through this literature is employing dialogue as a tool and mindset to help researchers negotiate subjectivity and power dynamics in the research process, particularly in interviews.

Caution is needed to avoid thinking that all interviews will automatically be dialogical as argued by Kvale (2006) who is critical of common conceptions of interviews as dialogues. A dialogue, according to Kvale, is a "joint endeavour where egalitarian partners, through conversation, search for true understanding and knowledge" (p. 483). This rendering of dialogue, however, as an egalitarian search for truth is problematic in that any claims of equality and reciprocity between interviewers and interviewees can surely only be mythical, and that any search for truth is always contingent on the entangled and contextual social world of the subject(s). Kvale points out that power dynamics are always at play in the interview. Rather than being dialogical, he suggests most interviews entail a "hierarchical relationship with asymmetrical power distribution" (p. 484) where there is a "one-way dialogue, and instrumental and indirect conversation, where the interviewer upholds a monopoly of interpretation" (p. 484). With the provocations of Kvale (2006) in mind, let us explore a case study of an attempt to use dialogue in a qualitative interview and analysis process from one of the doctoral study by one of the authors (Hill, 2011).

Case study 1: Dialogical interviews in the re-envisioning of outdoor education

Hill's (2011) doctoral project worked collaboratively with a group of ten secondary school outdoor education teachers in Aotearoa New Zealand to critically reflect on their practice and take action to incorporate education for sustainability (EfS) in their programmes and practice. Based on a bricolage methodological framework (Kincheloe and McLaren, 2005), incorporating critical ethnography and participatory action research, the study utilised a dialogical interview process. The project was grounded on values of collaboration, reciprocity and reflexivity in an attempt to produced findings by mutual dialogue. Consequently, the research design sought to privilege dialogue in the planning, conduct and analysis of interviews.

The project was multi-phased and included individual interviews, group professional learning and development (PLD) workshops/interviews, informal conversations, action projects and critical reflections. In the first phase of the project, individual semi-structured interviews were conducted. These interviews did not vary much from standard semi-structured interviewing protocols as discussed in research methods texts such as Creswell (2012). Perhaps the only thing to distinguish these interviews was the mindset of the researcher to treat teacher-research-collaborators as colleagues and respected equals; that is, throughout the research process there was a keen awareness of the asymmetric power dynamics that often accompanied interviews. Certainly, as the interviews were conducted, there was a conscious effort to ensure they were underpinned by 'mutual dialogue'.

Yet there was a feeling of dissatisfaction with attempts to be dialogical through semi-structured interviews; after all it was mostly the researcher asking all the questions. So strategies were put in place to engage in more meaningful dialogue with and between the teacher-research-collaborators through the use of a sticky note interview activity in a workshop that followed the initial interviews. After some preliminary coding analysis of all of the interview transcripts, a series of themes were identified (e.g. cultural assumptions in outdoor education, connection to place, cross-curricular thinking). Selected quotations from the initial interviews were collated onto themed posters and placed around a room for part of the first workshop. Teachers were given two hours to circulate in small groups or pairs around the posters to read, discuss and place comments, questions and clarifications via sticky notes on the posters. This type of sticky note activity is not a documented or common form of 'interview'; however, variations are often used as teaching and learning methods in education.

The sticky note activity was followed by a group discussion. The teacher-research-collaborators and researchers circulated around each of the posters together and engaged in dialogue about what had been stimulated by the sticky note activity. These dialogues were recorded and selectively transcribed to add richness to the themed analysis from the initial interviews.

This approach encouraged collaboration and engagement with the interview analysis by teachers. What is important to highlight here is the use of innovative ways of both conducting 'interviews' and analysing interview data with participants in ways that were dialogic. In doing so, the boundaries of what constitutes an interview can be pushed, and knowledge construction can become participatory, emergent and collaborative.

Such boundaries can be pushed in other ways. What has been described above deals only with interactions between human subjects in interviews. Increasingly in outdoor and environmental education contexts, we are witnessing a greater commitment to inclusion of the more-than-human world in both pedagogical and research considerations. The possibilities that concepts of place and posthumanist theories can have for interviews in outdoor studies is profound. It is to this we now turn our attention.

How can interviews in outdoor studies take account of place and the more-than-human world?

Given the central role of 'place' in outdoor studies it is critical to understand the ways in which place might be positioned within interviews. Such considerations raise questions about the way in which places might be 'listened to' but also the centrality and positioning of the human gaze. Gough (2016), for example, highlights an apparent contradiction in relation to much outdoor and environmental education research:

> On the one hand, many outdoor and environmental activists, philosophers, and educators view anthropocentrism as an undesirable ethical position and valorise conceptual alternatives signified by terms such as 'biocentrism' and/ or 'ecocentrism'. On the other hand, many reports of outdoor and environmental education research privilege an anthropocentric gaze, which assumes autonomous human subjects as starting points for knowledge production and the focus of attention for data production and analysis.
>
> *pp. 58–59*

Decentring of the human in the practice of outdoor studies research (via interviews and other methods) should not be underestimated; indeed, it is a difficult task. Gough's (2016) provocation also suggests further possibilities that move away from traditional methods such as interviews; however, we highlight here, that when considered critically and in light of such provocations, interviews still have much to offer. So how might interviews shift away from human-centredness? How might we design interviews to go beyond a primary focus on human interactions? Here we can draw inspiration from posthuman theorists, who help us find new ways of imagining and providing agency to the more-than-human world.

By rejecting the essential individual at the heart of humanist research, posthumanist approaches allow the reimagining of research methods and methodologies. Taylor (2016), for example, suggests that "posthumanism opens ways of researching that seek to undo tired binaries such as theory/practice, mind/body, emotion/ reason, human/nature, human/animal producing instead heterogeneous knowledge pathways" (p. 7). By adopting a posthuman perspective towards interviews, ideas of studying phenomena (human or non-human) in isolation are unsettled; instead, awareness emerges that phenomena are multiple, subjective, and arise from complex relations (Ulmer, 2017). Interviews, then, shift from traditional modes of inquiry – that connect causality/reliability and validity – to inquiry that asks what is knowledge and how does knowledge come to be? Adsit-Morris (2017) suggests a level of humility is required to make this epistemological and ontological shift:

> It is what Barad calls the "irreducible entanglements of response- ability". It requires being attentive to who we gather up, live-with, think-with, become-with; it requires being response-able to the beings/figures/narratives we use to narrate and live our collective lives with.
>
> *p. 106*

A posthuman approach shifts the focus from an interpersonal interaction to a "more mangled orientation between bodies, things and sensations" (Springgay, 2014, p. 79), generating a potentially broader ethical research practice. In this way, the focus shifts from interpersonal interactions to entanglements with and between humans and the more-than-human. For Ulmer (2017), a posthuman approach radically alters what is possible in research methods, suggesting that if humans alone are not the only subjects/objects worthy of study, then an abundance of possibilities arises. With this in mind, and in light of considering interviews, we must think hard about the status of human subjectivity, ethical entanglements, values and cultural norms that are suitable for our times.

Case study 2: Interviews with(in) and through Kooyoora National Park

In the initial stages of a doctoral research project, one of the authors (Morse) explored possibilities for imagination in outdoor environmental education through a posthuman lens. In doing so, she set out to consider how pedagogical events might contribute to practices that decentre humans and respond to human exceptionalism. The early research involved exploring events involving Kooyoora National Park (central Victoria, Australia), pre-service teachers and primary school students on a four-day outdoor education programme.

Interviews were conducted in a conversational manner (short 2–10 minutes) where opportunities arose and in the contexts in which they occurred. Crucially, an alternative focus on the complexity and disruptions operating within messy entanglements of human and more-than-human encounters was undertaken. Working with posthuman concepts, and based on the Deleuzo-Guattarian (1987) concept of *assemblage*, interview methods were employed that attempted to explore beyond traditional discursive data. Imaginative arts-based education sessions were developed by pre-service teachers to engage learners, and possibilities for understanding complex and messy entanglements involving the material surrounds were included and implicated in interviews.

As an example, at one stage a group of primary school students encountered a pair of diamond firetail finches; incredibly eye catching with their brilliant red beaks and white polka dots, the birds were observed carefully, quietly and purposefully building their nest. One of the primary school students asked, "how was this possible?" This led to a collective conversation, trying to understand what it was to be a finch in this place. As they watched, students were drawn into the careful weaving of the finch/grass/hedge wattle assemblage. The students then went on to explore various materials and methods that finches might use to construct their nests to create their own art-based response – in some cases a nest (Figure 5.1).

During this arts-based session, a number of short interviews were undertaken that attempted to draw the surrounding environs and the material artefacts into the conversation – to search for ways in which the place itself acted/emerged as a point of departure to drive not only the conversation but further explorations; to be attentive to the way in which the material surrounds were active constituents in the interview event. In this way, the encounter and material making of the bird nests brought into being the interviews with no basis for one to necessarily hold

FIGURE 5.1 Fairy grass nest

Source: Image taken by Morse (2017) during Diamond Firetail Finch encounter at Kooyoora State Park

a privileged status. What, then, might happen if equal attention was paid to the material or sensuous interactions that guided the conversation? How might more-than-human surrounds invite attention and guide events through their very history, texture, sound or appearance?

Such an approach to interviews required two key undertakings. First, it required continuously thinking *with* and *through* posthuman concepts such as, in this case, Deleuze and Guattari's (1987) *assemblage* and Bennet's (2010) *vibrant matter*. Second, it involved approaching interviews with an attentiveness to the possibilities for more-than-human affect; to be implicated within the place and actively open to its agential nature. As Bennet (2010) suggests, "the capacity to detect the presence of impersonal affect requires that one is caught up in it. One needs, at least for a while, to suspend suspicion and adopt a more open-ended comportment. If we think we already know what is out there, we will almost surely miss much of it" (p. xv). The way in which the interviews unfold or themselves might act as points of departure in their own right, cannot be predicted, yet the very act of undertaking interviews in this way offers an exciting set of possibilities.

A posthumanist approach to research methods also demands alternative approaches to analysis. In this case, rather than a traditional coding approach, a mapping of messy entanglements of encounter was attempted in a way that was not about defining what the entanglements *were*, but, rather, what they might *produce* or *do*. Thus, the unit of reference for thinking and analysis becomes an assemblage or entanglement rather than the individual self, so that "rather than analyzing data in a search for meaning or attaching data in the form of interview transcripts to a specific subject who can speak the essence of experience, this methodological approach positions data and voice as agents in their own right" (Mazzei, 2013, p. 739).

Conclusion

By this point the reader may be asking – where to from here? We recognise that researchers approaching the use of interviews in their research design will be informed by a variety of theoretical and methodological ideas or frameworks. Through this chapter we are not arguing that researchers should adopt a dialogical or posthumanist approach to interviews. We do, however, encourage readers to more critically examine the role of interviews in their research and return to the questions that we asked in the introduction. We hope that the theoretical insights and examples of different ways of engaging with interviews stimulate thinking around these questions. As research continues to evolve, it is incumbent on researchers to continually reflect on the assumptions and practices that inform their research. We believe it is no longer appropriate to take for granted the role that interviews play in research. Interviews are not a straightforward data collection method. When used, interviews are inevitably entangled with and in the social and more-than-human worlds. Teasing out such entanglement in different ways for different inquiries and contexts is an exciting challenge that opens possibilities for outdoor studies researchers.

References

Adsit-Morris, C. (2017). *Restorying environmental education*. New York, NY: Palgrave Macmillan.

Bennet, J. (2010). *Vibrant matter: A political ecology of things*. Durham, NC: Duke University Press.

Carspecken, P.F. (1996). *Critical ethnography in educational research: A theoretical and practical guide*. New York, NY: Routledge.

Creswell, J.W. (2012). *Educational research: Planning, conducting, and evaluating quantitative and qualitative research* (4th ed.). Upper Saddle River, NJ: Pearson Education/Merrill Prentice Hall.

Deleuze, G. & Guattari, F. (1987). *A thousand plateaus: Capitalism and schizophrenia*. Minneapolis, MN: University of Minnesota Press.

Denzin, N.K. & Lincoln, Y.S. (2011). *The SAGE handbook of qualitative research* (4th ed.). Thousand Oaks, CA: SAGE.

Fontana, A. & Frey, J.H. (2005). The interview: From neutral stance to political involvement. In N.K. Denzin & Y.S. Lincoln (Eds.) *The SAGE handbook of qualitative research* (3rd ed. pp. 115–159). Thousand Oaks, CA: SAGE.

Gough, N. (2016). Postparadigmatic materialisms: A "new movement of thought" for outdoor environmental education research? *Journal of Outdoor and Environmental Education, 19*(2), 51–65.

Hill, A. (2011). *Re-envisioning the status quo: Developing sustainable approaches to outdoor education in Aotearoa New Zealand* (PhD thesis). Dunedin: University of Otago.

Kincheloe, J. & McLaren, P. (2005). Rethinking critical theory and qualitative research. In N.K. Denzin & Y.S. Lincoln (Eds.) *The SAGE handbook of qualitative research* (3rd ed., pp. 303–342). Thousand Oaks, CA: SAGE.

Kvale, S. (2006). Dominance through interviews and dialogues. *Qualitative Inquiry,* 12(3), 480–500.

Mazzei, L. (2013). A voice without organs: interviewing in posthumanist research. *International Journal of Qualitative Studies in Education, 26(6)*, 732–740.

Miller, J. & Glassner, B. (2016). The "inside" and "outside": Finding realities in interviews. In D. Silverman (Ed.) *Qualitative Research* (4th ed.). Thousand Oaks, CA: SAGE.

Peraklya, A. & Ruusuvuori, J. (2011). Analyzing talk and text. In N.K. Denzin & Y.S. Lincoln (Eds.), *The SAGE handbook of qualitative research* (4th ed., pp. 529–544). Thousand Oaks, CA: SAGE.

St Pierre, E.A. (2011). Post qualitative research: The critique and the coming after. In N.K. Denzin & Y.S. Lincoln (Eds.) *The SAGE handbook of qualitative research* (4th ed., pp. 611–626). Thousand Oaks, CA: SAGE.

St Pierre, E.A. (2014). A brief and personal history of post qualitative research. *Journal of Curriculum Theorizing, 30*(2), 2–19.

Scheurich, J.J. (1995). A postmodernist critique of research interviewing. *International Journal of Qualitative Studies in Education, 8*(3), 239–252.

Springgay, S., (2014). "Approximate-rigorous abstractions": Propositions of activation for posthumanist research. In N. Snaza, J. Weaver (Eds.) *Posthumanism and educational research* (pp. 90–102). New York, NY: Routledge.

Sullivan, P. (2012). *Qualitative data analysis: Using a dialogic approach*. Los Angeles, CA: SAGE.

Taylor, C.A. (2016). Edu-crafting a cacophonous ecology: Posthumanist research practices for education. In C. Taylor & C. Hughes (Eds.) *Posthuman research practices in education* (pp. 5–24). Hampshire: Palgrave Macmillan.

Ulmer, J.B. (2017). Posthumanism as research methodology: Inquiry in the Anthropocene. *International Journal of Qualitative Studies in Education (QSE), 30*(9), 832–848.

6

METHODS AND TECHNIQUES FOR CAPTURING EMPIRICAL MATERIAL FROM EXPERIENCES AND STORIES IN OUTDOOR SPACES AND PLACES

Heidi Smith

Introduction

The gathering and creation of physical data in outdoor contexts challenges the researcher in terms of accessibility of individuals/groups, capturing informal spontaneous conversations as empirical material, structured vs. unstructured document collection, and maintaining the physical technology to record the data in the field often, although not always, for extended periods of time in a range of environments and conditions. Rationales for each method, their distinguishing features, strengths and weaknesses, along with benefits and a guide for undertaking each approach are elucidated. Methods are presented in such a way that readers can replicate or modify them to suit their own research contexts. Tried and tested techniques for recording data in the wilds, in terms of technology, where battery life and weather conditions challenge the researcher, are of particular focus given the outdoor context. The range of data collection techniques and methods, challenges, issues and tension, provide the researcher with several options when considering empirical material generation from experiences and/or participants' stories in the outdoors.

Underpinning paradigm

While paradigmatic differences across qualitative research exist, collecting data through observations, intentional conversations and document collection are similar (Hatch, 2002). Underpinning the methodological approaches recommended in this chapter lies the interpretative paradigm (Sparkes, 1992) and suits researchers concerned with understanding the "feelings and world views" (Neumann, 1997, p. 73) of participants through their experiences and stories shared (Freedman, 2006). The interpretative paradigm allows for asking how meaning is constructed and social interactions negotiated in outdoor programmes and associated activities/

learning experiences (Scott & Usher, 1999). Participant observation, intentional conversations and document collection explore the participants as they are (Punch, 2009; Sparkes, 1992) in context (Creswell, 2007; O'Donoghue, 2007) and take into account social interactions with peers, students and the participant observer in the construction of knowledge and understanding (Denzin & Lincoln, 2018; Mason, 2002; Neumann, 1997).

In the outdoor context, researchers often take on the role of participant researcher and directly take part in the activities/programme/learning experiences outdoors, while collecting empirical data at the same time as observer. The number of days spent in the field and the natural environment together often play a significant role in determining the materials that are collected, and not collected (Hatch, 2002, Simons, 2009; Wolcott, 1995). While in the field, the researcher makes decisions about "who to talk to, where to be, and when to be in certain places" (Sparkes, 1992, p. 33), as well as when to be absent. This chapter presents rationales, distinguishing features, strengths and weaknesses, along with benefits and lessons learned for undertaking: participant observation, intentional conversations and document collection. We begin with dominant ethical considerations that underpin these data collection methods.

Ethical considerations

Before moving to each of the data collection methods specifically, it is important to address the dominant ethical considerations which inform participant observation, intentional conversations and document collection. Prior to entering the outdoor research context, significant attention to ethical considerations of the individual context, its participants and the personal biases you bring with you into the field is essential (Takyi, 2015). While it is not possible to address all ethical considerations, I acknowledge the need for thoughtful considerations of ethical considerations specific to each outdoor context. The three presented here have been found to be essential when conducting participant observation, intentional conversations and document collection in the outdoor learning research context.

Expertise and experience

When an outdoor educator as researcher enters the field to observe an outdoor learning programme, they come to the research context with particular biases from personal experience and expertise, and these impact how they see the outdoor learning context, including its actors. They need to be aware that their own presence, no matter how engaged and involved they are as a participant, remains fraught to some degree as their main role is to observe and gather data from the participants, the place and the interactions that occur (Hatch, 2002). One way to contend with such reflexivities is to keep a reflexive journal in an attempt to bracket such perceptions and subjectivity (Cope, 2014). Similarly, the researcher needs to be aware of personal judgements they bring with their expertise.

Timing of data collection

Collecting data in the outdoor learning context, more often than not, occurs over a number of days dependent on programme length and place. For example, if the programme or people under study are on an extended journey, the researcher will most likely need to be present for the whole journey as exiting may not be possible due to place. While this is not always the case, since much outdoor learning happens locally and in easier to access and exit locations, participants may prefer that the researcher is present as a participant, however that role is outlined at the outset. Therefore, it is important to be clear to participants and/or the group you are observing how you will go about your observation. This could include whether you will be observing all of the time or some of the time and, if this changes, how you will let them know.

Impact of researcher on participant experience in an outdoor programme

While the presence of the researcher and the impact this has on the participant experience cannot be ignored, their presence ideally does not necessarily result in negative outcomes for participants – rather by being a part of the research process, there is potential for the participants' experience to be enhanced. For the researcher, being mindful of your impact, and taking time to be absent from the group to allow participants to be unobserved for a while ensures that the research does not overshadow the participant experience of the outdoor programme. It is essential that you answer all questions truthfully and maintain an open line of communication to all participants. When designing consent, consider the use of verbal consent as you go through the observation period, giving individuals the opportunity to provide or retract consent as you go. Regular checking of transcripts and field notes can be conducted as you go or after the observation period is concluded (Mertens, 2010). With these ethical considerations in mind, we turn now to the three specific methods of data collection: participant observation, intentional conversations and document collection.

Participant observation

Observation as a method requires the researcher to systematically look at and take note of people, places, events, behaviours and routines (Angrosino, 2007; Cohen, Manion & Morrison, 2011). It also provides the researcher with a comprehensive "sense of settings" and documented events that allow the researcher to develop "rich descriptions" (Simons, 2009, p. 55). Essentially, observation involves observing what people do, listening to what they say, and asking clarifying questions (Gillham, 2000) to understand the phenomena under study (Denscombe, 1998; Simons, 2009) in a specific context, over a period of time (Takyi, 2015). Seeing what would normally go unnoticed, observation provides the researcher with first-hand data instead of relying on second-hand accounts: "what people do may differ from what they say they do" (Cooper & Schindler, 2001, p. 374). Observations are systematic, explicitly guided by a theoretical or conceptual framework (Bechhofer & Paterson, 2000) where observation schedules and checklists are used to maintain the focus of the observation (Mertens, 2010; Smith, 2011).

In the participant observer role, the researcher aims to develop an "intimate familiarity with each participant and the related setting" (Smith, 2011, p. 86) by engaging in active participant observation, doing "what the others do" (Mertens, 2005, p. 382) in the outdoor setting, "without blending in entirely" (Smith, 2011, p. 86). In order to be successful, observers need to be able to participate in the outdoor activities/programme under study while simultaneously maintaining a professional distance to allow for observation and recording data (Smith, 2011; Takyi, 2015; Yin, 2014). The empirical data collected is dependent on the outdoor context, the outdoor programme, the amount of time spent observing, how many instances of observation there are in the research project and how the researcher looks or sees.

Strengths of participant observation as a method include its sensitivity to context, how it moves beyond perception-based data that interviews provide and captures what might otherwise be unconsciously missed (Denzin & Lincoln, 2018). It is highly flexible, provides access to interactions within a social context and records these. Participant observation complements other forms of data collection such as intentional conversations, interviews and document collection which together provide possibility for triangulation of data (Cohen, Manion & Morrison, 2011). The prolonged engagement of the researcher often results in a high trust relationship between researcher and participants (Lincoln & Guba, 1985), with participants acting 'normally', in a way that may not be possible if the researcher were to sit back from the group and observe (Takyi, 2015). Through this intimate engagement with the group, the participant observer becomes privy to conversations and actions that may not be heard or witnessed if not ensconced within the group. The researcher can be flexible in terms of when to be present and when to be absent. Not always being present collecting observation data helps to reduce and avoid excessive data collection and provides the researcher with time to consider materials collected and what is still required (Sparkes, 1992).

Fallibility of the researcher, the potential to be highly selective in the observation process, and getting caught up in the group process, the activity and or teaching/leading in the outdoor programme being observed, have all been identified as potential weaknesses of this method (Denzin & Lincoln, 2018; Hatch, 2002, Takyi, 2015). In the outdoor context it can be difficult to observe groups in their entirety if the activities mean that the group breaks apart, separates, or is difficult to access. Depending on whom the researcher is observing and for what purposes, the decision on where to place themselves is significant as it may not be possible to change this, perhaps for the whole day or whole outdoor programme being researched.

As participant observer, it is important the researcher is well practiced in the skills of observation prior to entering the field. Hatch (2002) provides a list of skills, which include building skills in paying close attention to what is happening, developing memory skills to retain what has been observed until you are able to write it down/record it, practicing suspending judgement and remaining open to what is observed, developing writing skills, and learning to be aware of your own biases and how these impact what you observe. In addition, when undertaking participant observation in the outdoor context it is important to consider: the activity, the place, access to technology to record observations, timing and your role as researcher within the group. An approach of being seen but not heard can be effective, whereby the researcher blends into the group as a participant observer. This provides another way of lessening researcher impact on participants.

When conducting participant observation in the outdoor context, much of the research space (e.g. in terms of environment and people) is out of the control of the researcher. While the researcher may enter the outdoor context with a specific plan for data collection, it is essential that the researcher is prepared with several back up plans to ensure the time is not wasted for either participant or researcher. The following highlights a few areas that need to be considered to ensure that effective modifications and problem solving occur seamlessly in the field.

The first thing to be clear on is the research question(s). It is recommended that the researcher have a few copies of their research question(s) printed and laminated for use in the field. This helps to keep the researcher focused and, at any moment, they can remind themselves what it is they are trying to answer. This will become instrumental in the decisions made about how long to observe, when to come and go from the context, and when to stop. Equally this will affect where you situate yourself, who you listen to or observe, how long you observe for, and when you choose to leave the field. Recording the reasons why you stopped, or began observing, are indispensable for your research. Prior to entering the research context, carefully think through the process of observation, from a clear introduction to participants/groups covering all aspects of the participant observation through to identifying challenges the researcher may face, and solutions to potential barriers to observation. Be open to changes, flexible, and willing to forego data if the group/individual learning requires. Intentional conversations support participant observation by enabling data collection from unstructured conversations that occur in the outdoor context.

Intentional conversations

Hatch (2002) describes informal interviews as "unstructured conversations" that while "*informal* are not without purpose nor undertaken randomly" (Hatch, 2002, p. 92). They include conversations with participants to discuss what is happening, as it happens (Sparkes, 1992), without disrupting the flow of what is being observed. The term *intentional conversations* has been chosen for this chapter as it best reflects what happens during an observation in the outdoor context. Intentional conversations may result in thick descriptions (Gillham, 2000) or "accounts" (Cohen, Manion & Morrison, 2011) which can be later crossed checked during interviews with participants for accuracy. Similar to observations, the researcher needs to be able to pay close attention to what is happening in the research context, listen deeply and ask pertinent questions there and then. Note taking is completed away from the intentional conversation, and audio recording is only done in cases where it will not disrupt the conversation (Hatch, 2002).

Intentional conversations advocate a two-way interchange instigated by either researcher or participant. They are not the primary source of data, rather they complement the participant observation and can provide direction for structured or semi-structured interviews that may follow the observation period. As purposeful conversations they take advantage of "the immediate context" and encourage participants to "reflect on what they have said, done or seen" (Hatch, 2002, p. 92) in the moment, or shortly following. Intentional conversations provide the researcher with the opportunity to invite participants to explain why they did what they did and an opportunity to share thoughts, ideas and motivations for what is happening as it happens. These intentional conversations provide the space to collect big and small stories (Phoenix & Sparkes, 2009) of experience in authentic ways.

In outdoor programmes, intentional conversations provide salient opportunities to collect rich data from the context as it is happening. They appear to the researcher as 'teachable moments' might appear, and can be easily missed if the researcher is not paying close attention. Therefore they go hand in hand with participant observation and act to further enrich the empirical data collected. The very nature of the intentional conversations makes them appear similar to the facilitation and teaching/learning process usually adopted in outdoor programmes and thus they appear a 'natural' part of the programme to participants. Intentional conversations provide a form of triangulation for the researcher as they help to confirm or disrupt what the researcher has observed. In the introduction of the researcher to the context, this is an opportunity to explain and clarify these intentional conversations so all participants are clear what they might look like, and their ability to say, "I do not wish that conversation to be a part of the research".

Due to their organic nature, intentional conversations pose few instances for modification or problem solving in the field. Rather they require skills in paying close attention and being able to record accurate field notes about what has transpired. Recording needs to occur as soon after the incident as possible; however, it also needs to be completed so that it does not disrupt the learning process. This is where skills in memory recall are essential. If you are sitting on a raft and you engage in a rich conversation that you wish to record, it is unlikely that you will be able to record it there and then. Instead you will need to retain the conversation for recording later. While you are retaining this conversation, more observations and conversations may well occur. As the researcher, you will need to decide what is most important and record later in as much detail as possible.

Before engaging in and recording intentional conversations, it would be ideal if you could practice recording intentional conversations and taking notes later to build your skills in this area. Practising your observation skills, paying close attention so that you are ready for intentional conversations when they arise and asking guiding questions when they do, is also helpful. It is important to let go of any conversations that you feel you missed, or notes you were not able to recall clearly.

Document collection

The collection of documents is usually undertaken to aid in the understating of the everyday functioning and dynamics of the research context (Mertens, 2010). Document collection can occur in both a structured and unstructured way. For example, documents that may be collected in relation to research conducted in outdoor contexts include: programme information/outlines, slogans, values, goals, posters and images that adorn walls in outdoor centres, and websites. While some of these may directly support the research questions and result in a structured approach to document collection, if the outdoor context is at a distance to the researcher it is recommended to engage in unstructured document collection whereby you collect more than you feel you may need at the time, and make decisions about their inclusion later during data analysis. Overall, the researcher needs to maintain sensitivity with regards to the content of documents collected, and ensure ethical consent is gained for their use (Boeije, 2010; Mertens, 2010). Document collection contributes to the triangulation of other data sets, such as interviews, intentional conversations and participant observations (Punch, 2009). Now that each method has been elucidated, the focus now shifts to the physical needs of collecting data in the outdoor context.

Physical methods for data collection

One of the greatest challenges once methodology and methods have been chosen is to physically record the data in the field. The technology used needs to be "cost effective, durable, light and small" (Stonehouse, 2007). The last decade has witnessed an exponential shift in options for technology (e.g. GoPro™, audio recorders) and its relative affordability, including increased battery life and recharging in the form of power banks and solar/wind energy. Ideally, not having to recharge is the best outcome, maintaining the focus on the data collection, and not on charging and recharging. When exploring recording options for my own research, I found some of Stonehouse's (2007) recommendations useful, while others were less effective. In the end it came down to personal preference, resulting in my inclination for technology that requires little to no battery power and is as robust as possible. A combination of hand written field notes, typed records, audio and video recorded data is recommended to support the data collection of the methods outlined in this chapter.

Hand written field notes are useful for key words, general comments, and drawing contexts/maps. However, for verbatim conversations/discussions or more detailed notes, I found a keyboard to be essential as I am able to type much faster than I am able to hand write. It also meant that I did not need to then type the observations from hand written notes to create research protocols for analysis (Hatch, 2002). I used voice recorders that are at the lower end of the technology spectrum that utilise AAA batteries, which last a very long time (with easy access to spare batteries) and are able to record many hours of content. As Stonehouse (2007) also found useful, I have repeatedly used the Alphasmart Neo2 keyboard, which was extremely light (less than 1kg), intentionally designed for robustness, and easy to keep dry in a dry bag or transport in a backpack. With a very small screen, it sat flat in my lap, and while it did draw some curiosity initially, this soon subsided as it did not attract the interest more modern devices have. The keyboard had a battery life of 700 hours on three AAA batteries. It was not possible to edit as you went, as the screen only showed two lines of text. The downloaded text was simple, but easily transferable to more complex word processing tools for editing and revision. The benefit of recording verbatim conversations quickly through my ability to touch type with the Alphasmart Neo2 was invaluable.

In terms of video I have utilised body-worn cameras on the researcher (e.g. GoPro™) with helmet/head/chest attachments, depending on the research, participants and activities. Similar to participants using body-worn cameras, they allow access to "insights that traditional observations" simply do not afford (Lloyd, Gray & Truong, 2018, p. 53). Once participants are used to the wearer having the camera on them, participants tend to forget they are there, and the researcher is able to record data in an unobtrusive way, while still being visible. The main limitation with the body-worn cameras is the battery life, particularly in cold environments. Access to spare batteries is recommended to ensure they do not need to be charged during the day. A static video camera can also be useful to be placed off to the side, running while an activity is underway, and the researcher can then engage with the participants and still record what is happening. This also allows the researcher to see the impact of their presence with the group as well.

Conclusion

When engaging in research in an outdoor context, attention to ethical considerations, the underpinning paradigm, research questions and methodology require the researcher to identify their personal biases, understand their place in the world and remain focused on the purpose and practice of the data collection. When selecting methods for data collection in the outdoor context, the unpredictability and changeability of the context itself, activities (typically) undertaken and the potential remoteness of the research context require a flexibility and creativity from the researcher that may not be required in other contexts. Access to a combination of methods facilitates the collection of big and small stories, told and retold through participant observation, intentional conversations and document collection, which together triangulate data collected and support other methods in qualitative research (e.g. interviews). The technical challenges of physically recording data in the outdoor context require a thoughtfulness and thoroughness in preparation for the research, and a flexibility once in the research context. Simple robust technology with long battery life and a range of collection tools supports the researcher in the ever-changing outdoor context.

References

Angrosino, M. (2007). *Focus on observation*. London: SAGE.

Bechhofer, F. (2000). *Principles of research design in the social sciences*. London: Routledge.

Boeije, H. (2010). *Analysis in qualitative research*. London: SAGE.

Cohen, L., Manion, L., & Morrison, K. (2011). *Research methods in education* (7th ed.). London: Routledge.

Cooper, D.R., & Schindler, P.S. (2001) *Business Research Methods*. London: McGraw-Hill Higher Education.

Cope, D.G. (2014). Methods and meanings: Credibility and trustworthiness of qualitative research. *Oncology Nursing Forum, 41*(1), 89–91.

Creswell, J.W. (2007). *Qualitative inquiry and research design: Choosing among five approaches* (2nd ed.). Thousand Oaks, CA: SAGE.

Denscombe, M. (1998). *The good research guide for small-scale social research projects*. Philadelphia, PA: Open University Press.

Denzin, N.K. & Lincoln, Y.S. (Eds.). (2018). *The SAGE handbook of qualitative research* (5th ed.). Thousand Oaks, CA: SAGE.

Freedman, M. (2006). 'Life "on holiday"? In defence of big stories. *Narrative Inquiry, 16*(1), 131–138.

Gillham, B. (2000). *Case study research methods*. London: Continuum.

Hatch, J.A. (2002). *Doing qualitative research in education settings*. Albany, NY: State University of New York Press.

Lincoln, Y.S. & Guba, E.G. (1985). Naturalistic inquiry. Newbury Park, CA: SAGE.

Lloyd, A., Gray, T. & Truong, S. (2018). Seeing what children see: Enhancing understanding of outdoor learning experiences through body-worn cameras. *Journal of Outdoor Recreation, Education and Leadership, 10*(1), 52–66.

Mason, J. (2002). *Qualitative researching* (2nd ed.). London: SAGE.

Mertens, D.M. (2005). *Research and evaluation in education and psychology* (2nd ed.). New York, NY: SAGE.

Mertens, D.M. (2010). *Research and evaluation in education and psychology* (3rd ed.). Thousand Oaks, CA: SAGE.

Neumann, W.L. (1997). *Social research methods: Qualitative and quantitative approaches* (3rd ed.). Needham Heights, MA: Allyn & Bacon.

O'Donoghue, T.A. (2007). *Planning your qualitative research project: An introduction to interpretivist research in education*. Abingdon: Routledge.

Phoenix, C. & Sparkes, A.C. (2009). Being Fred: Big stories, small stories and the accomplishment of a positive ageing identity. *Qualitative Research, 9*(2), 219–236.

Punch, K.F. (2009). *Introduction to research methods in education*. London: SAGE.

Scott, D. & Usher, R. (1999). *Researching education: Data, methods and theory in educational enquiry*. London: Cassell.

Simons, H. (2009). *Case study research in practice*. London: SAGE.

Smith, H.A. (2011). *Extraordinary outdoor leaders: An Australian case study* (Doctoral dissertation). Retrieved from http://ro.uow.edu.au/theses/3551.

Sparkes, A. (1992). *Research in physical education and sport: Exploring alternative visions*. London: Falmer Press.
Stonehouse, P. (2007). Recording in the wilds: A reflection on research-technology needs on an expedition. *Australian Journal of Outdoor Education, 11*(1), 47–49.
Takyi, E. (2015). The challenge of involvement and detachment in participant observation. *The Qualitative Report, 20*(6), 864–872.
Wolcott, H.F. (1995). *The art of fieldwork*. Walnut Creek, CA: Altamira.
Yin, R.K. (2014). *Case study research: Design and methods* (5th ed.). Thousand Oaks, CA: SAGE.

7

MOBILISING RESEARCH METHODS

Sensory approaches to outdoor and experiential learning research

Sue Waite and Phil Waters

Introduction

One of the challenges of researching outdoor and experiential learning is that much of this learning occurs on the move. Fincham, McGuiness and Murray (2010, p.170) compiled a selection of ways in which researchers have sought to accommodate mobility through appropriate dynamic methodologies that often involve 'being there' and sharing the particular contexts of movement, but this is not without attendant problems. 'Mobilising' in our title both acknowledges this movement in and through outside spaces and attempts to marshal some potential methods to meet challenges in capturing the multisensory and material contexts of places with(in) which such learning takes place without destroying the fluidity that marks such experiences. It also refers to mobility across time frames as experience may be interpreted with immediacy and on reflection, so that time is crosscut in the representations that various research methods allow. As a vehicle for considering this mobilisation, we offer five examples of sensory approaches from our own research. Our methods seek to recognise the heightened importance of sensory experiences in natural environments where children and young people construct meaning using all the senses: touch, smell, taste, sight and hearing. The first example considers how very young children can be helped to express the meaning of experiences through props to support articulation. The second focuses on how adult/child researcher/researched relationships may distort the object of study by unintended influence. The third notes how 'capturing' perspectives can never be straightforward, and demonstrates that robust research inevitably shines a light on this process of layering and positioning. The fourth and fifth examples with young people aged 16–25 illustrate how tensions in intentions for participatory methodologies may be exacerbated by financial or pragmatic research constraints. All the research showcased here attempts to capture participants' perspectives on their own terms, deeply embedded in the physical circumstances of their experience, and acknowledges the ethical implications for active methodologies. We note in conclusion material agency in the co-construction of possibilities for outdoor learning, troubling taken-for-granted anthropocentric perspectives and that this turn poses additional questions about how to research in and across places.

Challenges of research with very young children outside

In a series of studies on Forest School for undergraduate dissertations (Davis & Waite, 2005), student researchers used various methods to explore Forest School processes. Carole Britton (Appendix 3A) asked children directly whether they liked being outside and what sorts of activities they do when outdoors, but adult prompts may have steered what children mentioned. Questioning also took place inside the classroom after experiences so children had to reconstruct what had happened, spatially and temporally shifting the locus of experience.

Niki Paulin (Appendix 3D) found 20 children out of 32 participating said they were excited and happy about Forest School, while two felt sad; others did not know. In Forest School, children's enjoyment of the session is often gauged either by thumbs up or down or showing happy or sad faces on 'tree cookies' (sawn slices of branches), but as only negative responses are usually followed up for explanation, this might encourage positive evaluations. Similarly, when children were asked what they had learned, they said to make things, to use knives and to saw or that they didn't know. Practical learning goals such as, 'today we are going to learn how to use a saw safely' were more frequently stated by staff and this may have led children to report these rather than tacit personal and social outcomes. The way in which questions are framed has a big influence on children's responses. Nishiyama (2018) found that developing a community of inquiry could help to deflect pressure to conform to adult expectations.

In another Forest School project, the researcher used simple maps that she chatted through with children and this helped them to link sites within the outdoor environment to how they prompted certain play and thereby access some of the multisensory affordances of the landscape. Mudholes, for example, were a prominent tactile affordance for play that emerged through an open-ended exploration of the space (Davis and Waite, 2005).

Niki also analysed 34 children's drawings of forest school and their most common features were: people (24); trees (20); grass (12); weather (9); camp (6); tree house (5); rope activity (4); flowers (4). However, children's inclusion of features may be limited by what they can draw and by a generic representation of 'nature', asking children to explain their drawings through 'talk and draw' (Yuen, 2004) or co-constructing images with them can help access their intended meanings (Tay-Lim and Lim, 2013).

These instances illustrate how accessing children's voice can be challenging, particularly in the early years when language is still developing and where they may be more inclined to seek a socially acceptable response (Clark, 2005). Another student, Julie Winsor reviewed techniques for encouraging children's voice and decided on the following methods for her study:

> The multiple methods I used included individual semi-structured interviews referring to a class book of photos for prompts. The children also participated in focus groups prompted by watching a video made of them in the woods. Finally, during a large group discussion I used a teddy as an independent focus to which the children could describe their experiences of the forest school project. These three data collection methods were tape-recorded, after clearing this with the children and explaining that I would not be able to remember what they had said. The children were also asked to draw what they remembered about their forest school experience as an alternative form of narrative and the older children annotated their pictures.
>
> *Winsor, Appendix 3H: Davis and Waite, 2005*

Providing multiple methods, memory aids and props can help young children to express their thoughts and feelings more freely (Clark, 2005; Due, Riggs and Augoustinos, 2014).

Protecting children's free outdoor play in research

We consider the next study from the point of view of how the focus of attention can be fundamentally altered by research methods. It was funded by the Economic and Social Research Council (ESRC) over 29 months to look at opportunities afforded by outdoor spaces to smooth transition between early years provision and school (Waite, Rogers and Evans, 2013). We selected an urban and a suburban school in the UK with outdoor space that had two Foundation Stage and two Year 1 classes, so we could compare different teachers in the same setting and across schools in eight classes with children aged between 4 and 6 years of age.

Our focus for data collection was on the "micro level of peer relations" (Vandenbroeck & Bouverne-de Bie, 2006: 128) in social interactions outdoors. We selected four target children from each class to ensure mixed-sex and teacher-assessed attainment levels. We followed 15 children from Foundation Stage into Year 1 (one child left the school). In total, 192 audio recordings were captured during summer and autumn 2009 and summer 2010, using the following methods:

- Observations to provide contextual field notes;
- Audio recordings to gather natural conversation by highly mobile children wearing digital recorders in small brightly coloured bags;
- Staff semi-structured interviews and informal conversations to explore reasons for pedagogical decisions;
- Images to understand contexts and provide a stimulus for reflection and analyses.

The detail of interactions during children's play was contextualised by the observers' field notes taken from different vantage points, visual data and by staff views about their pedagogical decision-making. We used a systems framework (Neill, 2008) to take account of what was offered, the programme and activity and the facilitator's role within those; and what was received, the culture, group and individual (Waite, Nichols, Evans and Opie, 2009).

In the Forest School research, we had found that following children around with a clipboard impacted on what children did (Waite & Davis, 2007). We wanted to capture unstructured free outdoor play without destroying it in the process. Adults can distort children's free play direction and duration (Thomson, 2014). However, simply observing at a distance tends to compromise detail in social and material interactions. We experimented with Geographic Positioning Systems and head-cams worn by children to video-record activity from their perspective, but the GPS definition was insufficient to link activity and particular places and children's rapid head turns resulted in dizzy-making footage so these methods were not pursued (Waite et al., 2009).

Getting framed and reframing

In the next example, Phil extends the sensory approach so that other senses can be explored more fully by the inclusion of more detailed videos from young children's perspectives and 'objective' distances. In a study exploring the impact of a story-based intervention on children's physically active play compared to free play, 57 year olds were equipped with GoPro chest-harness cameras and asked to play in the natural spaces of their school grounds (Waters, Waite & Frampton, 2014). In addition to the close-quartered cameras capturing a first-person view, a third-person viewpoint was being recorded by supplemental cameras located on landscape features and buildings around the school site. Using Rose's sites of *production*, *image* and *audiencing* (2012) combined with the autoethnographic method (see Chapter 10), restorying (Ollerenshaw & Creswell, 2002), a framework for audio-visual analysis was conducted at specific points during the research project.

Most surprising was the analysis of film taken from close-quartered cameras, which tended to reposition the researcher as a child, sometimes as a female, and locate him within scenarios that were both time and place separated. Analysis of children's films would often happen months after they were collected, yet the multisensory richness of being immersed in a child's play-world by proxy of their films, had the researcher responding to voices, gestures, and other signifiers as if being present in real time. The same could not be said from analysing films taken from the supplemental third-person cameras. Not only did they provide for some unusual perspectives – three being high up on poles – but they offered the standard practitioner viewpoint; outside, looking on at children as they play.

The potency of children's first-person viewpoints was demonstrated at a conference, when a member of the audience expressed that she was uncomfortable watching the sequence of a child who, having wrestled a play object through the school grounds, ended up sat beneath it for quite some time. Naturally concerned for the child's safety, thinking she could not breathe, a second sequence showed the same play but from the child's perspective. The child was in no danger and as the film demonstrated, all her scraping, scuffling, wriggling, breathing, sighs and giggles, signalled that she was not only safe, but enjoying the experience.

Analysing children's first-person films requires a whole new appraisal within visual methodologies. The almost transcendental experience of being an observer without body, yet still feeling, empathising and sensing what children experience directly, asks for an analysis that goes beyond the visual. Likewise, listening to children as they wrestle, carry, drag and drop objects, goes beyond the audio. Their sounds of effort and physicality ask more of researchers than just the coding of behaviours, or the objective representation of action. It asks for a subjective positioning that embraces a sensory tableau. In this study the bringing together of visual and narrative methods has helped to frame both objective and subjective inferences made during analysis. Restorying asks the researcher to recast what is being observed through a subjective lens, and like all good storytelling, multisensory narrativity paints a richer and deeper text. This is not to deny that recording behaviours, dialogues or monologues at face value is not equally important. It is, and they are. Objectivity does, after all, bring validation and companionship to the subjective stance utilised in restorying methods. Instead, it asks us to recognise that children are multisensory beings playing in a sensory-rich world, and if we wish to explore this, we ought to select methods that are appropriate and representative of children and their worlds.

Images, reflections and shifting identities

A temporal aspect through restorying may be more successful with older children and young people and we now turn to a study of social and economic aspirations of young people aged between 16 and 24, living on or near Exmoor National Park within North Devon and East Somerset (Merchant, Quinn & Waite, 2013). The funding from the National Park Authority was insufficient to accompany young people as they carried out their daily business across the Park so we had to devise other methods to capture roaming experiences.

Within a participatory research methodology, research methods were chosen through consultation with young people in a pilot initial focus group, which informed how we sought further participants and provoked use of photographs as prompts and data sources to expand data from surveys and focus groups. From June to October 2012, 109 qualitative questionnaires distributed to 16–24 year olds who lived and/or grew up in or near to Exmoor National Park were collected online or as hard copy. The field researcher was a similar age to our target group. This encouraged relaxed 'banter' in focus groups of 5–10 people in the pubs and cafes that the young people themselves suggested, further levelling power relations between researcher and researched (Leyshon, 2002). Focus group members were mainly recruited via existing community groups and through 'snowballing' (where a participant recruits someone they know), so most participants were familiar each other and felt comfortable expressing their views. However, quieter and more isolated individuals might be less well represented in this context; those approached through the Youth Offenders Team and Children's Services did not take part in focus groups. Survey methods provided an alternative way to contribute but accessing a representative sample of young people through snowballing is problematic because it favours homogeneity.

As the initial focus group had talked a lot about traditional farming and the heritage of Exmoor, in addition to predetermined questions, we showed photographs depicting Exmoor rural life in the 1960s to stimulate discussion of what the images meant to the young people, if/how they related to them and whether they regarded them as valid depictions of rural life now. Participants were also given a digital camera to capture a 'typical day in their life' for the research.

Visual prompts are valued to help reflection, particularly in relation to embodied experiences (Merchant, 2011; Pink, 2006). MacDougall (1997) argues visual representation can offer pathways to the other senses and help to address challenges in communicating emotions, time, body, senses and identity. As Rose (2007) points out, rich contextual information is instantly accessible in pictures compared with written descriptions and can provide hooks to highlight significant sensations and emotions. We hoped that the photos would "carry and evoke ... information, affect and reflection" (Rose, 2007, p. 238). To our surprise, given their earlier emphasis on maintaining tradition, the two focus groups shown photographs felt that these historical images cast them as 'backward'. They chose not to participate in this element of the research. They were more enthusiastic about taking photos but ultimately only two returned photographs, so it was not possible to interpret them reliably. For this reason, they were excluded from data analysis. We had developed these methods to allow for more sensory representations to be available but, although initially appealing, the methods were not eventually useful for eliciting data, illustrating a gap between intended and received meanings and reinforcing that affective responses can be stimulated by the research process in unforeseen ways.

Deepening relationships and meaning

In our final example of research methods employed to try to capture mobile multisensory experiences in outdoor learning, we turn to an evaluation of engagement of young people aged 16–25 with National Parks across the UK (Waite, et al., 2016). The Campaign for National Parks' Mosaic project (2013–2016) was designed to build youth skills, citizenship, health and confidence through volunteering and encouraging other young people to enjoy National Parks. Criteria for volunteers, dubbed Young Champions, included being Not in Employment, Education or Training; limited access to public transport and services; not knowing what career path to follow; being the main carer in the home; needing to improve physical or mental health or suffering from stress.

We explained our intention to use pictures as well as conversations to find out how they felt about their involvement in the Mosaic project. We wanted to access the affective element of the programme. Participants were loaned cameras and invited to select three images that encapsulated their experiences with brief explanatory notes; however, participants were slow to respond, caught up with aesthetics of image production or reluctant to share feelings.

Our photo-voice methodology (Sutton-Brown, 2014) was supported by an artist-researcher who held workshops to explore the meanings of their photographs with young people. Gradually the relationship with the researchers grew and the young people shared more of their stories and feelings, even suggesting that this method may have contributed to positive impacts by making them more aware of their emotional responses. One participant included two pathways in his chosen photos, representing movement through the landscapes of Exmoor and his life. They suggested a process of transitional becoming. One track was an icy roadway framed by the car windscreen. The young man had stated that he valued unmediated nature; that cameras filtered his direct engagement. His framing with the windscreen on an organised 'photosafari' perhaps suggests that some Young Champion experiences may have been too structured for his preferred engagement with nature, but he also acknowledged that creating lasting images had enabled reflection on what places meant to him.

He captioned another photograph of sun bursting through dark menacing clouds as 'Explosion', a word that on one hand summoned images of his earlier military career but also seemed to reflect the vibrancy of the sky and sudden opening to positive ideas and opportunities through his Mosaic involvement.

> if I didn't do this, I'd be working is how I'd be, doing everything the same except from I wouldn't have a place where I could go to relax ... have a day off ... and have fun ... Like all those things where you have memories of a positive thing, because life is just so negative in every aspect, so if I didn't have it [Mosaic experience] I'd be a lot more negative.

Through their photos, we glimpsed meanings and opened conversations that helped us explore relationships between places and affect. Time to reflect on the images afterwards allowed multi-layered understandings of motivations and feelings to emerge that may not have been possible without this shared focus. "Multisensoriality is integral both to the lives of people who participate in our research *and* to how we ethnographers practice our craft" (Pink, 2009, p. 1).

Unfortunately, funded research timescales and budgets often do not permit gradual relationship building, yet this can be vital to sharing of embodied and affective experiences. There is a risk that visual methodologies are employed for images' powerful immediacy, but literal 'snapshots' may come at the expense of deeper meaning making which benefits from trust over time.

Ethics in mobilised multisensory research

These various examples are underpinned by knotty ethical dilemmas, which may not be recognised in standard ethics procedures within controlled research conditions (see Chapter 2). Ebrahim (2010, p. 290) argues that a preconceived document alone cannot account for the complexity of lived experience, and that a "flexible stance where the reflexive researcher develops through relationships and shapes research outcomes through engagement with ethics in practice" enables "situated ethics" (Simons and Usher, 2000) that are responsive to circumstances using multiple methods to co-construct meanings. The outdoor and experiential context is inherently unpredictable; so, although possible eventualities are considered and included in informed consent procedures, the acid test of ethical research is if the researcher is primed to respond ethically in the moment. This requires them to be attentive to the promises made to those taking part in the research and balancing different responsibilities and rights so that they can respond appropriately and report any diversions from anticipated procedures (Christensen and Prout, 2002; Waters and Waite, 2016).

For example, in the ESRC study, we sought consent from the head-teachers, staff, children and their parents in the classes of both schools. We used brightly coloured bags to remind children of their choice in taking part and help them remain conscious of the research when they wore them, but it was clear from recordings that children had not always understood or remembered that their voices would be captured even when no adult was present. We promised that non-participants would not be disadvantaged, but the bags conferred an unanticipated status, so to avoid non-participants feeling left out, we allowed all children to wear them if they asked but either kept the recorder turned off or immediately deleted recordings if there was no consent.

Methods were developed in consultation with young people in the study of social and economic aspirations, but those not well received in practice were abandoned. An ethics protocol is not a dead document to be left on a shelf. It is the means to make explicit ethical commitments and prepare for possible scenarios in research for use in the field as the touchstone for active ethical research. This living quality is especially important in unpredictable contexts like natural environments. An example of a responsive and responsible approach to ethical decision-making in practice is included in Chapter 2 and is discussed more fully in Waters and Waite (2016).

Concluding thoughts

The examples we have given here suggest that we need to mobilise a range of research methods to do justice to research involving outdoor and experiential learning with children and young people. As with all research, alignment of aims with appropriate theory and methods is key to generating robust and trustworthy research data. The examples here have been largely child- or young people-centred capturing their movement through and sensory engagement with landscapes. Recent thinking stresses the agency of the material world in shaping outdoor experience (Taylor, 2013; Quinn, 2016). Decentring does not necessarily require new methods but demands changes in analysis approaches away from paradigmatic silos towards complex new materialist perspectives (Gough, 2016; Hultman and Lenz Taguchi, 2010) (see Chapter 1) where boundaries between the subjective and 'other' are contested, complicated and indistinct. Nevertheless, we suggest that creative methods that admit and capitalise on multisensory dimensions (e.g. Wright, Goodenough & Waite, 2015) are more likely to support fuller understandings in research undertaken outdoors.

References

Christensen, P. & Prout, A. (2002). Working with ethical symmetry in social research with children. *Childhood, 9*, 477–497.

Clark, A. (2005). Ways of seeing: Using the Mosaic approach to listen to young children's perspectives. In A. Clark, A. Kjørholt & P. Moss (Eds.) *Beyond listening: Children's perspectives on early childhood services* (pp. 29–50). Bristol: Policy Press.

Davis, B. & Waite, S. (2005). *Forest Schools: an evaluation of the opportunities and challenges in Early Years*. Plymouth: Plymouth University.

Due, C., Riggs, D. & Augoustinos, M. (2014). Research with children of migrant and refugee backgrounds: A review of child-centered research methods. *Child Indicators Research*, 7(1), 209–227.

Ebrahim, H. (2010). Situated ethics: Possibilities for young children as research participants in the South African context. *Early Child Development & Care, 180*, 289–298.

Fincham, B., McGuinness, M. & Murray, L. (Eds.). (2010) *Mobile Methodologies*. Basingstoke: Palgrave.

Gough, N. (2016). Postparadigmatic materialisms: A "new movement of thought" for outdoor and experiential education? *Journal of Outdoor and Environmental Education, 19*(2), 51–65.

Hultman, K. & Lenz Taguchi, H. (2010). Challenging anthropocentric analysis of visual data: a relational materialist methodological approach to educational research. *International Journal of Qualitative Studies in Education, 23*(5), 525–542.

Leyshon, M. (2002). On being "in the field": Practice, progress and problems in research with young people in rural areas. *Journal of Rural Studies, 18*(2), 179–191.

MacDougall, D. (1997). *The corporeal image: Film, ethnography, and the senses*. Princeton, NJ: Princeton University Press.

Merchant, S. (2011) The body and the senses: Visual methods, videography and the submarine sensorium. *Body & Society, 17*(1), 53–72.

Merchant, S. Quinn, J. & Waite, S. (2013). *Social and economic aspirations of young people living in or near Exmoor National Park*. Plymouth: Plymouth University.

Neill, J. (2008). *Enhancing life effectiveness: The impacts of outdoor education programs*. (PhD thesis draft). University of Western Sydney. Available at https://researchdirect.westernsydney.edu.au/islandora/object/uws:6441

Nishiyama, K. (2018). Using the community of inquiry for interviewing children: theory and practice. *International Journal of Social Research Methodology, 21*(5), 553–564.

Ollerenshaw, J.A. & Creswell, J.W. (2002). Narrative research: A comparison of two restorying data analysis approaches. *Qualitative Inquiry, 8*(3), 329–347.

Pink, S. (2009). *Doing sensory ethnography*. London: SAGE.

Quinn, J. (2016). Theorising learning and nature: Post-human possibilities and problems. In C.A. Taylor & G. Ivinson (Eds.) *Material Feminisms: New Directions for Education* (Chapter 5). London: Routledge.

Rose, G. (2007). *Visual methodologies: An introduction to the interpretation of visual materials*. London: SAGE.

Rose, G. (2012). *Visual methodologies: An introduction to the interpretation of visual materials* (3rd edn.). London: SAGE.

Simons, H. & Usher, R. (2000). Introduction: Ethics in practice of research. In H. Simons & R. Usher (Eds.) *Situated ethics in educational research* (pp. 1–11). London: Routledge.

Sutton-Brown, C.A. (2014). Photovoice: A methodological guide, photography and culture, 7(2), 169–185.

Tay-Lim, J. & Lim, S. (2013). Privileging younger children's voices in research: Use of drawings and a co-construction process. *International Journal of Qualitative Methods, 12*, 65–83.

Taylor, A. (2013). *Reconfiguring the natures of childhood*. London: Routledge.

Thomson, S. (2014). "Adulterated play": an empirical discussion surrounding adults' involvement with children's play in the primary school playground. *Journal of Playwork Practice, 1*(1), 5–21.

Vandenbroeck, M. & Bouverne-De Bie, M. (2006). Children's agency and educational norms: A tensed negotiation. *Childhood, 13*(1), 127–143.

Waite, S. & Davis, B. (2007). The contribution of free play and structured activities in Forest School to learning beyond cognition: An English case. In B. Ravn & N. Kryger (Eds.). *Learning beyond cognition* (pp. 257–274). Copenhagen: Danish University of Education.

Waite, S. Nichols, M., Evans, J. & Opie, M. (2009). Methods for exploring how pedagogies are shaped in outdoor contexts. Fourth International Outdoor Education Research Conference, Beechworth, Australia, April 15–18, 2009. www.latrobe.edu.au/education/assets/downloads/2009_conference_waite_nichols_evans_opie.pdf

Waite, S., Rogers, S. & Evans J. (2013). Freedom, flow and fairness: Exploring how children develop socially at school through outdoor play. *Journal of Adventure Education and Outdoor Learning, 13*(3), 255–276.

Waite, S., Waite, D., Quinn, J., Blandon, C. & Goodenough, A. (2016). *MOSAIC MATTERS: External Evaluation of the Mosaic project*. Plymouth: Plymouth University.

Waters, P. & Waite, S. (2016). Towards an ecological approach to ethics in visual research methods with children. In D. Warr, M. Guillemin, S. Cox & J. Waycott (Eds.) *Ethics and visual research methods: Theory, methodology and practice* (pp. 117–127). New York, NY: Palgrave Macmillan.

Waters, P., Waite, S.J. & Frampton, I.J. (2014). Play frames, or framed play? The use of film cameras in visual ethnographic research with children. *Journal of Playwork Practice, 1*(15), 23–38.

Wright, N., Goodenough, A. & Waite, S. (2015). Gaining insights into young peoples' playful wellbeing in woodland through art-based action research, *Journal of Playwork Practice, 2*(1), 23–43.

Yuen, F. (2004). "It was fun … I liked drawing my thoughts": Using drawings as a part of the focus group process with children. *Journal of Leisure Research, 36*(4), 461–82.

8

CAPTURING COMPLEXITY AND COLLABORATIVE EMERGENCE THROUGH CASE STUDY DESIGN

An ecosocial framework for researching outdoor sustainability education practice

Alison Lugg

Introduction

Freebody (2003, p. 9) asserted that educational practices "stand at the meeting place of a society's contesting instincts for stasis and change". The focus of this chapter was, largely, about how this "meeting place" played out for the pre-service teachers participating in a teaching practicum project referred to as SOIL (Sustainability through Outdoor Integrated Learning). Central to the study are the experiences of pre-service teachers being inducted in the practice of teaching in a secondary school while simultaneously developing a new experiential sustainability curriculum and team-based pedagogy in that professional setting. These dual processes necessitated bridging: two institutions (the school and the university); several discipline areas (outdoor education, science and humanities); differing teaching and learning environments (classroom, urban and outdoor). The notion of bridging underlines pre-service teacher participation as negotiating boundary spaces (Akkerman & Bakker, 2011; Edwards, 2009, 2011; Engeström, 2008; Wenger, 2009). Work in boundary or liminal space is often characterised by uncertainty, complexity and the potential to drive change (Engeström, 2008). These characteristics are key features of sustainability education and, depending how it is practised, outdoor education.

Sustainability education in teacher education and outdoor education

To learn to live sustainably on Earth is, arguably, the greatest challenge of our time. Education is one means of addressing issues arising from unsustainable human activity that, if well supported by government policy and resources, has high potential for instigating long-term change. The potential of education to instigate change in this arena was acknowledged by the UNESCO Decade for Education for Sustainable Development (UNDESD 2005–2014) and, currently, by goal four in the United Nations Sustainable Development Goals.[1]

The SOIL study was undertaken in Bendigo, Victoria in south-east Australia, during 2009–2010, mid-way through the UNDESD. Then, the Australian government had produced a new National Action Plan, *Living Sustainably* (Commonwealth of Australia, 2009), that highlighted the need for the education sector to address sustainability issues in schools and universities and for teacher education providers to prepare pre-service teachers for teaching sustainability across the curriculum. Sustainability is now an emerging theme in Australian and international education research in teacher education, and also in outdoor and environmental education, where the possibilities for holistic, collaborative and critical pedagogy are recognised (Stevenson, Ferreira, Davis & Evans, 2014; Ross, Christie, Nicol & Higgins, 2014). However, despite these policy and research directions, Australian teachers are only minimally prepared for the challenges of sustainability education, and sustainability in teacher education programmes is still on the curriculum fringe (Stevenson et al., 2014).

Within Australian outdoor education curricula, the Victorian Certificate of Education (VCE) Outdoor Environmental Studies for senior secondary students is a lone example of a coherent school subject focusing on sustainability education. This subject is offered as an elective option in approximately one third of Victoria's secondary schools but, apart from the VCE study, outdoor education curriculum and practice in Australian schools shows only minimal evidence of reflecting national or global sustainability imperatives (Gray & Martin, 2012). In this context then, how might pre-service outdoor education teachers respond to the challenge of developing and teaching a sustainability education programme in a secondary school and how might this experience impact their professional learning? These issues provided the impetus for the longitudinal case study that was the focus of the investigation discussed in this chapter.

The SOIL study

With the convergence of the sustainability policies and initiatives outlined, the interdisciplinary Year 9 SOIL practicum was instigated in 2009 as a pilot partnership between a Year 7–10 secondary college and La Trobe University. It was designed to meet the needs of a local school to engage Year 9 students in a new sustainability education unit and to provide outdoor education pre-service teachers with a unique opportunity to design, teach and evaluate an outdoor sustainability curriculum. The SOIL programme continued for five years and in each iteration pre-service teachers worked in cross-disciplinary teams with peers, supported by teacher mentors from humanities and science, to design a four-week unit of work focused on sustainability education for Year 9 students. The unit was to involve experiential approaches to sustainability integrating humanities and science, culminating in an overnight camp.

This case study was conducted in the pilot year when the SOIL practicum was first implemented and in 2010 during the 12 months after participants had graduated. The emergent nature of this SOIL practicum meant that there was considerable uncertainty about how it would work, especially as the school had not included a sustainability unit before, nor had the teacher mentors been involved in team teaching, interdisciplinary curriculum or outdoor education. Therefore, as the following comment illustrates, the pre-service teachers, along with their mentors (and myself as a university staff mentor), were entering unknown territory:

> … it was tough because it was the first time it had been run and we were like the crash-test dummies, that's cool! I got that in my head pretty early that it won't progress smoothly, there will be teething problems, so … I figured out ways to work around it.
>
> *Barry, Focus Group, September 2009*

The study investigated how the pre-service teachers responded to this novel situation and how it impacted their professional development. The pre-service teachers engaged in this practicum as members of *teams* and in relation to school and university cultures and policies, not as isolated individuals. Therefore, I was primarily concerned with how they *collectively* interpreted and navigated an evolving, complex practicum programme. This chapter focuses on the challenges of designing a research methodology for an emergent project where the unit of analysis was an activity system and in which I, as the researcher, was immersed as part of that system.

Methodological considerations

The multi-layered interactions under investigation called for a research design and analytical framework that would facilitate investigation of both *collective* and *individual* perspectives, changing *relationships* and *conditions*. Consequently, I was mindful of what Thomas (2010) referred to as symmetries between research content and research process. In Thomas' research this symmetry was unexpected but, in my study, it was intentional. I wanted the research methodology to reflect the messy nature of the SOIL project itself so that the interpretive framework could be both flexible and reliable enough to capture the complexity while enabling meaningful analysis as suggested by Davis, Sumara and Luce-Kapler (2008).

Freebody (2003) argued that one of the main contributions of qualitative research in education is its ability to illuminate indigenous or local knowledge, cultures and issues, in ways that support and validate situated knowing and complex relationships. A qualitative methodology enabled close examination of the pre-service teachers' interactions, conceptualisations and professional learning outcomes in the local context as explicated through the research questions:

1. How did the pre-service teachers' participation in an innovative sustainability education practicum programme impact their professional development as outdoor education teachers?
2. How did the pre-service teachers conceptualise and practice their outdoor education teaching in relation to the sustainability curriculum?
3. In what ways did the collaborative structure of the SOIL practicum impact the pre-service teachers' practice?
4. In what ways did this new practicum enhance or hinder their professional learning and transition to teaching?

The *how, why, what* and *who* questions outlined above, are typical of case study research where complex phenomena or events are studied in their naturalistic form and the researcher has minimal control (Corcoran, Walker & Wals, 2004; Yin, 2003). Yin's (2003) longitudinal case study methodology was apt for this purpose.

Case study research

The strength of case study for *education* research is its "closeness to real-life situations" (Flyvbjerg, 2006, p. 223), its ability to embrace the particularities and dynamics of everyday teaching and learning processes, potentially revealing insights that are inaccessible via more generic research methodologies. Case study is a common form of educational research because teachers and students deal with real people, subject matter and places where "conditions are not 'background' variables, but rather lived dimensions that are indigenous to each teaching-learning event" (Freebody, 2003, p. 81).

Case study research is particularly useful for investigating sustainability and environmental education innovations because it is grounded in practice with all its complexity, its focus on context and on practitioner actions and beliefs (Corcoran et al., 2004; Lotz-Sisitka & Raven, 2004). The attention to the particular is a strength that counters common criticisms that case study methodology findings are non-generalisable and lack wider relevance. Corcoran et al. (2004), Lotz-Sisitka and Raven (2004) and Yin, (2003) argued that questions and issues emerging from case study analyses arise from the *particular* activities and relationships inherent in the case and are not necessarily solved by studying more cases in other contexts or by changing the research design because such modifications would generate a *different set of relationships* and issues. However, they asserted that, in each case study, some elements will be more generalisable (to practice or theory) than others, and knowledge generated can offer useful insights into similar situations.

A key reason for the choice of case study methodology for the SOIL study was its capacity to examine the multifarious interactions involved in teacher preparation, sustainability and outdoor education. I adopted Yin's (2003) view that a case study, is "an empirical inquiry, that investigates a contemporary phenomenon within its real-life context when the boundaries between phenomenon and context are not clearly evident" (p. 13). This notion of blurred boundaries and the emphasis on contextual conditions, were significant factors in identifying case study methodology, because the conditions for pre-service teachers' professional learning were important and the boundaries between subjects, context and object in the SOIL project were fuzzy and evolving.

Case study design

The research was designed as a longitudinal case study, investigating an identifiable group of pre-service teachers' experiences and changing conceptions over a period of "two or more different points of time" (Yin, 2003, p. 42). To this extent it contributed to an identified gap in outdoor education research where most case studies offer snapshots of programmes or individual outcomes over short time spans (Preston, 2008).

Both Yin (2003) and Freebody (2003) suggested that case study design may require multiple levels of analysis where each level is embedded in a nested analytical system. In this study, pre-service teachers' views and practices were considered *in relation to* contextual elements such as the views of their peers, the teacher mentors, the practicum structure and local education and sustainability policies. Yin's (2003) *embedded case design* was used and adapted to account for the multiple components of the *activity system* (Engeström, 2001) that formed the primary basis for analysis of data. Analytical 'layers' required different sources of evidence to enable a richer understanding of the case. "Converging" or cross-checking and synthesising information from different data sources provided a more reliable interpretation of the phenomenon under study (Yin, 2003).

While the SOIL case study was guided by a predominantly qualitative methodology, it also included quantifiable survey questions to provide initial baseline data. Qualitative data were then generated primarily via pre-service teacher interviews and focus groups at different stages of the research project. Additional 'layers' of data were obtained via participant observation on camps and in classes as well as document analysis of unit plans, meeting notes and emails. Mentor teachers were also interviewed prior to and after the SOIL programme to provide further important contextual information. Figure 8.1 shows a convergent approach to data generation that accounts for the whole group of pre-service teachers as well as teaching and camp teams and individual perspectives.

This is an example of Yin's (2003) criterion for case study where the boundaries between subjects and context are blurred.

Ecosocial conceptual framework

The theories comprising the ecosocial conceptual framework developed for this study take into account the "meeting place" (Freebody, 2003) between the dual processes of induction into established practice and innovation and change. This framework, represented in Figure 8.2, allowed for dynamism in its structure by enabling detailed analysis of the pre-service teachers' activity without losing sight of the broader social, cultural and environmental milieu. It supported a view of

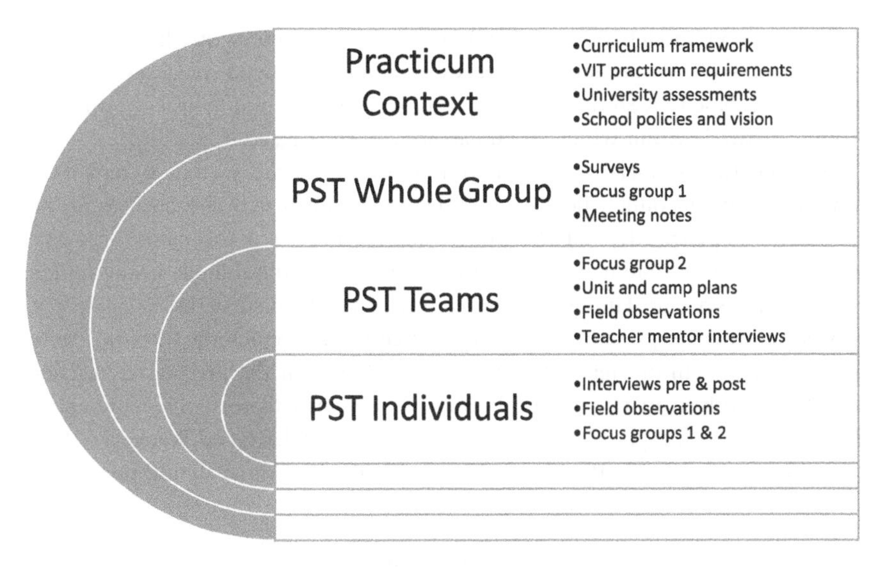

FIGURE 8.1 Embedded case study design: Analytical layers and methods

Note: PST refers to pre-service teachers; VIT refers to Victoria Institute of Technology

experience and learning that is holistic, situated and sufficiently detailed to consider the nature of interactions between individuals, groups and the learning contexts.

The ecosocial framework adopted Bronfenbrenner's (1994) notion of nested systems that represents human activity as structurally coupled with systems; evolving and responding in a *mutually constitutive* relationship (Davis et al., 2008; Opfer & Pedder, 2011; Sterling, 2004). Whether or not phenomena were considered as distinct (parts) or collectives (wholes) depended on the level of observation and the focus of investigation (Davis et al., 2008). Although it is not possible to investigate all systems with equal depth, it is possible to foreground one while taking into account other system levels and relationships between them (Rogoff, 2003). The representation of nested activity shown in Figure 8.2 underlines that pre-service teachers' interpretations and enactment of outdoor sustainability education practice cannot be studied in isolation but needs to be looked at in relation to activity systems, communities of practice and broader social, cultural and material systems. The theories that contributed to the framework all share a concept of human activity:

- Ecological theory – complexity, nested systems, embodied experience (Bronfenbrenner, 1994; Davis et al., 2008; Sterling, 2004);
- Communities of practice (CoPs) –legitimate peripheral participation, professional development (Lave & Wenger 1991, Wenger, 2009);
- Cultural Historical Activity Theory (CHAT) – mediation, expansive learning, knotworking (Engeström, 2001, 2008);
- Relational agency, relational expertise, common knowledge (Edwards, 2009, 2011).

Unlike most models of nested systems, Figure 8.2 represents the *macrosystem* as the *biophysical* environment and processes rather than broad social systems. This concept was based on Sterling's (2004) argument that the biophysical system is the "superstructure" within which all human activity takes place.

In using the ecosocial framework for data analysis, the various system levels were taken into account where relevant. However, the analytical focus was on the activities and perspectives of the pre-service teachers thus the *microsystem* level within the activity system, represented in Figure 8.2 by the triangle in the middle of the nested systems. The SOIL activity system was formed at the *boundary zone* between intersecting Communities of Practice (CoPs) represented by the circles in the microsystem of Figure 8.2. It was anticipated that the pre-service teachers would be involved in negotiating boundaries between these CoPs, representing Freebody's "meeting place" between stasis and change. Lave and Wenger's (1991) theory of learning in CoPs was used as a general means of analysing pre-service

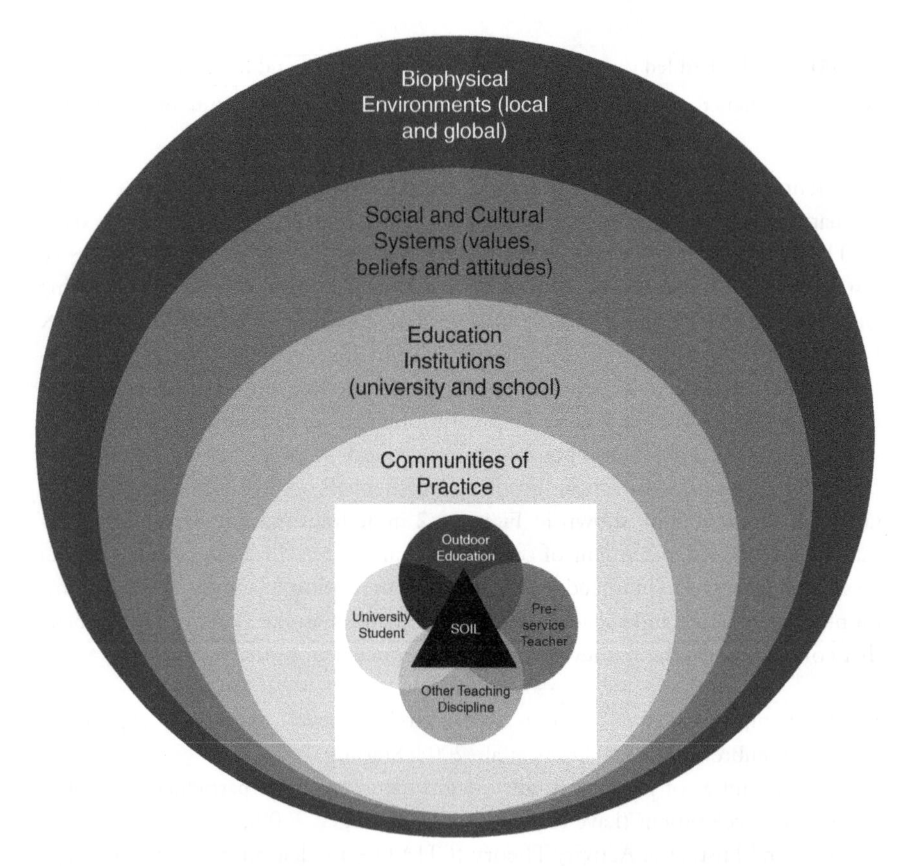

FIGURE 8.2 Ecosocial framework

Source: Adapted from Bronfenbrenner, 1994; Davis, Sumara & Luce-Kapler, 2008; Engeström 2001; Lave & Wenger, 1991; Lim, 2002; Sterling, 2004

teachers' induction into established teaching practice, while Engeström's (2001, 2008) Cultural History Activity Theory (CHAT), represented by the triangle, was used to investigate pre-service teachers' activity in more detail with an emphasis on their *joint* activity and changing perspectives.

Learning, according to Lave and Wenger (1991), is about becoming a member or participant in social communities where cultural patterns of activity are adopted and reproduced by learners such as apprentices or, in this case, pre-service teachers. CoPs are dynamic, self-organising systems with diverse membership, changing roles and complex relationships (Wenger, 1998). They differ from communities in general, in that membership of CoPs involves "sustained pursuit of a shared enterprise" through *mutual engagement*, a *joint enterprise* and a *shared repertoire* of practice (Wenger, 1998, p. 45). This theory highlights the significant role of teacher mentors in the development of pre-service teachers.

Cultural Historical Activity Theory

There is a strong epistemological alignment between CHAT theory, sustainability, outdoor education and pre-service teacher preparation as all are based in dialectical, contingent, notions of learning and development. 'Activity' such as teaching or learning, from a CHAT perspective is "a matter of what, why and how people do things together, either cooperatively or conflictually, over time; mind as a thoroughly *social* and *material* as well as historical phenomenon" (Fenwick, Edwards & Sawchuk, 2011, p. 56). Based on Vygotsky's theory of *mediation,* CHAT is concerned with how people interpret, respond to, think/feel about and transform the artefacts with which they interact. As they interpret and respond to the possibilities created by these artefacts, the nature of their activity changes, producing new techniques, rules, relationships and, over time, new object or motive of activity (Engeström, 2001).

As a theoretical framework CHAT enabled a socio-cultural account of individual *and* collective learning while putting boundaries around analysis of the context of activity without necessitating examination of limitless socio-cultural factors. The subjects in this study were the SOIL group, the teams and individuals at different stages of the research. The context from a CHAT perspective is the web of relationships between subjects, mediating artefacts, rules, structures and social norms shaping the activity system. The interactions between the subject(s) and activity system elements were dialectical.

Using Engeström's (2001, 2008) representation of a second-generation activity system, Figure 8.3 is a visual representation of the SOIL activity system at the final stage of the study where pre-service teachers had graduated and, in most cases, were working as teachers. The object of activity had evolved significantly since the beginning of the programme as had the outcome. The darker, arrowed lines indicate the key relationships within the system at that stage. These were determined by the frequency of references to or the importance placed on these interactions by the pre-service teachers. The text in italics indicates the system

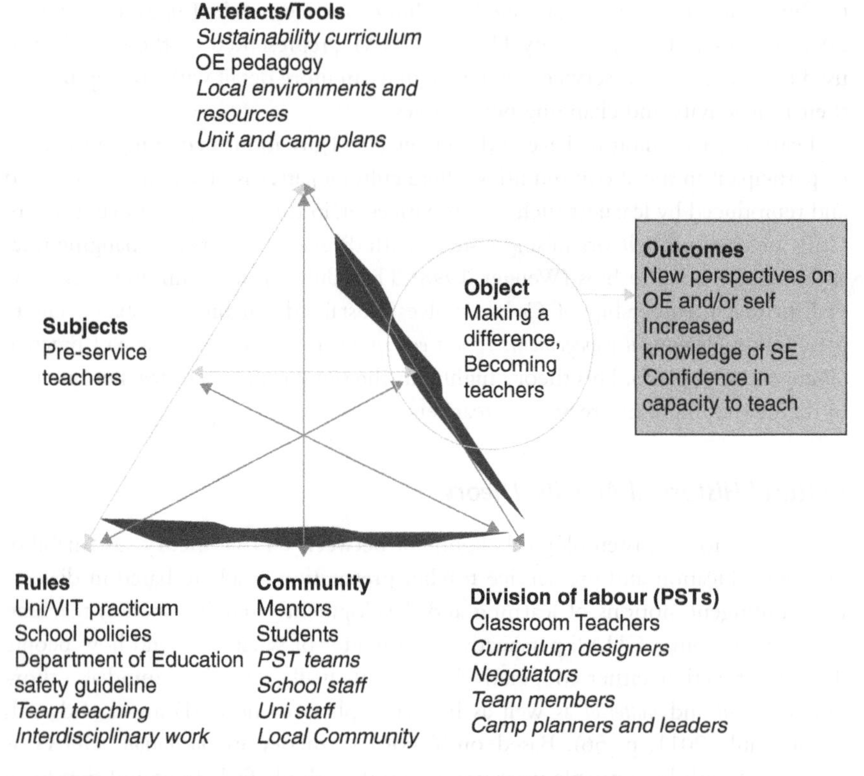

FIGURE 8.3 The human activity system as it related to the SOIL practicum
Source: Adapted from Engeström, 2001, p. 135

elements that were different in the SOIL practicum to the traditional practicum model, thus emphasising the increased complexity of the SOIL practicum. According to Engeström (2008), dissonance or "contradiction" between elements within an activity system are essential drivers of system change because participants attempt to resolve these problems and, in so doing, create change. The systemic contradictions in the SOIL programme are represented by the thick, black zig zag lines. These indicate where dissonance was evident between particular elements within the activity system where pre-service teachers identified an inconsistency or problem arising from dissonance.

Data analysis

Because the data in this case study were generated via an inductive process over an extended period of time using multiple data sources, data analysis was necessarily simultaneous and iterative (Cresswell, 2008). Iterative cycles of data generation and analysis enabled me to modify research instruments as interesting information or issues were revealed. An interpretive approach to data analysis was adopted since the purpose of the study was to *understand* the participants' experiences rather than to deconstruct them. Interpretation in this case study occurred on two levels – (1) the pre-service teachers' or teacher's interpretations of events or learning; and (2) my interpretation of what they said and the meanings I ascribed to their words. NVIVO was used as a data management programme to assist thematic coding.

Coding *is* analysis according to Miles and Huberman (1994, p. 56), because it involves making sense of data by differentiating between segments based on their relevance to the research questions, while at the same time being mindful of the relationships between the segments. Coding in this study, meant assigning tags or labels to paragraphs, sentences or phrases in the data source texts, according to ideas/concepts or themes that were evident in those text segments. Initially I took a semi-grounded theory approach to coding the interview and focus group data, according to the themes/ideas I could identify as prevalent on first reading. Some codes related to words/phrases that were regularly repeated (for example, 'environment'), while others related to broader pedagogical concepts that dominated the pre-service teachers' discourse, such as 'hands-on' learning, or teaching as 'facilitation'. These codes were identified by single words with a description of each. At that initial stage, 85 codes were generated, clearly more than were useful.

As suggested by Cresswell (2008, p. 251), this initial coding was reduced during the second stage of coding by deleting redundant codes and by identifying or adjusting codes to address key issues as informed by the literature (see Figure 8.4). Through this reduction process, I was gradually able to develop a sense of what mattered in this study. Once the initial codes were refined, they were collapsed to form broad themes (step 4 in Figure 8.4) in relation to: the research questions; ideas and issues arising from the literature; and to the ecosocial framework. The coding model depicted in Figure 8.4 portrays an orderly, linear process, but this was not so in reality. Rather, data interpretation, coding and thematising were iterative processes including checking the original data and revisiting the research questions at each stage in order to mitigate distortion of the original data.

Researcher roles – juggling multiple hats

Apart from the usual ethical and operational considerations in ethnographic research, a particular challenge for me was that I had multiple roles relating to the SOIL practicum: school liaison, outdoor education method lecturer, practicum supervisor (for students not involved in the study) and researcher. I was therefore enmeshed in the processes of planning and supporting the SOIL practicum throughout the research period. This involvement provided a strong basis for naturalistic enquiry; as in insider, I had a natural legitimacy for participating in both the university and school settings, enabling me to investigate phenomena with which I was familiar (Creswell, 2008). This insider view afforded personal experience of the processes, structures and relationships that allowed me to act as a "broker" (Akkerman and Bakker, 2011) across boundaries of the university and school. I could readily empathise with the

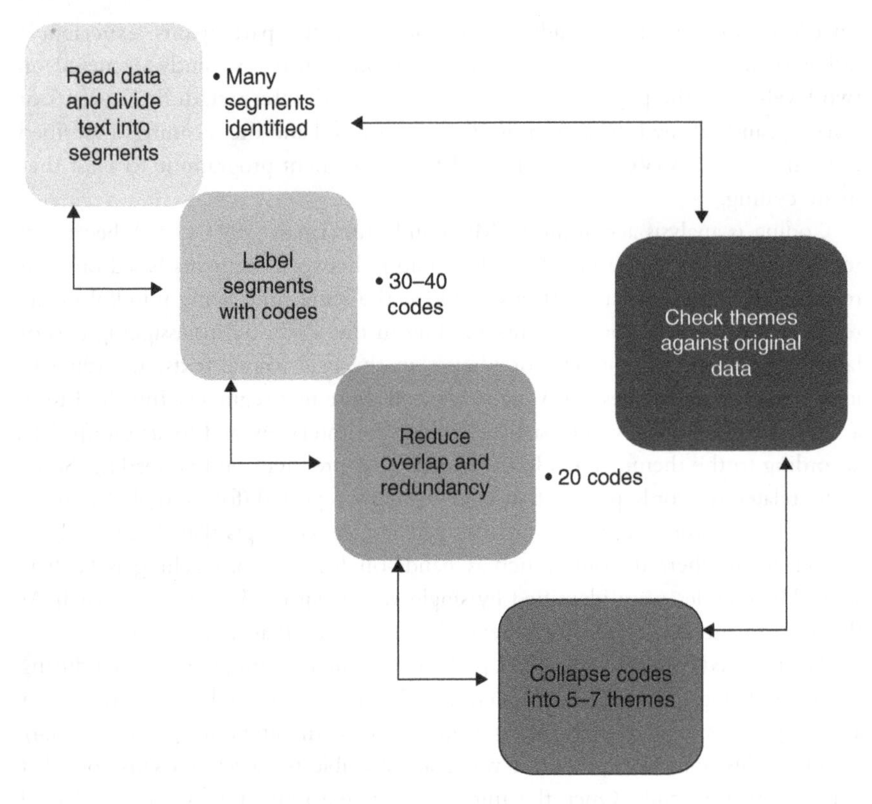

FIGURE 8.4 Data coding process
Source: Adapted from Creswell, 2008, p. 251

pre-service teachers' concerns and also be responsive to "subtle details or unanticipated events" (Wong, 1995, p. 22).

However, tensions between the research, liaising and teaching roles were unsettling when conflicting moral and ethical responsibilities arose. I needed to be particularly vigilant about maintaining confidentiality and trust with all concerned with the SOIL practicum. I had to be clear about my role in the school on any given day (for example: researcher/practicum supervisor/teacher/school-university liaison) and whether information provided was on or off the record (Cresswell, 2008). I also needed to be aware of particular ways in which I conceived the programme, including my *affective* involvement in its unfolding events. In my teaching role I prepared the pre-service teachers for participation in the SOIL practicum but as researcher, I wanted to minimise my influence over participants' conceptualisations of their SOIL work. This was quite a tricky balance as, in both roles, I was a significant mediator of pre-service teachers' SOIL experiences.

Like Wong (1995), I determined that we have a moral responsibility to be fair to students and that educational research must always serve the interests of learning and teaching in school contexts. Research opportunities needed to take a back seat during the SOIL practicum so as not to detract from or interfere unnecessarily with the pre-service teachers' professional development or student learning (Cresswell, 2008). Wong recommends a change in the culture of teaching and learning towards negotiated activity where the boundaries between learner, teacher and researcher are blurred and levels of engagement are more fluid.

Conclusion

This chapter has focused on the challenges of investigating complex professional learning activity where roles and boundaries are blurred and conditions are dynamic. It highlighted the importance of taking into account the conditions and context for learning or activity in the analytic process and of developing a research framework that aligns with the nature of the research object or content. The ecosocial conceptual framework used to frame the investigation offers outdoor educators, sustainability educators and teacher educators a tool for analysing complex educational practice on multiple levels. Ultimately it was the "meeting place" between stasis and change (Freebody, 2003) that proved both challenging and incredibly empowering in terms of the pre-service teachers' professional development. Engeström's (2001) CHAT provided a useful means of understanding the dynamic relationships between elements in the SOIL activity system and of identifying key relationships and conditions that hindered or helped the pre-service teachers' professional learning.

Note

1 See www.un.org/sustainabledevelopment/education/

References

Akkerman, S.F. & Bakker, A. (2011). Boundary crossing and boundary objects. *Review of Educational Research, 81*(2), 132–189.

Bronfenbrenner, U. (1994). Ecological models of human development. In T. Husén & T. N. Postlethwaite *International encyclopedia of education* (2nd ed., pp. 1643–1647). Oxford: Elsevier.

Commonwealth of Australia, Department of the Environment, Water, Heritage and the Arts. (2009). *Living sustainably: Australia's national action plan for education for sustainability*. Canberra. Retrieved from: www.environment.gov.au/sustainability/education/publications/living- sustainably-national-action-plan.

Corcoran, P.B., Walker, K.E. & Wals, A.E.J. (2004). Case studies, makeyourcase, and case stories: A critique of casestudy methodology in sustainability in higher education. *Environmental Education Research, 10*(1), 7–21.

Creswell, J.W. (2008). Educational research: Planning, conducting, and evaluating quantitative and qualitative research (3rd ed.). Upper Saddle River, NJ: Pearson Education.

Davis, B., Sumara, D.J. & Luce-Kapler, R. (2008). *Engaging minds: Changing teaching in complex times*. New York, NY, & London: Routledge.

Edwards, A. (2009). Becoming a teacher: A sociocultural analysis of initial teacher education. In H. Daniels, H. Lauder & J. Porter (Eds.) *Educational theories, cultures and learning: A critical perspective* (pp. 153–164). Oxford, New York, NY: Routledge.

Edwards, A. (2011). Building common knowledge at the boundaries between professional practices: Relational agency and relational expertise in systems of distributed expertise. *International Journal of Educational Research, 50*(1), 33–39.
Engeström, Y. (2001). Expansive learning at work: Toward an activity theoretical reconceptualisation. *Journal of Education and Work, 14*(1), 133–156.
Engeström Y. (2008). *From teams to knots: Activity – theoretical studies of collaboration and learning at work.* Cambridge: Cambridge University Press.
Fenwick, T., Edwards, R. & Sawchuk, P. (2011). *Understanding CHAT: Emerging approaches to educational research: Tracing the sociomaterial.* Oxford: Routledge.
Flyvbjerg, B. (2006). Five misunderstandings about case study research. *Qualitative Inquiry, 12*, 219–244.
Freebody, P. (2003). *Qualitative research in education: Interaction and practice.* London: SAGE.
Gray, T. & Martin, P. (2012). The role and place of outdoor education in the Australian national curriculum. *Australian Journal of Outdoor Education, 16*(1), 39–50.
Lave, J. & Wenger, E. (1991). *Situated learning: Legitimate peripheral participation.* Cambridge: Cambridge University Press.
Lim, C.P. (2002). A theoretical framework for the study of ICT in schools: A proposal. *British Journal of Educational Technology, 33*(4), 411–421.
Lotz-Sisitka, H. & Raven, G. (2004). Learning through cases: Adopting a nested approach to case-study work in the Gold Fields participatory course initiative. *Environmental Education Research, 10*(1), 67–87.
Miles, M.B. & Huberman, A.M. (1994). Qualitative data analysis (2nd ed.). Thousand Oaks, CA, London: SAGE.
Opfer, V.D. & Pedder, D. (2011). Conceptualising teacher professional learning. *Review of Educational Research, 81*(3), 376–407.
Preston, L. (2008). *Becoming green: The formation of environmental ethics in outdoor education* (Unpublished PhD thesis). Melbourne Graduate School of Education, University of Melbourne.
Rogoff, B. (2003). The cultural nature of human development. Oxford: Oxford University Press.
Ross, H., Christie, B., Nicol, R. & Higgins, P. (2014). Space, place and sustainability and the role of outdoor education. *Journal of Adventure Education and Outdoor Learning, 14*(3), 191–197.
Sterling, S. (2004). *Sustainable education: Re-visioning learning and change.* Foxhole: Green Books.
Stevenson, R.B., Ferreira, J., Davis, J. & Evans, N. (2014). *A statewide systems approach to embedding the learning and teaching of sustainability in teacher education.* Sydney: Australian Government Office for Teaching and Learning.
Thomas, G. (2010). Doing case study: Abduction not induction, phronesis not theory. *Qualitative Inquiry, 16*(7), 575–582.
Wenger, E. (1998). *Communities of practice: Learning, meaning and identity.* Cambridge: Cambridge University Press.
Wenger, E. (2009). A social theory of learning. In K. Illeris (Ed.) Contemporary theories of learning: Learning theorists … in their own words (pp. 209–218). London: Routledge.
Wong, E.D. (1995). Challenges confronting the researcher/teacher: Conflicts of purpose and conduct. *Educational Researcher, 24*(3), 22–28.
Yin, R.K. (2003). *Case study design and methods* (3rd ed.). Thousand Oak, CA: Sage.

9

ETHNOGRAPHIC RESEARCH IN OUTDOOR STUDIES

Ina Stan

Introduction

This chapter highlights the process of ethnographic research in conducting research on group interaction in the outdoor classroom. Ethnography is a methodology which is sensitive to participants and can uncover teacher-pupil interactions (Davies, 1984), enabling a holistic understanding of such processes (Fetterman, 1989). This chapter discusses the suitability of an ethnographic approach in outdoor education, examining the rationale for using participant observation, semi-structured interviews and a reflective diary in order to collect rich and diverse data. It also explores the significance of reflexivity within research and the importance of being flexible while conducting fieldwork, drawing on examples from research undertaken at an outdoor centre involving groups of primary school children, their teachers and outdoor facilitators. The chapter also explains the benefits and challenges of participant observation, focusing particularly on the issues that need to be considered when studying children. The chapter considers the appropriateness of using semi-structured interviews and explains the rationale for choosing who should be interviewed and the importance of building rapport. Ethnography requires a continuous engagement in a process of data interpretation during the data collection, culminating in a systematic stage of data analysis after the completion of fieldwork, which is also briefly explained.

Theoretical framework

Ethnography is a research methodology that belongs to the qualitative/interpretative paradigm (Kumar, 2014), aiming to study the culture of a group of people over an extended period of time to gain a holistic insider's understanding of that culture by participating in the group's experiences (Kramer & Adams, 2017; Bloome & Green, 2018). Qualitative research, unlike its positivistic counterpart, does not rely on prediction and measurable proof, but rather it produces 'subjective' data and findings, and this has been criticised for lacking scientific rigour (see Hammersley & Atkinson, 1995). Nevertheless, the idea that quantifiable data and objective observation can lead to verifiable truth is challenged by feminists for providing patriarchal world views (Keller, 1985). Thus, naturalism emerged as a way to study the social world in its 'natural' state, without, where possible, influence from the researcher, arguing that, since human behaviour is not caused in a mechanical manner, it cannot be analysed causally and manipulated through variables, typical of quantitative research (Hammersley & Atkinson, 1995).

Traditional ethnographers' claim to neutrality and objectivity was challenged by Kuhn (1970), who argued that all knowledge is influenced by paradigmatic presuppositions, which means that the paradigm in which researchers situate themselves will have an impact on any validity of scientific claims that are made. Denzin and Lincoln (2000), when presenting a history of qualitative research divided into seven moments (initially five, see Denzin & Lincoln, 1994), talk about a crisis of representation in the fourth moment, which took place in the middle of the 1980s, caused by questions around how social reality should be appropriately represented. Nevertheless, Atkinson, Coffey and Delamont (2010, p. 466) argue that there have always been a multitude of perspectives within ethnographic research and that "it has never been totally subsumed within a framework of orthodoxy and objectivism". They critique Denzin and Lincoln's (1994) linear development of ethnography, emphasising that this development has encompassed many different aesthetic and interpretative viewpoints, arguing that "these so-called trends actually reflect long-standing tensions, rather than constituting a new unique moment in ethnographic research" and that these tensions still continue in the present (Atkinson et al., 2010, p. 466). Similarly, Sparkes (2002) and Sparkes and Smith (2014) argue that there is still a crisis of representation and legitimation (see Chapter 10).

What is ethnography and why use it in outdoor education?

The main aim of ethnography is gaining a holistic understanding of the culture of a group, community or organisation (Bloome & Green, 2018; Kramer & Adams, 2017). Wolcott (1990, p. 188) describes ethnography as providing a "picture of a way of life of some identifiable group of people", while Tedlock (2000) argues that it encompasses a continuous process of placing particular situations, encounters and events into a meaningful context. Since an aim of ethnography is to have an understanding of the culture of a specific group from the point of view of the members of that group (Wolcott, 1990), there is an expectation that ethnographers immerse themselves within settings for extended periods of time (Cushion, 2014). The favoured data collection method, participant observation, involves ethnographers conducting fieldwork in the groups/communities they are observing and actively and attentively participating in their practices (Delamont, 2004; Kramer & Adams, 2017). This enables an insight into how those cultures 'work', uncovering the meaning behind people's beliefs, day-to-day activities, feelings and relationships (Mills & Morton, 2013), and identifying the complex patterns of social life (MacPhail, 2004). Researchers build rapport with the participants and gain knowledge about their language and experiences through regular, in-depth and prolonged contact (Cushion & Jones, 2012). To gain an 'insider perspective', in my research, I immersed myself in the outdoor centre during two phases of data collection, lasting for two months each. I participated in their activities, had lunch with children and staff, gaining an understanding of the culture of the place and how things were done and why.

Ethnography does not have a predetermined course, and researchers use their judgement in the field to determine the direction of their research. Nevertheless, having a clear and developed research problem prior to conducting any fieldwork, or a 'well-developed theory', can be beneficial for the progress of the research (Hammersley & Atkinson, 1995). Medawar (1994) argues that the step-by-step model of the research process perpetuates the myth of objectivity, which is unrealistic, as even in science, 'discoveries' are actually the result of a complex process of social construction (Latour & Woolgar, 1979). My research adopted an emergent design, which meant that the research evolved as my knowledge and understanding increased.

As ethnographers themselves are the principal sources of data (Woods, 1994), subjectivity is inevitable in the research process. Therefore, researchers need to be explicit about their involvement in the research (Mills & Morton, 2013), keeping an open mind about what they are observing and the most appropriate way of writing about it, finding a balance between "suspending preconceptions and using one's present understandings and beliefs to enquire intelligently" (Walford, 2001, p. 9). Researchers need to be aware of any preconceived ideas (Dey, 1993). Achieving this awareness involves engaging in a constant review of the development of one's own ideas, considering carefully why certain decisions are made, particular questions asked and why data are obtained in one way and not another (Walford, 2001). Moreover ethnographers must make explicit any assumptions or beliefs that they may have and how these may impact their research, since the researcher is a part of the research act.

Ethnographers are the research tool, as they are the only instruments flexible enough to have the ability to grasp the complexity, subtleness and unpredictability of human experience (Lincoln & Guba, 1985). Fetterman (1989, p. 41) reinforces this view when stating: "Relying on all its senses, thoughts and feelings, the human instrument is a most sensitive and perceptive data gathering tool". Being empathetic is useful when conducting ethnographic research, but it is also important to for the researchers to use their own judgement to interpret social situations, body language and emotions (Mills & Morton, 2013).

The idea of indwelling is seen by Arendt (1958) as closely linked to the concept of human plurality, according to which all humans are equal, but different from one another. Furthermore Arendt (1958) argues that by embracing this diversity and accessing each other's worlds and experiences as equals, human beings can become aware that one's own understanding of a situation may not be the same as someone else's (Arendt, 1958). Thus human plurality only makes it possible to approximate such understanding, without guaranteeing it.

Reflexivity acknowledges that researchers' views will be influenced by their socio-historical background, rejecting the idea that social research can be conducted separate from the wider society and from the researchers (Hammersley & Atkinson, 1995). Delamont (2004) highlights the importance of reflexivity in ethnography in achieving credibility. Moreover, Humberstone (1997, p. 200) points out that reflexivity enables the reader to gain an insight into "the ways in which webs of power work in both the culture under exploration and within the particular research process". Hence, reflexivity in ethnography develops the researchers' understanding of their own research process.

Ethnography is influenced by social presence (Patton, 1990), as the data collected is mediated by the researcher's theoretical and social background (May, 1999). According to Fetterman (1989, p. 11), ethnography is "the art and science of describing a group or culture". Thus, ethnography allows researchers to use their creativity in the research process, uncovering new knowledge, which can be considered the 'art' in ethnography. However, undertaking an ethnography entails being disciplined (Walford, 2002), as well as having a strong theoretical background (Delamont, 2002; Malinowski, 1922), which can be seen as the 'science' in ethnography.

The focus of my research was to explore interactions within groups of primary school children and their teachers/facilitators while taking part in outdoor activities. Previous research on groups in outdoor education had largely used traditional group concepts, which I considered limiting (see Stan, 2009). Moreover, groups are rarely the focus of research in outdoor education (Stan, 2010, pp. 10–23), even though they are used extensively in outdoor activities. I used ethnography to study groups involved in outdoor activities to gain an insight into teacher-pupil and pupil-pupil interaction.

Rickinson et al. (2004) call for a more in-depth evidence-based understanding of the outdoor educational process. Hence, I believed that an ethnographic exploration of interactions in outdoor classrooms would lead to a deeper understanding of the outdoor educational process. Ethnography is used extensively in education (Mills & Morton, 2013). A key strength of ethnography is its emphasis on understanding people's perceptions and cultures (Walford, 2002). Ethnography tends not to focus on the outcome of the educational process, but rather on the "customs and behaviour of the group and, in particular, the members' *understanding* of the world in which they operate" (Denscombe, 1983, p. 107). Delamont (2002) highlights the importance for researchers not to limit themselves to one type of research setting, but to also explore different sites where learning may occur. Thus, an ethnography of outdoor classrooms contributes to expanding the boundaries of education research.

Being in the field

Fieldwork is essential in ethnography. Pens and notepads, which I considered unobtrusive, practical and inexpensive, were used. Ethnographic methods are eclectic, and I used participant observation, semi-structured interviews and a reflective diary to collect rich and diverse data. Collecting a variety of data, which may range from chance conversations, informal interviews, overheard remarks, observational notes and even some quantitative data, can help ethnographers to better understand and construct the world investigated (Kramer & Adams, 2017). Outdoor researchers like myself can draw upon field notes and reflective diaries that contain an abundance of records of chance conversations with teachers, pupils and facilitators. Public documentation, such as outdoor centres' notice boards and websites can be examined to understand the philosophy of an organisation. However, before collecting any of the data, the researcher needs to negotiate access to the centre.

Choosing a setting requires careful consideration of practical issues, such as travel and relevance. I was introduced to the centre by a former student at my university who worked there and who was an invaluable go-between (Fetterman, 1989). The centre was formally contacted by my supervisor over the telephone and in writing, after which I visited personally. I was well received by the two gatekeepers of the centre, who were the centre's administrator and the deputy director. Gatekeepers have control over key resources.

Gaining access is a continuous process, and does not stop once the researcher enters the setting. There is also a long process of extracting information about the place and the people while developing trusting relationship with the participants, avoiding the withdrawal of access, which can take place at any time (Walford, 2001). Throughout my fieldwork, I came into contact with many different school groups, thus I had to negotiate access with the teachers accompanying the groups. However, having "a strong recommendation and introduction" can strengthen the ethnographer's ability to work in the community, improving the data quality (Fetterman, 1989, p. 44). The centre informed the schools of my presence before their arrival, explaining my role there, which helped my being accepted more readily by the teachers.

Nevertheless, I also had to negotiate access with the children themselves, as simply having the access from the teachers does not guarantee that the children would be ready to accept the researcher (Boyle, 1999). In reality, children are gatekeepers themselves, and it is essential to negotiate access with them to be able to learn about their world (Mandell, 1988). Consequently, I was honest with the children and answered all their questions, which helped me build a trusting relationship based on respect.

As my fieldwork involved being in close contact with children, I needed to be subject to a Criminal Records Bureau Check (now known as the Disclosure and Barring Service), which is a requirement for any adults working in an organisation where there are children. Thus gaining access is a complex issue that does not simply entail having the permission to enter the research setting, but rather it also involves ethical considerations around consent.

As ethnography requires an immersion into a group's culture with the goal of producing new knowledge, researchers need to ensure that no harm is caused to the participants, that there is no deception or invasion of privacy. Ethics need to be considered throughout the whole research process, and ethnographers reflect upon the sensitivity of the topic and how confidentiality and anonymity will be ensured (Mills & Morton, 2013) (see Chapter 2). (It is worth noting that ethnographers put ethical considerations at the centre of their research, arguably long before universities required 'tick box' ethical proposals.) Hence, I informed all participants about my research and my role there. However the extent to which I explained the research varied in accordance with the interest of the person asking for the information. Fetterman (1989) argues that researchers should only go into detail if the person manifests further interest, so I only presented a general idea of my research, but was happy to provide further explanation when this was solicited.

I ensured that I obtained verbal and written consent from the outdoor centre and the teachers. Before observing any activities, I would ask the children for their permission to accompany them and take notes of what they were saying and doing. The children appeared to be positive and unencumbered by my presence. I also asked the centre staff if they consented to me observing their activities, which was always granted. Moreover I used pseudonyms in my research in order to protect the confidentiality of the participants (Christians, 2005).

I used participant observation to collect data during my fieldwork, which entails observing everything, writing detailed field notes, taking the time to expand on them, reflecting on them outside the field, seeking explanations from the participants and collecting any relevant documents, pictures or other artefacts (Delamont, 2004). Ethnographers aim to produce what Geertz (1973) calls thick description of the setting and the participants, which provides sufficiently rich detail to allow the reader to visualise that setting without infringing the principles of the setting. Thus participant observers have to listen carefully and keenly observe what is happening around them to develop a deep understanding of the participants in the social situation observed. Consequently, researchers engage in a learning process of which they must be explicitly aware.

There are various degrees of participation, ranging from "complete participant", "participant-as-observer", "observer-as-participant" and "complete observer" (Junker, 1960) and most researchers assume roles somewhere between the participant-as-observer and observer-as-participant. It is important not to be too rigid and fixed in one role, but rather to be flexible and adapt, which I had to do in order not to compromise my position in the field. The challenge for ethnographers is to combine participation and observation to understand the setting and the people as an insider, while describing them for outsiders (Patton, 1990).

I took on the role of Schutz's (1964) "stranger" on entering, since I was in an unfamiliar setting within a new culture, i.e. a Romanian former teacher in a British outdoor centre. Conducting fieldwork in an unfamiliar culture can be more difficult, but it can also be easier, as everything is strange and the researcher will be able to express that, while researching a familiar culture may lead to failure to report many aspects of the setting (Delamont, 2004). I began my fieldwork with a survey period (Fetterman, 1989), by getting acquainted with the setting and the people. Everything seemed important at first, and I tried to record everything I could, by observing more and participating less so that I would be able to make my observations. I still took part in activities, but only when I thought it appropriate.

It is difficult to assume a fully participative role when researching children, due to the physical size and perceived power of the researcher (Fine & Glassner, 1979). Thus I adopted the role of least-adult (Mandell, 1988), which allowed me to build rapport with the children. I did this by waiting for the children to approach me first, rather than always being the one initiating conversations. As a researcher, I was aware that respecting children, suspending adult judgements and recognising that they hold knowledge themselves would enable me to better understand their world (Boyle, 1999).

Establishing 'normal' social interactions by finding a neutral ground with participants allows a space for informal conversations. Doing this enabled me to build rapport with the participants. However, it was important to allow the participants to have their own space, so during breaks I would sit a little further from them, but gradually participants started to approach me and we engaged in informal discussions.

These informal discussions or "interviews as participant observation" (Hammersley & Atkinson, 1995) became part of my field notes, which often happens in ethnography (Kramer & Adams, 2017). I supplemented these with formal interviews with the staff at the centre, who I considered as key individuals, since they were present throughout the duration of my fieldwork. Teachers and pupils, on the other hand, would be different every week and they also had limited time at the centre, and I felt that interviewing them would have been too intrusive. Moreover, the interviews took place towards the end of each fieldwork stage, by then I had built a strong rapport with the staff at the centre and they appeared to be open to me interviewing them.

I conducted semi-structured interviews, which involved having an interview guide to ensure that all the topics needed were covered, and using open-ended questions to give the opportunity for participants to express their own feelings and thoughts (Sparkes & Smith, 2014). The interviews were tape recorded to avoid missing out valuable data. Participants were able to choose the location of the interview, to make them feel comfortable and, at the beginning of each interview, I explained the purpose of the interview, and reassured them about confidentiality and their right to refuse to answer any questions (Fetterman, 1989).

After the first stage of data collection, I engaged in data analysis as I formulated interpretations of my observations, which I recorded in my reflective diary and which helped narrow down the focus of the next stage of fieldwork. Data analysis is not the final stage in ethnography, but rather an ongoing process since ethnographers reflect on the data collected each day by expressing their thoughts and feelings in a fieldwork diary, which is an early stage of data analysis. However this was not sufficient, and when my fieldwork was completed, I engaged in a more systematic process of data analysis (Bhatti, 2002).

Ethnographic analysis entails undertaking an intensive analysis of the data to discover patterns, which can help to explain the culture studied. I did this by reading carefully all the field notes, the diary and the interviews transcripts looking for patterns of participant behaviour, repetitions in the way the activities were conducted, comparing and contrasting different kinds of interactions between participants. I thus identified cultural domains, which are important basic units in every culture (Spradley, 1980). I also conducted a taxonomic analysis by examining how the cultural domains were organised. Taxonomies reveal the relationship between all the included terms in a domain and identify subsets and the way they are related to the whole (Spradley, 1980). I then carried out a componential analysis by searching for attributes of terms in each domain. Finally, I conducted a theme analysis by searching "for the relationships among domains and for how they are linked to the cultural scene as a whole" (Spradley, 1980, pp. 87–88).

Conclusion

In the words of Delamont (2004, p. 215): "ethnography is hard work: physically, emotionally and mentally exhausting. The research does not proceed in a straight line, but in a series of loops, because each step leads the researcher to reflect upon, and even revisit, earlier steps". This quote describes my ethnographic journey accurately, and this chapter shows the complexity of this methodological approach, by first discussing the theoretical framework underpinning ethnography and how ethnographers do not sit in a comfortable position, removed from their research, but are rather part of the research. I explain what ethnography entails and why it is a useful approach for a study in outdoor education. The complexities of conducting ethnographic fieldwork are highlighted. These include careful consideration of the researcher's role in the field, interaction with the participants, types of methods and data, and ethical issues. The data analysis I conducted enabled me to gain an understanding of the cultural scene studied.

References

Arendt, H. (1958). *The human condition.* Chicago, IL: University of Chicago.

Atkinson, P., Coffey, A. & Delamont, S. (2010). Ethnography: post, past, and present. In Atkinson, P. & Delamont, S. (Eds.) *SAGE qualitative research methods* (pp. 461–471). Thousand Oaks, CA: SAGE.

Bhatti, G. (2002). On the doctoral endeavour. In G. Walford (Ed.) *Doing a doctorate in educational ethnography, Vol.* 7 (pp. 9–27). Oxford: Elsevier.

Bloome, D. & Green, J.L. (2018). In Frey, B. (Ed.) *The SAGE encyclopedia of educational research, measurement, and evaluation* (pp. 612–622). Thousand Oaks, CA: SAGE.

Boyle, M. (1999) Exploring the words of childhood: The dilemmas and problems of the adult researcher. In A. Massey & G. Walford (Eds.) *Studies in educational ethnography, Vol. 2, Explorations in methodology.* Stamford, CT: Jai Press Inc.

Christians, C.G. (2005). Ethics and politics in qualitative research. In N.K. Denzin and Y.S. Lincoln (Eds.) *The SAGE handbook of qualitative research* (3rd ed., pp. 187–205). Thousand Oaks, CA: SAGE.

Cushion, C. (2014). Ethnography. In L. Nelson, R. Groom & P. Potrac (Eds.) *Research methods in sport coaching* (pp. 171–180). Abingdon: Routledge.

Cushion, C.J. & Jones, R.L. (2012). A Bourdieusian analysis of cultural reproduction: Socialisation and the "hidden curriculum" in professional football. *Sport, Education and Society, 19*(3), 276–298.

Davies, L. (1984). *Pupil power: Deviance and gender in school.* London: Falmer Press.

Delamont, S. (2002). *Fieldwork in educational settings: Methods, pitfalls and perspectives* (2nd ed.). London: Routledge.

Delamont, S. (2004). Ethnography and participant observation. In Seale, C., Gobo, G., Gubrium, J. F. & Silverman, D. *Qualitative Research Practice* (pp. 205–217). London: SAGE.

Denscombe, M. (1983). Interviews, accounts and ethnographic research on teachers. In M. Hammersley (Ed.) *The ethnography of schooling* (pp. 105–128). Driffield: Nafferston Books.

Denzin, N.K. & Lincoln, Y.S. (Eds.). (1994). *The SAGE handbook of qualitative research.* Thousand Oaks, CA: SAGE.

Denzin, N.K. & Lincoln, Y.S. (Eds.). (2000). *The SAGE handbook of qualitative research* (2nd ed.). Thousand Oaks, CA: SAGE.

Dey, I. (1993). *Qualitative data analysis: A user-friendly guide for social scientists.* London: Routledge.

Fetterman, D.M. (1989). *Ethnography – step by step.* London: SAGE.

Fine, G. & Glassner, B. (1979) Participant observation with children. *Urban Life, 8*, 153–174.

Geertz, C. (1973) Thick description: Towards an interpretive theory of culture. In C. Geertz (Ed.) *The interpretation of cultures: Selective essays* (pp. 3–30). New York, NY: Basic Books.

Hammersley, M. & Atkinson, P. (1995). *Ethnography: Principles in practice.* London: Routledge.

Humberstone, B. (1997). Challenging dominant ideologies in the research process. In G. Clarke & B. Humberstone (Eds.) *Researching women and sport* (pp. 199–213). London: Macmillan.

Junker, B. (1960). Field *w*ork. Chicago, IL: University of Chicago Press.

Keller, E. (1985). *Reflections on gender and science.* New Haven, CT: Yale University Press.

Kramer, M.W. & Adams, T.E. (2017). Ethnography. In Allen, M. (Ed.) *The SAGE encyclopedia of communication research methods* (pp. 458–461). Thousand Oaks, CA: SAGE.

Kuhn, T.S. (1970). *The structure of scientific revolutions* (2nd ed.). Chicago, IL: University of Chicago Press.

Kumar, R. (2014). *Research methodology: A step-by-step guide for beginners* (4th ed.). London: SAGE.

Latour, B. & Woolgar, S. (1979). *Laboratory life. The social construction of scientific facts.* London: SAGE.

Lincoln, Y.S. & Guba, E.G. (1985). *Naturalistic inquiry.* Beverly Hills, CA: SAGE

MacPhail, A. (2004). Athlete and researcher: undertaking and pursuing ethnographic study in a sports club. *Qualitative Research, 4(2)*, 227–245.

Malinowski, B. (1922). *Argonauts of the Western Pacific.* London: Routledge & Kegan Paul.

Mandell, N. (1988) The least-adult role in studying children. *Journal of Contemporary Ethnography, 16*(4), 433–467.

May, T. (1999). *Social research: Issues, methods and process.* Buckingham: Open University Press.

Medawar, P. (1994). Feminism and the possibilities of a postmodern research practice. *British Journal of Sociology of Education, 14*(13), 327–311.

Mills, D. & Morton, M. (2013). *Research methods in education: Ethnography in education.* London: SAGE.

Patton, M. (1990). *Qualitative researching.* London: SAGE.

Rickinson, M, Dillon, J., Teamey, K., Morris, M., Young Choi, M., Sanders, D. & Benefield, P. (2004). *A review of research on outdoor learning.* London: National Foundation for Educational Research & King's College London.

Schutz, A. (1964). The stranger: An essay in social psychology. In A. Schutz (Ed.) *Collected Papers, Vol. II.* The Hague: Martinus Nijhoff.

Sparkes, A.C. & Smith, B. (2014). *Qualitative research methods in sport, exercise and health: from process to product.* Abingdon: Routledge.

Sparkes, A.C. (2002). *Telling tales in sport and physical activity: A qualitative journey.* Champaign, IL: Human Kinetics.

Spradley, J.P. (1980) *Participant observation.* New York, NY: Holt, Rinehart and Winston.

Stan, I. (2009) Recontextualising the role of the facilitator in group interaction in the outdoor classroom. *Journal of Adventure Education and Outdoor Learning, 9*(1), 23–43.

Stan, I. (2010) *Group interaction in the 'outdoor classroom': The process of learning in outdoor education.* Saarbrücken: VDM Verlag Dr. Müller.

Tedlock, B. (2000). Ethnography and ethnographic representation. In N.K. Denzin & Y.S. Lincoln (Eds.) *The SAGE handbook of qualitative research* (2nd ed., pp. 455–486). London: SAGE.

Walford, G. (2001). *Doing qualitative educational research: A personal guide to the research process.* London: Continuum.

Walford, G. (Ed.) (2002). *Doing a doctorate in educational ethnography, Vol.* 7. Oxford: Elsevier.

Wolcott, H. (1990). Ethnographic research in education. In R. Jaeger (Ed.) *Complementary methods for research in education* (pp. 187–206). Washington, DC: American Educational Research Association.

Woods, P. (1994). Collaborating in historical ethnography: researching critical events in education. *International Journal of Qualitative Studies in Education,* 7(4), 309–321.

10

AUTOETHNOGRAPHY

Creating stories that make a difference

Barbara Humberstone and Robbie Nicol

Introduction

In 2009, the authors were attending the European Outdoor Network conference in Uekermunde, Germany. Robbie was presenting *Canoeing Around the Cairngorms: A Circumnavigation of my Home* (Nicol, 2016). Afterwards Barbara commented "that sounded a bit like autoethnography". Since that encounter, we have both adopted autoethnographic onto-epistemological positioning and published a number of narratives based upon our research. Autoethnography is a significant research approach that has been adopted within the social sciences at varying times (Jones, Adams & Ellis, 2016).

This chapter examines autoethnography as methodology and method, its development, its critiques, and highlights its potential for understanding lived sentient experiences and evoking social and environmental awareness. The chapter provides (re)-presentations of autoethnographic approaches for exploring the senses in embodied movement in nature and memory in connecting with place.

In search of meaning

Autoethnography is a holistic interpretative qualitative research approach that can uncover unique insights into the embodied experiences of the life-worlds of being and becoming within social, cultural and political contexts. It has been referred to as introspective ethnography, personal narrative and narrative of self in the social sciences. Past interrogation of dominant social science research drew attention to major contradictions/ambiguities in how research could/should be represented and judged (Sparks & Smith, 2014). Two of these are identified as firstly, the crisis of representation: How researchers write, explain and describe the social world needed to take account of the complexities of human actions; and secondly, the crisis of legitimation: Whether the researcher can speak the 'truth' for 'others'. Consequently, narrative scholars drawing upon their embeddedness in particular cultures through ethnographic methods began to story their research writing evocatively and with rich description, beginning with "impressionistic tales" (Van Maanen, 1988) and moving into other forms of writing research (Sparkes, 2000).

This crisis of legitimation and representation in the social sciences thus prompted new ways of judging and representing research. Such methodologies as ethnography and autoethnography and the like, which are underpinned by phenomenological perspectives (Hockey & Allen-Collinson, 2007), provide unique methodologies for exploring human and non-human interactions in the outdoors. New forms of social science were developed in response to these crises. Impressionistic tales of social analysts' own cultural experiences, focusing on specific cases, rich description and experiential knowledge emerged (Denzin, 2001). Narrative scholars began to 'story' their understanding of themselves and their social worlds, developing a new 'moment' in social research (Denzin & Lincoln 2000, Coates, 2010). Autoethnography is a qualitative methodology connected to storied writing.

Autoethnographers' actions, feelings and emotions are part of the research process. Autoethnographic research is self-reflexive and embedded in the field of narrative research. It is an increasingly popular approach to uncovering and generating knowledge. The term consists of 'auto', referring to self, 'ethno' referring to culture and human interaction and 'graphy' referring to the process of doing research. Hammersley and Atkinson (1995) posit it is similar to reflexive ethnography. Reflexive ethnographies build upon personal experiences. Exploring situated experiences, reflexive ethnographers attempt to make sense of their feelings and selves, and their personal experiences and relations with other participants in the research context to understand the participants' life-world and experiences. In autoethnography, the self is positioned at the centre of the research, rather than centring other participants. Critiquing this centring of self, Anderson (2006) argues that autoethnographic research falls into the trap of self-absorption. Delamont (2007) reasons that autoethnography is problematic, maintaining that it cannot be published ethically, it is experiential not analytic, it focuses on the wrong side of the power divide and is self-indulgent. Likewise Atkinson (2006, pp. 402–403) argues that autoethnography is based on "its experiential values, its evocative qualities and its personal commitment rather than its scholarly purpose, its theoretical bases and its disciplinary contributions".

We, however, agree that autoethnography is a form of reflexive ethnography. It goes beyond the self, engaging with social, cultural and interactional relations and issues. It is the way in which researchers draw upon their feelings, emotions and experiences, which legitimates and gives credibility to autoethnography as a significant means of inquiry. The critiques above do not fully interrogate or do justice to the notion of an 'embodied' situated researcher (Olive 2016). Ellis and Bochner (2000, p. 739) argue that autoethnography is personal narrative, "writing and research that displays multiple layers of consciousness, connecting the personal to the cultural". Different and varying emphases are placed on the varied aspects of the narrative in different situated research. In this way, drawing upon personal stories and researcher feelings, connections between personal experience and particular life-worlds and places unfold which provide the basis for critical reflexive accounting of the social and cultural. Humberstone (2013, p. 499) suggests that "there is a combination of autoethnography with ethnography so that 'others' and relations with others move in and out of the centre of the research as the self moves fluidly from centre to periphery of the research".

Anderson (2006, p. 78) supports an analytical or "realist" form of autoethnography that constitutes, "1. complete member research status, 2. analytical reflexivity, 3. narrative visibility of the researcher's self, 4. dialogue with informants beyond self, and 5. commitment to theoretical analysis". Analytical autoethnographers aim for theoretical understanding whilst artistic autoethnographers write evocative stories. Burnier (2006) and Ellis (2004) argue for both analytical and artistic forms of autoethnography and the term creative analytical process (CAP) (Richardson, 2000) combines research that is both creative and analytical.

Clough (cited in Goodley et al., 2004) raises critical concerns regarding such ways of writing research:

> If we are arguing for the use of narrative in research reports there has to be a critical questioning of the purpose of such accounts. We must ask: Who are these stories for? What effects do the stories have on whom?
>
> *p. 184*

These are some of the questions that need to be asked, and answered, when undertaking autoethnographic research in outdoor studies.

Like ethnography (see Chapter 9), in autoethnography the researcher is the research instrument and data are collected through the experience of the researcher. The autoethnographer is not restricted to any particular method. Through methodological pluralism the authoethnographer investigates interactions with non-human and humans in situated contexts (Chang, 2008; Edensor, 2000). In autoethnographic research current phenomena, informed by past and present experiences and how they might affect the future, are central to the analyses. What are the smells, sounds, bodily feelings i.e. rain on skin or balance sensations; what are/were the emotions (Humberstone, 2019)? These provide for sensuous analyses. Our senses tell us a great deal about the 'atmosphere', the subtleties and nuances of the situated research (see Ellis, 2004; Pink, 2009; Sparkes 2017).

Autoethnographic analysis is ongoing, reflective and depends on the researcher and their experience perspective. Since the human condition and human interactions are complex we would argue that theoretical frameworks will always be partial but consistent with a process of change and becoming. For example, Brown, Collins and Humberstone (2018, pp. 489–502), in order to share their lifelong involvement in the outdoors within changing times, used topical life stories in the form of co-autoethnographies. In these narratives they chose to examine their lives, "through a specific lens, that of engagement with nature/the nonhuman world, rather than other lenses such as our experiences of being women in a male-dominated field". They propose that if they had, "focused through the latter lens, the narratives ... would have been very different. These narratives may have spoken to inequality and silencing" (Humberstone, 1998).

Memory, thus, has a powerful influence on autoethnographic processes. Giorgio (2016, p. 406) asserts that, "as autoethnographers, we use memory for much of our data; through memory we ground our analyses; our memories inform our epistemologies and methodologies". She continues "I re-live and re-imagine, shaping my memories into autoethnography, a suturing of lived experience with theory, (and) memory with the forgotten" (pp. 406–407). Autoethnographic research can not only enable exploration of cultural contexts of life span, but enable experiences of place and embodiment to be memorialised, sensorialised and (re) presented in various ways.

The following section shows one process of ethnographic methodology in exploring relations to place, bringing memory to the fore.

Memories, place and process

July 2018

Jane and I (Robbie), our two Labradors, Farril (14) and Cuillin (1), holidayed this summer in Scotland's Hebridean Islands. We stayed a few nights on the Island of Eriskay, parking our campervan near the pub called *Am Politician* popularised through Compton Mackenzie's book *Whisky Galore.* The age gap between Farril and Cuillin means different exercise routines and so we often walk them separately. In this late summer evening I am alone with Farril.

Meandering slowly along a sandy beach overlain with the fragments of broken mollusc shells both of us following different, sightless paths with no destination in mind. Farril is enjoying the smells and tasty morsels of decaying shellfish and I am left with my thoughts. These thoughts are playing heavily on my mind because by the time I get home this article needs to be written – but I need to remain a full participatory member of this family holiday. A wry smile forms as I think of my academic friends and colleagues who all seem to end up with this issue.

In that moment I was reminded that choices are necessary to adopt appropriate methods to conduct an enquiry, and the rigour of these methods, and the data they elicit, should be considered through some form of methodological analysis. Autoethnography is all of these things but a lot more too. It is not just a method or methodology, but a way of life – a life characterised by the need for "creating change within selves, communities and cultures" (Jones et al., 2016 p. 39). For me this translates into the pursuit of social and ecological justice (Nicol 2013, 2014, 2018). There's the irony. Holidays are supposed to provide quality time with loved ones away from work yet inequalities know no such boundaries. A perfect recipe for writer's paralysis and doing nothing, hence my moody reflections. Jane asked earlier "what's wrong?" and I explained. She encouraged me to write. With four weeks ahead we have lots of time, she reasoned.

But what to write about?

Paws and boots shuffle along the debris of shells and sand. My eyes are cast downward in search of inspiration amongst the disaggregated assortment of tide and time. Something, perhaps a change of light, or a synapse transfer, results in a subconscious shift in my attention as I look from the beach to the horizon where Beinn Sciathan (185m) breaks the skyline, and in that moment I know that I want to walk up it. Simultaneously, I know what I will write about because I feel my autoethnographic self mirrored back to me. A mere whim, flight of fancy, change of mood, a lift of the head, whatever, has just provided me with a writing goal and I am not entirely sure why this has happened. I do know that it has happened before and will happen again.

Like most people interested in outdoor studies I am fascinated by outdoor places, what makes them special and why some seem more special to me/us than others. An autoethnographer, like any other researcher, needs to relate their enquiries to the background literature(s) into which their own research will make a contribution. I had recently been reading and writing about Place-Based Education (PBE). I am drawn to this literature /theoretical framework because of its explicit, and sometimes implicit, focus on social and ecological justice. Wattchow and Brown (2011, p. 54) explain "how people live within subjective lifeworlds that not only influence their experience of life, but which direct much of their actions", and I have argued that the moral significance of our relationship with places is based on the attention we pay to them (Nicol, 2014).

Cultural geographers have informed this literature. Tuan (1974) and Relph (1976) pointed out how human experience is affected by *dwelling* in places and spaces. Gruenewald (2003, p. 8) has suggested the need for "place-consciousness" which is related to Seamon's (2014, p. 11) term "place attachment" which he defines as "the emotional bonds between people and a particular place or environment". These ideas of dwelling, consciousness and attachment and the emotions they invoke provide an interesting tripartite synthesis to better understand my own fascination with Beinn Sciathan. Yet there is more.

Making meaning from memories

Schama (2004) explores the storied landscape. He shows how memory is not simply individual but cultural and that layers of rock are related to layers of memory as both stretch back in time. There is potential for creative frisson when landscape and memory are combined providing rich research opportunities for autoethnographers. Chang (2008, p. 71) explains that: "personal memory is a building block of autoethnography because the past gives a context to the present self and memory opens a door to the richness of the past" (and, for reasons explained above, 'the future'). When I gaze upon Beinn Sciathan it is a place that can be described objectively. It is also a place that fills my subjective self with longing and desire. The source of that is found in my memories. This might at first sound odd since I have never before set eyes on Beinn Sciathan, never mind set foot within this more-than-human place. In many ways though I have been here before, and on many occasions, when I consider the memory-ridden longing and desire I have for places like this. Walking has been so much a part of my life, the Alps, the

Pyrenees, the Scottish Munros, Corbetts, lochs, sheilings, burns and glens. All these layers of memory are brought from the past through a stream of consciousness and fixated on this one place I experience in the present. Thus my memories infuse my current experience with levels of familiarity and intimacy that shape how I feel and think about Beinn Sciathan.

It was almost as though these memories had disappeared from my consciousness only to return with a reminder when faced with the immediacy, directness and anticipation while gazing upon Beinn Sciathan.

Autoethnography goes beyond the self, bringing together mind and world engaging social, cultural, spatial, environmental and interactional relations and issues. I have taken the experiences of Beinn Sciathan and memorialised them through the collective experiences of other places which in turn acts as a prism bringing history into the present and vice versa. But it can go further. For Jones et al. (2016, p. 21) autoethnography "creates a space for a turn, a change, a reconsideration of how we think, how we do research and relationships, and how we live". This change may be through our own reflexive analyses, but also through the ways in which narratives are represented, 'read' and evoked.

Narrative (re)-presentations

The question of representation argues for showing rather than telling. Consequently, autoethnographic narratives, the way that the research process is represented, are frequently written or presented as a story. Some autoethnographies may be written in relatively traditional form with a review of literature, methods, findings, discussion and conclusion. This is rare as findings and discussion are usually synthesised. Other more creative/artistic forms use storying to create an evocative narrative piece (see Chapter 14), others use creative nonfiction (see Chapter 13), whilst others may use poetic style text and art to bring the reader to the experience.

For example Brown represents his embodiment, senses and ocean sailing thus:

Seascapes, the sea, the horizon
Rhythm and movement, waves and wind,
Salty lips, cresting waves, blue sky,
feel the movement, side to side, up and down
forwards and backwards
Reflected sun on sea – glistening, sparkling – beauty in harmony
Seascapes, the smells, the motions
Dysrhythmia, jerky movements, waves and wind
Taste the bile, sloshing waves, blue sky,
feel the movement, side to side, up and down
forwards and backwards
Reflected sea on boat-banging, slopping-beauty in disharmony

Brown, in Humberstone, Fox & Brown, 2017, p. 82.
Reproduced with permission.

In the text above, Brown attempts to bring to the reader his sentient embodied experiences, both sublime and grossly uncomfortable, of ocean sailing through this 'poetic' narrative form. This provides something of the visceral sensations. lisahunter & emerald (2016, p. 39) propose and ask:

> *Using* the senses to create a text and creating a text that can be engaged sensorially: can it be touched, smelt, tasted, can a research text evoke pleasure or pain, where/when is it in place/space/time, how can a text capture me (turn me)?

And, we would add, the reader?

Technologies may provide another way of (re)-presenting autoethnographic experiences (see Chapter 20).

In the text focusing on "seeking the senses", Barbara draws upon her journal of an afternoon kayaking amongst some marshes in the Solent, UK, recalling (representing) her sentient interaction with sky, sea/land, birds and pollution.

> *2nd November 2015*
>
> *Last night's flight to Dublin for seminar cancelled due to fog throughout Britain. Today the fog is thick, drifty and it is drizzling, a dismal day. BUT there is a rumour that the coast has sunlight and the tide is up.*
>
> *Late afternoon and I drive down to the spit; within a mile the mist is swirling high and small clouds cling to the power station chimney. On and around the spit, the sky is clear blue, the sun shines across the calm still sea as we unlock the plastic kayaks. We carry our kayaks to the water's edge, I get into my yellow kayak and push into the creek, turn the kayak and paddle;*
>
> *left, right, left, right, left, right ...*
>
> *straight into the sun which sparkles from its low angle on the horizon. All around and above is blue, at the edges is a wall of mist which surrounds us like cotton wool sheets, I think strangely of 'worm holes' and 'parallel universes'. The Isle of Wight, usually clear and distinct, has vanished behind a misty veil.*
>
> *We paddle across the creek, past the mini spit, the sun now to my left. The small orange-hulled commercial fishing boat lies motionless in the calm water to my right, while two larger trawlers move across our path working the sea bottom. A passing ferry in the channel creates a set of small waves upon which the kayak surfs easily to the marshes. The marshes are full of geese, ducks and other birds resting on the marsh-grass and shore. As I paddle up to the marshy island they rise into the air as one ... geese vigorously honking, ducks yodelling and grunting. Wild fowl fly for a short distance, landing again further away, quiet descends once more. I push ashore and walk along the mud/shingle shoreline where broken bottles lie, worn smooth by the motion of the sea and the stones. Picking up a discarded plastic fishing twine and a large plastic bag, I put them in my kayak. The sun is lower and larger, a big red-orange orb balanced on the power station. I feel warm, happy and content. It is nothing like the damp, mist and greyness outside the cotton wool cloak.*

> *We make our way back, passing closer in shore using the small waterways that cut through the marshes. I sit tall and rotate from my waist pushing with my top hand and pulling with bottom hand. A rhythm which I remember from so long ago which is both mesmerizing and self-affirming:*
>
> *right, left, right, left, right, left …*
>
> *The water laps gently against the kayaks as we move smoothly along. We see a crested grebe but she dives vanishing under the water and we can't find her again. The sun, now an enormous red glowing orb to my right, sits on the horizon ready to disappear. We paddle back to the shore, getting out, stretching our legs as the sun disappears below the horizon. And the world changes back to a dismal grey.*
>
> *Humberstone, in Humberstone et al., 2017, pp. 91–92*

In the example above, Barbara attempts to show the feelings/emotions of her connections with the wider universe and the embodied, physical repetitive motion of paddling the plastic kayak set against the more-than-human wildlife and the human interventions of plastics and power-station. Does it touch/turn the reader?

Concluding reflections

We have shown that autoethnography can be adopted not only to explore emplaced, embodied contemporary phenomena in outdoor studies, but also that it is informed by former experiences, various theoretical frameworks and may extend thinking into the future. Autoethnographic narratives can be storied, documentary, 'poetic', arts-based, visual and so forth. They may evoke a variety of emotions in the reader. We should trust the text and its reader. Autoethnographers are often inspired to challenge injustices. Barbara is inspired by ecofeminisms' thought and praxis (Humberstone, 2018, 2019) and interactions with more-than-human. For Robbie it is the inequalities brought about by human-centredness, the domination of one species over others and developing action-orientated educational responses that challenge these inequalities. These issues shape who we are and so our pasts influence how we experience the present. Barbara's sea kayaking narrative links with the past and shows plastic pollution and industrial artefacts in the midst of natural beauty and the innocence of, and interconnection with, the more-than-human. For Robbie it is how the challenges outlined by PBE might be addressed through suitable methodological enquiry processes. Both of us are not only writing about the world, but in a sense creating narrative that might bring about change. This brings us to the future. Autoethnography has a performativity function through the consummation of some form of action and if not, then it justly deserves the critiques of self-absorption we have outlined above. Performativity has both implicit forms such as the exploration or creation of new knowledge (which could lead to change but may not be the initial intent) and more explict change functions. The key to autoethnography therefore is how experiences of the past are filtered through the direct experience of the present and then used to direct future actions to re-imagine and 'turn' a better world.

References

Anderson, L. (2006). Analytical auto-ethnography. *Journal of Contemporary Ethnography, 35*(4), 330–345.

Atkinson, P. (2006). Rescuing auto-ethnography. *Journal of Contemporary Ethnography, 35*(4), 373–395.

Brown, H., Collins, D. and Humberstone, B. (2018). Three women's co-ethnography of life-long adventures in nature. In T. Gray and D. Mitten (Eds.). *The Palgrave international handbook of women and outdoor learning*. (pp. 489–502). London: Palgrave.

Burnier, D. (2006). Encounters with the self in social science research: A political scientist looks at ethnography. *Journal of Contemporary Ethnography, 35*(4), 410–18.

Chang, H. (2008). *Autoethnography as method*. Walnut Creek, CA: Left Coast Press.

Coates, E. (2010). A personal journey through "moments": doctoral research into parents who rock climb. *Journal of Adventure Education and Outdoor Learning, 4*(2), 142–157.

Delamont, S. (2007). Arguments against auto-ethnography. *Qualitative Researcher, 4*, 2–4.

Denzin, N.K. (2001). *Interpretive interactionism*. Thousand Oaks, CA: SAGE

Denzin, N.K. and Lincoln, Y.S. (2000). Introduction. In N.K. Denzin and Y.S. Lincoln (Eds.) *The SAGE handbook of qualitative research* (2nd ed., pp. 733–768). Thousand Oaks, CA: SAGE.

Edensor, T. (2000). Walking in the British countryside: Reflexivity, embodied practices and ways to escape. *Bodies & Society, 6*(3–4), 81–106.

Ellis, C. (2004). *The ethnographer I: A methodological novel about autoethnography*, Walnut Creek, CA: Altamira Press.

Ellis, C. & Bochner A.P. (2000). Autoethnography, personal narrative, reflexivity. In N.K. Denzin & Y.S. Lincoln (Eds.) *SAGE h*andbook of qualitative research (2nd ed., pp. 733–768). Thousand Oaks, CA: SAGE.

Giorgio, G. (2016). Reflections on writing through memory in autoethnography. In S. Jones, T. Adams and C. Ellis (Eds.) *Handbook of autoethnography* (pp. 406–424). Oxford: Routledge.

Goodley, D. (2004). Approaching: methodology in the life story: A non-participatory approach. In D. Goodley, R. Lawthorn, P. Clough & M. Moore (Eds.) *Researching life stories: Method, theory, and analysis in a biographical age* (pp. 56–60). London: Routledge.

Gruenewald, D. (2003). The best of both worlds: A critical pedagogy of place. *Educational Researcher, 32*(4), 3–12.

Hammersley, M. & Atkinson, P. (1995). *Ethnography. Principles in practice*. London: Tavistock Publications.

Hockey, J. & Allen-Collinson, C. (2007). Grasping the phenomenology of sporting bodies. *International Review for the Sociology of Sport, 42*(2), 115–131.

Humberstone, B, (1998). Re-creation and connections in and with nature: Synthesizing ecological and feminist discourses and praxis? *International Review for the Sociology of Sport, 33* (4), 381–392.

Humberstone, B. (2013). Adventurous activities, embodiment and nature: Spiritual, sensual and sustainable? Embodying environmental justice. *Motriz, Journal of Physical Education, Rio Claro 19*(3), 565–571.

Humberstone, B. (2018). Foreword: Nourishing terrains-nurturing terrains: Network of connections. In T. Gray & D. Mitten (Eds.) *The Palgrave international handbook of women and outdoor learning* (pp. vii–xi). London: Palgrave Macmillan.

Humberstone, B. (2019). Bodies and technologies: Becoming a mermaid: Myth, reality, embodiment, cyborgs, windsurfing and the sea. In K. Peters, & M. Brown (Eds.) *Living seas* (pp. 183–195). London: Routledge.

Humberstone, B., Fox, K. & Brown, M. (2017). Sensing our way through ocean sailing, windsurfing and kayaking. In S. Sparkes (Ed.) *Seeking the senses in physical cultures: Sensual scholarship in action* (pp. 82–100). London: Routledge.

Jones, S. Adams, T. & Ellis, C. (2016). Coming to know autoethnography as more than a method. In S. Jones, T. Adams & C. Ellis (Eds.) Handbook of autoethnography (pp. 17–47). Oxford: Routledge.

lisahunter & elke emerald (2016). Sensory narratives: capturing embodiment in narratives of movement, sport, leisure and health. *Sport, Education and Society, 21*(1), 28–46.

Nicol, R. (2013). Returning to the richness of experience: Is autoethnography a useful approach for outdoor educators in promoting pro-environmental behaviour? *Journal of Adventure Education and Outdoor Learning, 13*(1), 3–17.

Nicol, R. (2014). Entering the fray: the role of outdoor education in providing nature-based experiences that matter. *Educational philosophy and theory, 46*(5), 449–461.

Nicol, R. (2016). *Canoeing around the Cairngorms. A circumnavigation of my home.* Aboyne: Lumphanan Press.

Nicol, R. (2018). "Dear oak": Showcasing the value of place-based education through local landscapes. Retrieved from: www.teaching-matters-blog.ed.ac.uk/?p=3069

Olive, R. (2016). Surfing, localism, place-based pedagogies, and ecological sensibilities in Australia. In B. Humberstone, H. Prince & K.A. Henderson (Eds.) *International handbook of outdoor studies.* (pp. 501–510). Oxford: Routledge.

Pink, S. (2009). *Doing sensory ethnography.* London: SAGE.

Relph, E. (1976). *Place and placelessness.* London: Pion.

Richardson, L. (2000). Writing: A method of inquiry. In N. Denzin & Y.S. Lincoln (Eds.) *The SAGE handbook of qualitative research* (pp. 923–948). London: SAGE.

Schama, S. (2004). *Landscape and memory.* London: Harper Press.

Seamon, D. (2014). Place attachment and phenomenology. In L. Manzo and P. Devine-Wright (Eds.) *Place attachment: Advances in theory, applications and methods* (pp. 11–22). London: Routledge.

Sparkes, A.C. (2000). Autoethnography and narratives of self: Reflections on criteria in action. *Sociology of Sport Journal, 17*(1), 21–43.

Sparkes, A.C. (Ed.) (2017). *Seeking the senses in physical cultures: Sensual scholarship in action.* London: Routledge.

Sparks, A.C. & Smith, B. (2014). *Qualitative research methods in Sport, Exercise and Health. From process to production.* London: Routledge.

Tuan, Y. (1974). *Topophilia: A study of environmental perception, attitudes, and values.* London: Prentice-Hall.

Van Maanen, J. (1988). *Tales of the field: On writing ethnography.* Chicago: University of Chicago Press.

Wattchow, B. & Brown, M. (2011). *A pedagogy of place: Outdoor education for a changing world.* Melbourne: Monash University Publishing.

11

THINKING THE SOCIAL THROUGH MYSELF

Reflexivity in research practice

Rebecca Olive

The challenge of reflexivity

Many lifestyle and action sports are about people engaging with nature through movement. Climbing, riding, flowing with, and being immersed in blue and green spaces – seas, rivers, forests, snowy mountains, soaring cliff faces – has shaped the cultures of activities like surfing, mountain-biking, snowboarding, wild swimming, rock climbing and base-jumping (Brymer and Gray, 2010). As these physical cultures and sports continue to grow in popularity in places around the world, researchers are working to understand how they impact individuals, communities, policies and the environments in which they are practised (Gilchrist & Wheaton, 2011). There are highly structured, competitive and professionalised aspects to many of these activities (Thorpe & Dumont, 2018). However, given that so many lifestyle and action sports are practised recreationally, outside of formal organisations and competition, it is not surprising that much of the research about action and lifestyle sports is based on participatory ethnographic approaches.

Ethnographic methods such as participation, observation and interviews, allow researchers to develop embodied understandings of how participants experience cultural practices, spaces and relationships. Action and lifestyle sport researchers often start out as, or become, embedded cultural insiders, empathetic to the cultural, spatial and historical nuances of the activities and participants in the field and online (Evers, 2006; Crockett, 2015; Laurendeau, 2011; MacKay & Dallaire, 2013; Olive, 2015; Roy, 2015; Thorpe, 2011; Wheaton, 2002). Clifton Evers (2006) argues that getting wet and getting involved in his research reveals a depth of complexity and connection that cannot be gleaned from cultural texts, such as books, magazines and films, and allows for the dynamic physicality and sensuality of, in his case, going surfing. As Cole (1994) has argued, this is important because bodies that are moving and performing are inseparable from the contexts they inhabit.

My research in this field has focused on women's recreational surfing. I have used ethnographic methods including interviews, participant observation and blogging as a form of cultural interaction. Using these methods, I seek to better comprehend how women understand, experience and negotiate what continues to be a male-dominated activity, but one in which the growth of women's participation is having significant cultural impacts. While interviews with women who surf are central, my own participation remains key in how I access participants and analyse the interviews. Going surfing situates me in the physical and cultural worlds of the women that my research focuses upon, so that instead of only talking about relationships, experiences and issues, I can place myself amongst them, sharing the experiences with other surfers in the water. Considering research in this way – as a collaborative process of mutual exchange – keeps the context, the research and the theory explicitly connected, and the analysis relevant to and reflective of participants' lives.

While ethnographic methods have driven new insights into the cultural politics of lifestyle and action sports, critics of the dominance of participatory approaches argue that researchers can struggle to fully account for their insider position (Donnelly, 2006; Wheaton & Beal, 2003). Researchers who are skilled at the activity and thus have high cultural status can be vulnerable to replicating existing cultural hierarchies in the field and in their analyses by keeping a focus on the core, dominant participants (Pavlidis & Olive, 2014; Wheaton, 2002). This remains an important concern: how do we critically analyse the practice, culture or community we are part of, that we love, that impacts our identity?

Developing an effective reflexive approach requires researchers to be engaged with their situatedness within their research, as an individual with their own subjectivity, history, knowledge, experiences and relationships. By developing an approach that makes visible the effects of their subjectivity, researchers can account for the influence of their own subjectivity over their research about cultural understandings and experiences. Collaborative research is an effective way to practise reflexivity (Olive et al, 2016; Pavlidis & Olive, 2014), yet the amount of time required in longer-term ethnographic projects means they are often carried out by individual researchers. Given the multiple connections that so many researchers have to their research fields, the task of locating and accounting for our own subjectivity in relation to our research fields by establishing a reflexive research practice is personally and methodologically challenging. In this chapter, I will discuss Elspeth Probyn's (1993) approach to reflexivity, which positions it as an ethical practice: "thinking the social through myself" (p. 3). Starting with a discussion of what reflexivity is, I will outline how I have used Probyn's work to develop a way of thinking through the effects of my own subjectivity on how I can understand various practices, contexts and relationships.

Reflexivity: Who is she and who am I?

Reflexivity refers to a process that all qualitative researchers "can and should use to legitimize, validate, and question research practices and representations" (Pillow,

2003, p. 175). At an individual and collaborative level, reflexivity helps us remain engaged with our subjective assumptions and experiences, and negotiate how the experiences, ideas, conversations, theories we have gathered through our research have all folded into and through us (St Pierre, 1997). It is about developing "situated knowledges" (Haraway, 1991), that reveal how researchers themselves shape projects and findings. Far from an effort to discredit the subjective-ness of research, developing a strong reflexivity is a strength. Rose (1997) reminds us that "all knowledge is marked by its origins", and is thus never universal, but that knowledges "are limited, specific and partial" (Rose, 1997, p. 307). Yet Rose also points out that these limits and partialities are of great value in how they highlight the diversity and possibilities of lived knowledges and experiences.

In this chapter, reflexivity is understood as more than a "methodological tool" (Pillow, 2003, p. 175) used during data collection or a theoretical concept that is applied during analysis, but as an essential part of doing research across all stages. In this understanding, reflexivity (and ethics) are not a separate and finalised part of data collection, but are continual, always already, becoming and being, and folding through the entire research project and process; the design, collection, analysis, writing, dissemination, etc. Taking this approach means reflexive approaches can be applied years after data collection to re-examine projects and to find new insights (Carrington, 2008; Thorpe, Barbour & Bruce, 2011). Reflexivity is key to effective participatory research but, clearly, it is a very difficult practice to establish.

I began my work about women's recreational surfing as a surfer and a member of the community I was reaching. I felt acutely aware of my insider status and sought to understand the implications of it. It was in feminist theory that I found direction, most significantly in Probyn's notion of "thinking the social through myself". Linking feminist and cultural studies theories, methods, ethics and questions of reflexivity, Probyn (1993) suggests that I may "conceive of thinking the social through myself" (p. 3) by using my own experiences as a relational point of understanding through which to understand the experiences of others as well. Probyn points out that while "the self here is constructed in the social formation that it seeks to transform", she is "convinced that … I can talk about my experiences of being in the social without subsuming hers" (p. 3). Probyn does not shy away from the subjectivity of participatory research; she embraces it as productive, and as a useful tool for finding our way into better understanding the lived experiences of others. This approach accounts for how lived experiences perform, challenge and reinscribe the discourses that shape us as researchers and participants.

This reflexive approach to research acknowledges that experiences are specific to individual subjectivities. When we think the social through ourselves, we allow for ways of knowing that are explicitly mediated through our researching subjectivity; our sex, gender, sexuality, race, ethnicity, dis/ability, age, class, skill level and more. Doing so reveals contextual specificity in terms of the cultures, times, places and networks we're inhabiting. At the same time, intersections of our subjective self with others and with space may produce understandings and experiences that are shared or unique, and which may reproduce, resist and disrupt normative understandings of cultures. Such intersections are key to understanding the embodied and relational nature of geographies, communities, cultures, histories, experiences and relationships.

We experience the effects of our subjectivities in terms of how we are positioned as contextualised subjects, highlighted by a range of voices of cultural authority, all of whom impact us to varying degrees, depending on our relationships to the people and contexts that are around us. As Probyn asks, "How can it be otherwise, given that our bodies and our sense of ourselves are in constant interaction with how and where we are placed?" (2003, p. 290). This way of thinking – through the researching self – implicates researchers in the messy interactions between spaces, places, cultures, bodies, discourses, and power (Ahmed, 1998; Probyn, 1993). It offers a theoretical framework through which to consider our own subjectivities not as an abstraction, but as lived and experienced, and as influenced, but not defined by, the discourses that surround us. Reflexivity is relational, ethical and productive.

Approaching research in this way has a risk of centralising the self over others. Importantly, as Couldry (1996) cautions, this approach "is *not* a licence for a subjective, overpersonalized form of writing; it should, rather, incite a re-examination of critical vocabulary" (p. 317, emphasis original). That is, I need to think through my subjective position without privileging it; the research is through me, not about me. Yet in Probyn's approach, being engaged with myself as a subject in my research is a constant reminder of the specificity of my lived experience and subjectivity, and thus, the limitations of my participation. In the context of women's recreational surfing, it reminds me that I am a researcher but that I am also a woman, daughter, sister, friend, surfer, nature-lover, feminist, and writer; that I am female, cis-gendered, heterosexual, white, able-bodied, middle-class, local, non-local, and a longboarder. In short, I occupy various positionalities that collectively impact me in different ways depending on the context.

> In bringing together the practices that we live and the problematizations of those practices, the self can provide a place to speak from. We can think of the "work" of the self; grounded in "the primacy of the real", the self must also be made to move analytically, revealing the character of the mediations between individuals and social formations
>
> *Probyn, 1993, p. 135*

As participatory researchers, we are thinking of the work of the self, as we do it; analysing as we swim, surf, skate, climb, dive, ski, paddle, play. We are listening and making notes and looking for patterns and blindspots, imagining how the cultures, practices and relationships we are interrogating are shaping and shaped by us. We are also listening, feeling, tasting, smelling, and touching, as well as responding, reacting, accommodating, negotiating, fearing, enjoying, and using all of our bodies and emotions to get a sense of a space, place, culture, community, network, knowledge and set of experiences (Evers, 2009). I'm using a lot of words here, but I do so to reflect the movement and interconnectedness of doing participatory research and to give a sense of how it feels when they, she, he, it, we, and I are tangled up in sets of experiences and relations. We might not understand or experience it all the same way, but we're all emplaced and implicated, and we better make sense of who she is if we understand who we are.

Researching surfing: Immersed, embodied, relational, subjective

So what does this look like in a practical sense? My own research uses participant observation (or what I call "going surfing"), as well as interviews and blogging as a research method. I have spent a lot of time in the water and in other spaces where surfing happens. I have surfed at a number of different places over the years, but I have spent the bulk of my time at the surfbreaks in and around Byron Bay, New South Wales, Australia where my access was easiest and most frequent. When I'm doing fieldwork, I go surfing as much as I can in all kinds of relevant spaces – amongst the waves, in beachside car parks, going to cafes, restaurants and parties with surfing crew, hanging out at each other's houses, cheering for friends at competitions, going to surfboard factories, films and discussions, as well as engaging with various social media, notably Instagram. I watch, listen, chat, laugh, argue, read, feel the sun on my back, the water on my skin, get sunburned, get cold, get stoked and catch as many waves as possible. Online, I behave similarly, posting, responding, liking, disagreeing, sharing and commenting. I share in the experiences of going surfing in all kinds of spaces and contexts, which helps me to develop a language for how surfing feels, as and how my friends and participants talk about surfing. These experiences become something mutual, rather than something they describe to me.

As well as other surfers, dolphins, whales, fish, birds and sharks, cultural and feminist theories, critiques and ethics become my constant companions, whispering in my ear, pointing to contradictions, questioning my assumptions, keeping me reflexive of my subjective relationships to places, people, communities, ideas and to surfing culture more broadly. I ask myself, *Who is she and who am I* (Probyn, 1993)?

For me, like so many others, making sense of my fieldwork looks a lot like writing as a form of praxis. I take this idea from many others, most notably from Richardson and St Pierre (2005), who suggest that we can use writing as a way of thinking through things and to help maintain a reflexive process. In this case, their interest was in how we can use appropriate language and writing techniques to discuss social and cultural ideas, but this builds on St Pierre's (1997) previous argument that we are always connected to our research, to participants and to the theory that arches over it all, and that we must reflect on how these impact each other in unavoidable and embodied ways. The reflexive tensions my fieldwork raises became apparent in my field notes and journals. Similar to Carrington (2008), as I developed methods for writing notes after various surfing experiences, I found it difficult to write observations and personal reflections as separate entries. Like my research experiences, my field notes showed the connections and contradictions I experienced between myself as a researcher, surfer and member of the community, with the spaces, cultures and theories blurring together in how I understood and recollected events. It became difficult to ignore my subjective position in recording the experiences and spaces as a participant.

When I go surfing, theory comes with me and I find that surfing and research blur into each other in complex ways and force me to face the limitations of my own subjectivity on how I can know a culture or place. There are days when my friends and I go for a barbeque at the beach – hanging out, telling jokes, teasing each other, surfing, sharing waves – and then something will happen or the conversation will turn and I remember I'm a researcher again, suddenly listening intently and making recordings in my head to replay and scrawl down later. In these ways, I am always surfing and always researching.

The participatory and embodied nature of going surfing as a research method meant that the entries I composed in my journals could not simply describe spaces and events to provide facts and data, but they also encompassed my own responses, experiences, agendas and interpretations as implicit in their recollection (Emerson, Fretz & Shaw, 2007). As Carrington notes,

> It could be argued that outside of qualitative research that explicitly locates itself within feminist scholarship, it is still rare to read in ethnographic accounts *genuine* attempts at narrating the author into the texts from the beginning and then using the self as site for analytical reflection. The problem, which is not often addressed is that it is incredibly difficult to write in this way as the ethnographer has to avoid using the detached, yet safe, third person pronouns, and acknowledge their own feelings and desires with no necessary safeguard that the approach will "work" (in terms of analytical insight generated) and not be dismissed, as is often the case, as narcissistic self-absorption
>
> *Carrington, 2008, pp. 432–433*

This is a long except from Carrington, but it makes an important point. Like Couldry, Carrington notes there can be a risk of self-indulgence or self-absorption, but this should not hold us back from putting ourselves in our own research picture. Having to account for myself in my field notes and journals helps me see more clearly where there are differences and similarities amongst participants (Olive, 2013). Writing about the impacts of romantic relationships on women's surfing lives brought to light my heteronormative assumptions and the blindness I had to LGBTIQ+ surfers. Writing about women's bodies forced me to admit I am a slim, white, cis-gendered woman too (Olive, 2015). Writing myself into critiques of 'localism' helped me think about where and how I was complicit in excluding 'non-locals' from surf spaces as well as from my research; in fact, I was replicating hierarchies, even without meaning to. Most significantly, it made clear my white settler identity in Australia (Olive, 2019, 2016).

More recently, researchers have begun using video as a way to objectively put ourselves in our own researching pictures. Evers (2015, 2018) takes his GoPro into the sea with him, to produce and explore engagements between the bodies of men who surf, as well as the water, waves, sand, weather, animals, and even pollution amongst which their bodily encounters are immersed (Evers, In Press). Palmer (2016) has used a GoPro video camera to capture her experience of training and running in "fitness philanthropy" (p. 226), using her own participation to record running movements and conversations within small groups of companions as well as in large crowds. I've not yet used video in my fieldwork, in part because the thought of seeing myself so explicitly in my research picture makes me nervous, but I'm convinced by Evers' and Palmers' arguments that the use of video offers new perspectives on our research fields, relationships and selves. We can see how we occupy, move through and impact space, time and culture; we can see ourselves in the social.

The use of body- or board-mounted cameras as an in-the-moment form of live field notes and interviews emphasises how important reflexivity is in research about outdoor spaces. In leisure and sport studies that take place outdoors, water, plants, rocks, animals, weather, climate, soil, and currents shape our participation, but are not adequately included in research. Lifestyle and action sport researchers have fairly recently begun genuine ethnographic engagement with more-than-human aspects of our sporting fields, and so we face new reflexive challenges. Not only are we cultural insiders, at risk of replicating existing hierarchies, we are also humans, at risk of replicating the same ecological assumptions of human entitlement that are the cause of the questions we are asking (Hammerton & Ford, 2018). As researchers attempt to decentre humans from research about the outdoors – nature, plants, animals, ecologies, blue and green spaces – we need new approaches for situating knowledges linked to our humanity. Evers and Palmer have both played with moving methods that capture their participation in sport as it happens and that put them, their perspectives and their experiences, in the picture. This is increasingly important as we include more-than-human aspects of lifestyle and action sports in our research.

As I negotiate these new challenges, my thinking continues to be inspired by Probyn, whose recent work, *Eating the Ocean* (2016), thinks the social through herself to understand the local, regional and global production and consumption of fish; socially, culturally, economically, environmentally, ecologically and politically. To do this, Probyn quite literally immerses herself in the world of fish, using "swimming as methodology of encounter" (p. 82) and using oceans as a method for understanding fish on their own terms. In forthcoming work, Evers (In Press) continues to use surfing as methodology of encounter to better understand how we surf in a world of "polluted leisure". Immersed in an ocean of plastic, chemicals and waste, Evers is not advocating for a return to nature, so much as rethinking nature in (and after) the Anthropocene.

To be reflexive is to constantly interrogate how our subjectivity impacts our research; from the questions we ask, the methods we choose, the theories we let guide us and the ways we publish our ideas. To do this means to constantly push ourselves in new directions, to ask new questions, to try new methods, to try new forms of communication. Thinking the social through myself as a reflexive practice has certainly led me into new territory; it has led me into rethinking what I know about the sea and how I apply that knowledge to green spaces. As I use swimming and surfing to understand human relationships to ecologies, I continue to find value in thinking the social through myself to recognise my own borders and boundaries, my own fears and longings and assumptions, my own joys and sensual experiences, my own sense of space. I use these to remind myself that the spaces themselves are more than how I see, feel and know them. By centring myself in my analyses of nature and the outdoors, I am able to recognise some of my boundaries, and think beyond them to new knowledges, directions, depths, volumes and possibilities.

References

Ahmed, S. (1998). *Differences that matter: Feminist theory and postmodernism*. Cambridge: Cambridge University Press.

Brymer, E. & Gray, T. (2010). Developing an intimate "relationship" with nature through extreme sports participation. *Leisure/Loisir, 34*(4), 361–374.

Carrington, B. (2008). "What's the footballer doing here?": Racialized performativity, reflexivity, and identity. *Cultural studies ⇔ Critical methodologies, 8*(4), 423–452.

Cole, C.L. (1994). Resisting the canon: Feminist cultural studies, sport, and technologies of the body. In S. Birrell & C.L. Cole (Eds.) *Women, sport and culture* (pp. 5–29). Champaign, IL: Human Kinetics.

Couldry, N. (1996). Speaking about others and speaking personally: Reflections after Elspeth Probyn's *Sexing the self. Cultural Studies, 10*(2), 315–333.

Crocket, H. (2015). Foucault, flying discs and calling fouls: Ascetic practices of the self in ultimate frisbee. *Sociology of Sport Journal, 32*(1), 89–105.

Donnelly, M. (2006). Studying extreme sports: Beyond the core participants. *Journal of Sport and Social Issues, 30*(2), 219–224.

Emerson, R.M., Fretz, R.I. & Shaw, L.L. (2007). Participant observation and fieldnotes. In P. Atkinson, A. Coffey, S. Delamont, J. Lofland & L. Lofland (Eds.) *Handbook of ethnography* (2nd ed., pp. 352–368). London: SAGE.

Evers, C. (2006). How to surf. *Journal of Sport and Social Issues, 30*(3), 229–243.

Evers, C. (2009). "The Point": surfing, geography and a sensual life of men and masculinity on the Gold Coast, Australia. *Social & Cultural Geography, 10*(8), 893–908.

Evers, C. (2015). Researching action sport with a GoPro™ camera: An embodied and emotional mobile video tale of the sea, masculinity and men-who-surf. In I. Wellard (Ed.) *Researching embodied sport: Exploring movement cultures* (pp. 145–163). London: Routledge.

Evers, C. (2018). The gendered emotional labor of male professional 'freesurfers' digital media work. *Sport in Society*, Online First. doi:10.1080/17430437.2018.1441009

Evers, C. (In Press). Polluted Leisure. *Leisure Sciences.*

Gilchrist, P & Wheaton, B. (2011). Lifestyle sport, public policy and youth engagement: Examining the emergence of parkour. *International Journal of Sport Policy and Politics, 3*(1), 109–131.

Hammerton, Z. & Ford, A. (2018). Decolonising the waters: Interspecies encounters between sharks and humans. *Animal Studies Journal*, 7(1), 270–303.

Haraway, D.J. (1991). *Simians, cyborgs, and women: The reinvention of nature*. London: Free Association Books.

Laurendeau, J. (2011). "If you're reading this, it's because I've died": Masculinity and relational risk in BASE jumping. *Sociology of Sport Journal, 28*(4), 404–420.

MacKay, S. & Dallaire, C. (2013). Skirtboarders.com: Skateboarding women and self-formation as ethical subjects. *Sociology of Sport Journal, 30*(2), 173–196.

Olive, R. (2013). "Making friends with the neighbours": Blogging as a research method. *International Journal of Cultural Studies, 16*(1), 71–84.

Olive, R. (2015). Reframing surfing: Physical culture in online spaces. *Media International Australia, 155*(1), 99–107.

Olive, R. (2016). Surfing, localism, place-based pedagogies, and ecological sensibilities in Australia. In B. Humberstone, H. Prince and K.A. Henderson (Eds.) *International handbook of outdoor studies* (pp. 501–510). Oxford, New York, NY: Routledge.

Olive, R. (2019). The Trouble with Newcomers: Women's Experiences of Localism in Surfing. *Journal of Australian Studies, 43*(1), 39–54.

Olive, R., Thorpe, H., Roy, G., Nemani, lisahunter, M., Wheaton, B. & Humberstone, B. (2016). Surfing together: Exploring the potential of a collaborative ethnographic moment. In H. Thorpe & R. Olive (Eds.) *Women in action sport cultures: Identity, politics and experience* (pp. 45–68). London: Palgrave Macmillan.

Palmer, C. (2016). Research on the run: Moving methods and the charity "thon". *Qualitative Research in Sport, Exercise and Health, 8*(3), 225–236.

Pavlidis, A. & Olive, R. (2014). On the track/in the bleachers: Authenticity and feminist ethnographic research in sport and physical cultural studies. *Sport in Society, 17*(2), 218–232.

Pillow, W.S. (2003). Confession, catharsis, or cure? Rethinking the uses of reflexivity as a methodological power in qualitative research. *Qualitative Studies in Education, 16*(2), 175–196.

Probyn, E. (1993). *Sexing the self: Gendered positions in cultural studies.* London: Routledge.

Probyn, E. (2003). The spatial imperative of subjectivity. In K. Anderson, M. Domosh, S. Pile & N. Thrift (Eds.) *Handbook of cultural geography* (pp. 290–299). London: SAGE.

Probyn, E. (2016). *Eating the ocean.* Durham, NC: Duke University Press.

Richardson, L. & St Pierre, E.A. (2005). Writing: a method of inquiry. In N.K. Denzin & Y.S. Lincoln (Eds.) *The SAGE handbook of qualitative research* (pp. 959–978). Thousand Oaks, CA: SAGE.

Rose, G. (1997). Situating knowledges: Positionality, reflexivities and other tactics. *Progress in Human Geography, 21*(3), 305–320.

Roy, G. (2015). Surfing friendships and encounters in the field. In I. Wellard (Ed.) *Researching embodied sport: Exploring movement cultures* (pp. 129–141). London: Routledge.

St Pierre, E.A. (1997) Methodology in the fold and the irruption of transgressive data. *Qualitative Studies in Education, 10*(2), 175–189.

Thorpe, H. (2011). *Snowboarding bodies in theory and practice.* Basingstoke: Palgrave Macmillan.

Thorpe, H., Barbour, K. and Bruce, T. (2011). "Wandering and wondering": Theory and representation in feminist physical cultural studies. *Sociology of Sport Journal, 28*(1), 106–134.

Thorpe, H. & Dumont, G. (2018): The professionalization of action sports: mapping trends and future directions, *Sport in Society*, Online First, DOI: 10.1080/17430437.2018.1440715

Wheaton, B. (2002). Babes on the beach, women in the surf: Researching gender, power and difference in the windsurfing culture. In J. Sugden & A. Tomlinson (Eds.), *Power games: A critical sociology of sport* (pp. 240–266). London: Routledge.

Wheaton, B. & Beal, B. (2003). "Keeping it real": Subcultural media and the discourses of authenticity in alternative sport. *International Review for the Sociology of Sport, 38*(2), 155–176.

12

FINDING MY PROFESSIONAL VOICE

Autobiography as a research method for outdoor studies

Mark Leather

This chapter is as much a story about my research journey as it is about the method of autobiography as part of my professional outdoor practice. If you have ever pondered the questions, who am I? And how on Earth did I end up here doing this? Then I suggest you have already started your autobiographical journey.

Why should you consider autobiography? In general terms, the personal and human decisions that are made about *all* research methods are inherently human, as Letherby (2015) emphasises:

> accepting that all research and scholarly writing takes place somewhere on an autobiographical continuum highlights how important it is to always keep a research diary within which the researcher(s) records connections to (or not), and personal, as well as intellectual, reflections on the research process
>
> *p.165*

There is great value in reflecting upon our teaching, learning and research journeys, both on the macro scale – as in the questions posed above – and on the micro scale. In respect of research, keeping a diary (reflective research journal or research log – there are many related terms) specifically for each research project undertaken can be highly beneficial. For a review of research logs see Fluk (2015) for as she highlights "they make students conscious of their research process" and "make students mindful of their research process" (p. 490). Research logs can help shape the complex and sometimes chaotic realities of academic research into coherent stories, and structure experiences so that we can make sense of them.

For me, a child growing up in the 1970s, in the suburban landscape of north London, how on Earth do I find myself so drawn to the sea, teaching sailing and place-responsive education on the south Devon coast (see Leather and Nicholls, 2016) and working at a small university[1] in Plymouth, UK, teaching outdoor education against the backdrop of a rich cultural seafaring heritage. I wrote about finding myself drawn to the sea in *Making connections with the sea: A matter of a personal and professional Heimat* (Leather, 2019) and used an autobiographical approach to do so. The sea is my *Heimat*: my homeland, my place, my sense of belonging in, by or on the sea. It is a place that is familiar, comfortable and comforting. I suggest it is useful to consider this autobiographical approach not just as a record of my life, "but as life itself" (Moss, 2001, p. 19).

So why use autobiography?

One of the modules I teach is Conceptual and Theoretical Discourses on a Master's in Research (MRes, Outdoor Education). This programme is designed for students with ongoing experience in an outdoor studies context. It aims to locate their professional practice within the wider theoretical and conceptual discourses of outdoor education. If we are to work well with others, which is a commonality across the diversity of outdoor studies contexts, then we first have to know and understand ourselves within the wider socio-cultural, political and historical discourse. For example, the questions I pose include: Who am I? Where have I been? Who are my role models? What 'critical incidents' have most influenced me? How did I end up here? How do I practice 'outdoor education'? What is the flavour of outdoor education I most enjoy? And most importantly, how does my experience relate to the wider educational, social, political, cultural and historical influences that have shaped and formed the wider outdoor studies community; locally, nationally and internationally? These questions are posed to help students feel comfortable in starting their research journey, situating the research process within them, and allowing them to position themselves within outdoor studies; in this case either identifying their practice as outdoor education or outdoor learning in whatever context they work. The answers to these questions are multitudinous and multifarious.

Gibson and Nicholas (2018) consider the relationship between autobiographical memories and outdoor activities. They explore how our autobiographical memories, our recollections of specific, personal events, are constructed through a personal narrative process; the way we choose to tell the stories of our lives, to ourselves, and crucially to others. Participants come to value memories of particular past events because they come to be seen as part of a shared history with others. Liddicoat and Krasny (2014) explore how memories are useful outcomes of residential outdoor environmental education. These programmes have been shown to yield lasting autobiographical episodic memories that reveal a variety of directive and social uses for their memories, including participating in outdoor recreation activities, being more knowledgeable about and appreciative of the local ecology, and engaging in environmentally responsible behaviours. Wright and Gray (2013) investigated how women achieve longevity in the outdoor learning profession through analysing the autobiographies of three experienced, successful Australian female outdoor educators.

Autobiography in education

My first encounter with autobiography was as one of Brookfield's (1995) four lenses for critical self-reflection (see Figure 12.1). In his classic text *Becoming a critically reflective teacher* he argues that inquiry as the pursuit of self-understanding is about knowing yourself as a practitioner. Figure 12.1 below shows how an autobiographical lens is useful as *one* perspective of this critical reflective process. I suggest that it is a vital part of the bigger picture.

A critical scrutiny of the self, using autobiography, is about 'consciousness-raising' and encapsulates a way of conceptualising as well as encouraging social change, as the product of rethinking the relationship between social and political structure, and human agency. Analysing our autobiographies as learners has important implications for how we teach and how we approach and conduct research. Bullough and Pinnegar (2001) provide insight as to when reflections upon the self, the regular practice that is familiar in a reflective learning cycle becomes research. They suggest that autobiographical researchers stand at the intersection of biography and history. The research questions come from a concern about, and interest in, the interaction of the self as an educator, in a specific context, over a period of time, with others whose interests represent a shared commitment to the development of students and the impact of that interaction on the self and the other. They argue that the ultimate "aim of this research is moral, to gain understanding necessary to make that interaction increasingly educative" (p. 15). Our autobiographies allow us to explore and acknowledge the experiences that have shaped our attitudes, beliefs and behaviours. This is because "our experiences as learners are

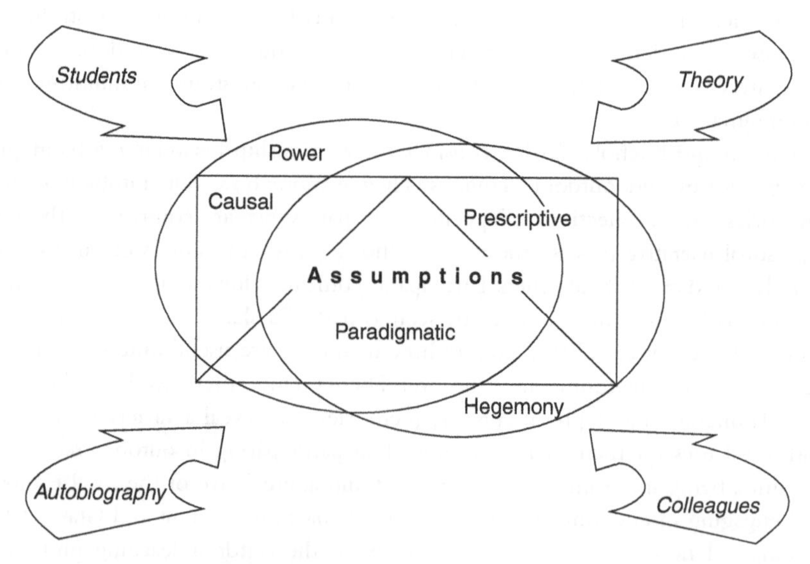

FIGURE 12.1 The critical reflection process

Source: Developed by Laurie Peterman in Brookfield, 1995, p. 30

felt at a visceral, emotional level that is much deeper than that of reason … and the insights and meanings for teaching that we draw from these deep experiences are likely to have a profound and long lasting influence" (Brookfield, 1995, p. 31). Autobiography is a good starting point to see yourself more clearly, and despite the intrinsic problems addressed below, associated with private self-reflection, "the critical journey has to start somewhere, and examining our autobiographies as learners and teachers is one obvious point of departure" (p. 33).

The usefulness of autobiography depends upon the reader, the telling of the story, and its comprehension as such, and we are reliant upon the human capacity of the reader to process knowledge in an interpretive way. This construction of my reality is what Bruner describes as "hermeneutic composability" (1991, p. 8) and is the characteristic whereby the narrative requires a negotiated role between the author and the reader, including the assigning of a context to the narrative. In doing so, what is said may resonate, at least in part, with the reader's experiences and context.

Autobiography in other disciplines

The wonderful aspect of working and writing about education is that it draws on multiple disciplines. In exploring my own story (Leather, 2019) and first engaging with autobiography as a research method, I drew upon geographical perspectives including the work of Buttimer (2001) and Moss (2001). Later, I have come to view autobiography from a sociological perspective, having listened to Letherby (2002, 2013) and also read Stanley (1993) as key influencers. As I consider below, autobiography has a long association with research in education, and the work of Pinar and Grumet (1976) on curriculum, although this work appeared to me later in my autobiographical research journey. Reflecting upon this interesting part of a research journey, exploring new approaches to research allows us to discover new work, see different perspectives and illuminate our thinking related to our research.

Moss (2001, p. 19) discusses from a geographical perspective how it is useful to consider an autobiographical approach not just as a record of my life, "but as life itself". It is useful to think of autobiography as a process, not only of recording – in the sense of documenting, orienting and analysing our lived experiences – but also of *becoming*, in the sense of our lives, subjectivities and identities. Philosophically this *becoming* provides us with the possibility of change; how have we become the outdoor studies practitioner we are today? While this critical self-reflection is not all-inclusive, or the only way of knowing ourselves and our outdoor practice, "it can be a helpful and workable approach in gaining insight into one's life as well as into the contexts within which one exists" (Moss, 2001, p. 20).

The reader's ability to interpret my narrative, my story and understand my *becoming* depends upon the themes they find in my story and their *hermeneutic composability*. These interpretive themes can be considered as the "meaning-metaphor-milieu"[2] of my story (Buttimer, 2001, p. 34); where *meaning* refers to my professional activity, *metaphor* to my cognitive style, and *milieu* to the environmental features of my childhood and formative years. In all three of these themes, there is scope for mutual understanding. In this sense then, autobiography is not only a data source or research approach, "it can also assist in critique and theory building" (Moss, 2001, p. 8), and here it is my intention that you (the reader) may find some resonance, insight and understanding in the *meaning, metaphor* and *milieu* presented in autobiographical research. The geographical perspective is enhanced if we also use a sociological lens with which to consider autobiography; these are not mutually exclusive perspectives on this research method.

Stephenson, Stirling and Wray (2015) used autobiographical methods to develop, within their students, a "sociological imagination"[3] that could then be used "to critically interrogate different cultural forms in order that they not only understood sociology as an academic discipline, but also possessed the tools with which to 'read' the wider social world" (p. 162). They highlight that a sociological perspective gives us the opportunity to understand our own lives and actions, as well as those of others in wider and interrelated contexts of group membership, institutions, hierarchies, ideologies, and material and social inequalities. Stanley (2005) also refers to the "sociological imagination" and highlights how "no study that does not come back to the problems of biography, of history and of their intersections within a society has completed its intellectual journey" (Mills, 1959, cited in Stanley, 2005). The sociological autobiography utilises sociological perspectives, ideas, concepts, findings, and analytical procedures "to construct and interpret a narrative text that purports to tell one's own history within the larger history of one's time" (Merton, 1988, cited in Stanley, 1993, p. 43). Letherby (2015, p. 129) highlights how her engagement with sociology "was from the beginning auto/biographical ... it remains so today". As such, "autobiographers are the ultimate participants in a dual participant-observer role, having privileged access to their own inner experience" (Merton, 1988, cited in Stanley, 1993, p. 43). These inner experiences and personal stories are given a valid voice and allow for the oppressed and lesser voices[4] to be heard. Consequently, autobiography emerged from feminist scholarship as a category of studying and practice, where individuals are compelled to display self-knowledge through the creation and presentation of stories about the self (Cosslett, Lury and Summerfield, 2000). Reflexivity in feminist research processes known as 'consciousness-raising' encapsulates a way of conceptualising as well as encouraging social change, as the product of rethinking the relationship between social and political structure and human agency. This reflexive understanding of the relationship between individual practice and social structure has been useful for outdoor studies. For example, the work of Humberstone (2000) allowed us to consider outdoor spaces as gendered and what this might mean in practice. The issues of equity, equality and women's stories of outdoor practice and social structure are to be found in Gray and Mitten (2018). According to Stanley (1993), reflexivity is located in treating one's self as subject for intellectual inquiry and "it encapsulates the socialised, non-unitary and changing self... Intellectual autobiography [can be] considered as a set of methodological practices, rather than just a dataset" (p. 44). Hollands and Stanley's (2009) discussion of the legacy of *critical sociology* highlights how these methodological practices include an analytical reflexivity that encourages reflection and a strong commitment, not to partisanship, but to moral accountability of ideas and knowledge claims.

Criticism of autobiography

Questions of validity inevitably arise with the use of autobiography due to the subjective nature of memory recall and the possibility of embellishment. Autobiography is to do with recovering a past "and depends on the deployment of an often shifting, partial and contested set of personal or collective memories" (Cosslett, et al., 2000, p. 4). Van Manen (1997) robustly defends the autobiographical approach, relating how it may include fictitious scenarios, emotions or moods within a life story, as this allows us to be imaginatively or emotionally involved. Additionally, autobiographical work is sometimes criticised for its so-called narcissism (Delamont, 2009), because it may assist the autobiographer to "sort myself out" (Rothman, 1986, cited in Letherby, 2015). The critiques made by Delamont (2009) are addressed in Chapter 10. Letherby addresses the criticism, arguing that *critical scrutiny* and the depth of reflexivity is vital, and stating:

> I have long argued that critical scrutiny of the self is completely different than mere navel gazing and that researchers and writers who do not engage in such self reflection are not undertaking the reflexivity needed to properly represent the significance of the complex relationship between the self and other
>
> *pp. 136–137*

In my own research journey, I uncovered the value of autobiography through writing of my personal connection to the sea (Leather, 2019). I had realised the value and usefulness of *autoethnography* in my practice, especially with Nicol's (2013) clear arguments for its appropriateness in outdoor education (see Chapter 10). So, my research journey wrangled with the question: what is the difference between autoethnography and autobiography?

Autobiography and autoethnography

Autobiography allows for my often shifting, partial and contested set of personal or collective memories and may also include partially fictitious scenarios, emotions or moods. I had always understood autoethnography to be a purposeful process of setting out to record these memories, emotions and incidents and to examine aspects of life within particular social, cultural, situational and ideological contexts (as Sparkes and Smith, 2014). For Sparkes and Smith (2014) autobiography features as part of "life history and narrative" approaches to research. These approaches seek to connect the private and the public, the personal and the social. This is done by connecting private, subjective perspectives to meanings, definitions, concepts and practices that are public and social. This type of research requires the researcher to "locate the life story and its events in the history and politics of the time so that the dialectical process between the agency of the individual and the constraints of social structure can be made evident" (Sparkes and Smith, 2014, p. 44).

One of the key authors on autoethnography for me (Chang, 2008), discusses reflexive self-narratives that can contribute to further developments of knowledge and greater understanding of specific topics or issues. And so, during a period of methodological wrangling and a little uncertainty, I was fortunate to attend a research seminar presented by Letherby, and so I raised my conundrum. As an experienced academic, she suggested that reflexive autoethnography was more of a North American term. This supported my reading of the literature for example – Chang at Eastern University (e.g. 2008) and Bochner and Ellis at the University of South Florida (e.g. Ellis and Bochner, 2006; Bochner and Ellis, 2016) have brought the term autoethnography to a wider audience. Autobiography was, perhaps, a more traditional and British term, with Stanley (e.g. 1993, 2005) at the University of Edinburgh and Letherby (e.g. 2002, 2013, 2015) at the University of Plymouth. I suggest that there may be a cultural bias against the term 'autobiography', given that sportsmen and women and celebrity media stars [sometimes ghost] write autobiographies, and these may not be considered as serious pieces. These factors may have resulted in a diminished use of autobiography as a common term in the research method literature. However, this appears to be changing with for example the work of Sparkes and Stewart (2016) as proponents of its use. My research journey in writing this chapter has been interesting here because the second edition of Brookfield's classic text (originally published in 1995 and revised in 2017) sees the term autobiography as notably absent. The label he now uses is 'personal experience', changed because the term autobiographical was a little confusing to some people; "I thought 'personal experience' might be more accurate. Just a different lexicon to refer to the same thing really" (Brookfield, 2019). Additionally, in exploring the literature, it was useful to read Davidson (2011), who uses both autobiography and autoethnography, making a clear distinction between the two. She argues that experience is best understood in retrospect and reflection, and for the importance of intellectual autobiography, as discussed above. She then revisited the account of her life experiences to express a more personal, vulnerable intimacy (Ellis and Bochner 2006), using autoethnography as a mode of enquiry.

However, whilst in this chapter I have outlined how autobiography has been useful as a research method to explore oneself with a critical scrutiny and a depth of reflexivity to make sense of our outdoor studies practice, there is another perspective to autobiographical research that is worth considering. This is articulated by Day (2016), Smith and Watson (2010) and Sparkes and Stewart (2016), who use autobiographies of others as serious analytical and pedagogical resources. For those readers interested in the use of autobiographies of *others* in outdoor studies, I recommend Smith and Watson's (2010) handbook, which provides a toolkit to help navigate the 'memoir boom' that has seen an exponential expansion in life writing in digital and visual media.

I encourage my students to engage in hermeneutic inquiry, as exemplified by Pinar and Grumet's (1976) approach to autobiographical curriculum inquiry. I contend that autobiographical research is useful for students of outdoor studies to, at the very least, get to know themselves, to give their own lived outdoor studies experience a voice that shapes their understanding and professional identity and impacts on their context. This hermeneutic inquiry, an exploration of one's own story, "is concerned with the ambiguous nature of life itself ... does not desire to render such ambiguity objectively presentable ... but rather to attend to it, to give it a voice." (Jardine, 1992, cited in Gough, 1999, p. 414). In attending to our own autobiographies, Gough (1999) provides us with a word of caution when working with others (clients, pupils, students etc.) that "we must take care that opening our autobiographies does not foreclose theirs [children's]" (Gough, 1999, p. 417). Whilst my autobiography provides me with the passion, excitement and enthusiasm for my outdoor studies context and practice, there must be space for my students to explore and tell their own stories. Their journeys have different starting points, routes and experiences to mine, that inevitably bring a richness, difference and diversity to the field.

Conclusions

For me, understanding autobiography was a research epiphany. There are those who devote their professional academic lives to one form of research method, one philosophical position or another. They understand the fine nuances of these approaches to research. What I have found, as an educational researcher who specialises in the theory, practice and research of outdoor studies, is that my students readily grasp the concept of autobiography and reflect on critical incidents from their past. My personal use of autobiography as a method continues to reveal insight and understanding and so informs my outdoor studies research practice. I have discussed in this chapter specifically how autobiography has a place in the outdoor studies research methods toolbox. It may appear that autobiography has been superseded by autoethnography. Nonetheless, it still has current value. For example, Petrov (2018) was amazed by the very experience and process of writing his personal autobiography whilst undertaking his PhD and wanted "to understand the 'magic' of the autobiographical experience" (p. 193). You may consider your personal autobiography as a useful approach to place yourself within your culture and position yourself within your own outdoor studies research. Stories of who we are, how we research and what questions we choose to ask, or not, are part and parcel of all our research in outdoor studies, whatever our approach to research, and whatever methodologies we employ.

Notes

1 University of St Mark and St John, Plymouth, Devon, UK – www.marjon.ac.uk
2 From mid-19th century French, from *mi* 'mid' + *lieu* 'place', meaning a person's social environment.

3 The term "sociological imagination" was first used by Mills (1959, cited in Stephenson et al., 2015) to demonstrate that the only way the individual can understand her or his place in society is through an understanding of the "intricate connection between the patterns of their own lives and the course of world history" (p. 4).
4 It is beyond the scope of this chapter to explore the discussions of hearing the people's voice. Paulo Freire's work is fundamental in exploring these ideas. For example, see Freire (1970/1996) *Pedagogy of the oppressed.*

References

Bochner, A. & Ellis, C. (2016). *Evocative autoethnography: Writing lives and telling stories.* Abingdon,: Routledge.

Brookfield, S.D. (1995). *Becoming a critically reflective teacher.* San Francisco, CA: John Wiley & Sons.

Brookfield, S.D. (2017). *Becoming a critically reflective teacher* (2nd ed.). San Francisco, CA: John Wiley & Sons.

Brookfield, S.D. (2019). *Personal email.*

Bruner, J. (1991). The narrative construction of reality. *Critical Inquiry, 18*(1), 1–21.

Bullough Jr, R.V. & Pinnegar, S. (2001). Guidelines for quality in autobiographical forms of self-study research. *Educational researcher, 30*(3), 13–21.

Buttimer, A. (Ed.). (2001). *Sustainable landscapes and lifeways: Scale and appropriateness.* Sterling, VA: Stylus Publishing.

Chang, H. (2008). *Autoethnography as method.* Abingdon: Routledge.

Cosslett, T., Lury, C. & Summerfield, P. (Eds.). (2000). *Feminism and autobiography: Texts, theories, methods.* London: Routledge.

Davidson, D. (2011). Reflections on doing research grounded in my experience of perinatal loss: From auto/biography to autoethnography. *Sociological Research Online, 16*(1), 1–8. Retrieved from: www.socresonline.org.uk/16/1/6.html

Day, M. (2016). Documents of life: From diaries to autobiographies to biographical objects. In B. Smith & A.C. Sparkes. (Eds.). *Routledge handbook of qualitative research in sport and exercise* (pp. 177–188). Abingdon: Routledge.

Delamont, S. (2009). The only honest thing: Autoethnography, reflexivity and small crises in fieldwork. *Ethnography and Education, 4,* 51–63.

Ellis, C.S. & Bochner, A.P. (2006). Analyzing analytic autoethnography: An autopsy. *Journal of Contemporary Ethnography, 35*(4), 429–449.

Fluk, L.R. (2015). Foregrounding the research log in information literacy instruction. *Journal of Academic Librarianship, 41*(4), 488–498.

Freire, P. (1970/1996). *Pedagogy of the oppressed* (M.B. Ramos, Trans.). London: Penguin.

Gibson, J. & Nicholas, J. (2018). A walk down memory lane: on the relationship between autobiographical memories and outdoor activities. *Journal of Adventure Education and Outdoor Learning, 18*(1), 15–25.

Gough, N. (1999). Surpassing our own histories: Autobiographical methods for environmental education research. *Environmental Education Research, 5*(4), 407–418.

Gray, T. & Mitten, D.S. (Eds.). (2018). *The Palgrave international handbook of women and outdoor learning.* London: Palgrave Macmillan.

Hollands, R. & Stanley, L. (2009). Rethinking "current crisis" arguments: Gouldner and the legacy of critical sociology. *Sociological Research Online, 14*(1), 1–13.

Humberstone, B. (Ed.). (2000). *Her outdoors: Risk, challenge and adventure in gendered open spaces.* Eastbourne: Leisure Studies Association (LSA) Publication 66.

Leather, M. (2019). Past and presents: Making connections with the sea: A matter of a personal and professional *Heimat*. In M. Brown & K. Peters (Eds.) *Living with the sea: Knowledge, awareness and action* (pp. 196–212). London: Routledge.

Leather, M. & Nicholls, F. (2016). More than activities: Using a "sense of place" to enrich student experience in adventure sport. *Sport, Education and Society, 21*(3), 443–464.

Letherby, G. (2002). Auto/biography in research and research writing. In G. Lee-treweek & S. Linkogle (Eds.). *Danger in the field: Risk and ethics in social research* (pp. 91–113). London: Routledge.

Letherby, G. (2013). Theorised subjectivity. In G. Letherby, J. Scott & M. Williams (Eds.) *Objectivity and subjectivity in social research* (pp. 59–78). London: SAGE.

Letherby, G. (2015). Bathwater, babies and other losses: A personal and academic story. *Mortality, 20*(2), 128–144.

Liddicoat, K.R. & Krasny, M.E. (2014). Memories as useful outcomes of residential outdoor environmental education. *Journal of Environmental Education, 45*(3), 178–193.

Mills, C.W. (1959). *The sociological imagination*. New York, NY: Oxford University Press.

Moss, P. (Ed.). (2001). *Placing autobiography in geography*. Syracuse, NY: Syracuse University Press.

Nicol, R. (2013). Returning to the richness of experience: is autoethnography a useful approach for outdoor educators in promoting pro-environmental behaviour? *Journal of Adventure Education and Outdoor Learning, 13*(1), 3–17.

Petrov, R. (2018). Autobiography as a psycho-social research method. In S. Clarke & P. Hoggett (Eds.) *Researching beneath the surface: Psycho-social research methods in practice* (pp. 193–213). Abingdon: Routledge.

Pinar, W.F. & Grumet, M.R. (1976). *Toward a poor curriculum*. Dubuque, IA: Kendall/Hunt Publishing.

Smith, S. & Watson, J. (2010). *Reading autobiography: A guide for interpreting life narratives* (2nd ed.). Minneapolis, MN: University of Minnesota Press.

Sparkes, A.C. & Smith, B. (2014). *Qualitative research methods in sport, exercise and health: From process to product*. London: Routledge.

Sparkes, A.C. & Stewart, C. (2016). Taking sporting autobiographies seriously as an analytical and pedagogical resource in sport, exercise and health. *Qualitative Research in Sport, Exercise and Health, 8*(2), 113–130.

Stanley, L. (1993). On auto/biography in sociology. *Sociology, 27*(1), 41–52.

Stanley, L. (2005). A child of its time: Hybrid perspectives on othering in sociology. *Sociological Research Online, 10*(3). Retrieved from: www.socresonline.org.uk/10/3/

Stephenson, C., Stirling, J. & Wray, D. (2015). "Working Lives": The use of auto/biography in the development of a sociological imagination. *McGill Journal of Education/Revue des sciences de l'éducation de McGill, 50*(1), 161–180.

van Manen, M. (1997). *Researching lived experiences: Human science for an action sensitive pedagogy* (2nd ed.). London, ON: The Althouse Press.

Wright, M. & Gray, T. (2013). The hidden turmoil: Females achieving longevity in the outdoor learning profession. *Australian Journal of Outdoor Education, 16*(2), 12–23.

PART III

Contemporary creative qualitative methods

13

CREATIVE NONFICTION IN OUTDOOR STUDIES

Ben Clayton and Emily Coates

What is creative nonfiction?

Creative nonfiction is arguably better shown rather than described, before it can be usefully explained. The below excerpt is taken from our research about the challenges of time for parents of young children trying to pursue their own interests as serious climbers:

> *Emma and Bill's mud-streaked Fiesta swings into the nearly full car park at the popular end of Stanage. Sam – wrapped in so many layers to be almost spherical – waddles quickly towards them as they exit the car. They had watched over Sam a number of times since Liz met them at a local climbing club shortly after Sam was born. Emma was as bubbly as Bill was quiet. Neither was a serious climber and instead preferred kayaking, but they loved the gritstone and were always keen to get out and were flexible on time and destination, which was just what Liz and Jack needed these days.*
>
> *"Hello you three", Emma greets Sam with open arms. "So nice to be getting out at last", she says, sweeping her long brown hair into a ponytail.*
>
> *"Oh, I know", Liz agrees. The rain and snow over the winter had led to many long hours and money spent at indoor bouldering walls around the Peak.*
>
> *The group begins the walk up the path to the crag. Sam at a half run doesn't take long to stumble and fall. He begins to cry.*
>
> *"It's alright, little man, you're okay", Jack helps him to his feet. "Just watch where you are going. Daddy will walk a bit slower". He engulfs his son's small hand and leads him slowly to the crag where Mike is warming up at the Goliath area,*
>
> *"Are you sleeping at the crag now?" Jack shouts from a distance.*
>
> *"I've only got this morning before I'm on double-trouble duty, so no wasting time for me", Mike replies.*
>
> *Clayton and Coates, 2015, p. 241*

What should be immediately clear is that this is written somewhat differently from the majority of academic works in the broad field of outdoor studies. We dispensed with more traditional modes of representation used in qualitative research, such as carefully selected verbatim quotes that unambiguously state the participants' points of view, and the immediate explanation, from some theoretical standpoint or another, of each and every extract of data. Instead, we used techniques for fictional writing to convey the findings of our fieldwork in a way that was more accessible and, we argued, more inclusive.

Creative nonfiction can be a vexed term. Most agree on the essential idea that it is an analytic practice that borrows from the literary arts, where empirical data are woven into a story akin to a fictional text (Cheney, 2001; Clayton, 2010; Gutkind, 2012). However, the full particulars of how this is done and especially the permissible scale of fictionalisation are subject to the reasoning of the individual author. That is to say that the 'creative' in creative nonfiction may refer simply to the substitution of traditional academic prose with that of the literary novel or to a more extensive use of dramatic licence.

Creative nonfiction, then, takes many forms. Leavy (2013) notes that it is such an expansive genre that it can be difficult to synthesise and delineate, and there may be any number of qualitative research projects published that we might call creative nonfiction, and perhaps several literary works that can be considered qualitative research projects. What they share in common is that they are "aesthetically *and* substantively impressive" (Barone, 2008, p. 107, emphasis not in the original). Creative nonfictions are literary essays and, whatever fiction is employed within them, the stories remain grounded in witnessed or experienced 'truth'. Sparkes (2002) highlights the importance of 'being there' and producing data about specific events using research protocols, to appeal to the same kind of authority and trust that are the measure of any social scientist. Ethnographic methods are perhaps the most commonly used by creative nonfiction writers, with many of these writers labelling their work 'ethnographic fiction' or 'ethnographic creative nonfiction' (e.g. Behar, 2001; Inckle, 2010; Smith, 2013). While a thoroughgoing ethnographic design is not requisite, it can be advantageous because it allows full immersion in the reality under study, which can aid in the creation of the kind of detailed scene, character development and empathy, and absorbing narrative that are the hallmarks of literary works.

Creative nonfiction in outdoor studies: Some examples of practice

Creative nonfiction has been usefully employed across many academic planes, including sport, health and recreation research, for some years and is becoming relatively commonplace. However, related fields of leisure, adventure education, outdoor studies and lifestyle sports, have been slower to adopt fictional techniques. There are a few noteworthy exceptions. Peacock, Carless and McKenna (2018) co-developed (with the participant) an evocative first-person account of one military officer's struggle with post-traumatic stress disorder (PTSD) and his experience, and the purported benefits, of an adapted adventurous training programme for recovering military personnel. Carless, Sparkes, Douglas and Cooke (2014) similarly used crafted first-person stories of adventure training experiences in the recovery of two soldiers with serious physical disabilities. In both papers, the soldiers' accounts from interview are woven into more fluid stories, but remain faithful to the experiences and emotions expressed by the soldiers themselves. In this way, the 'truth' is reported, but is more accessible, evocative and less disjointed, as the excerpt below demonstrates:

> *[Y]esterday I walked a fair bit – that hurt like hell. And the climbing, you know, I was bricking it beforehand to be honest. I'm not scared of heights, but I don't much like them. And, you know, with one leg that I can't move how am I always gonna keep three points of contact? Ha ha! That'd mean I just couldn't move! But once they got me up there, I kind of used that fear, I guess to help me focus – to think about what I had to do and could do to make the climb. I mean as a soldier, I always hoped for the chance to use my training in action – I'd have hated to have all that training and never see any action. And I think a lot of us are like that. I wanted the fear, the adrenalin of contact. In those situations, the sheer fear – of being shot, say – keeps you going. You just have to keep going. So the fear focuses you on what needs to be done. And it was a bit like that with the climbing.*
>
> *Carless et al., 2014, p. 128*

Further, Higgins and Wattchow (2013) use creative nonfiction to draw out the embodied and rational experiences of a descent down the River Spey in Scotland by outdoor education students. The writers here go beyond a process of enrichment of a first-person narrative, as in the examples above, and employ new levels of fiction in which the students' experiences are combined and represented within a fictionalised dialogue between tutor and student:

> *"OK. Take the Spey descent we did a few months ago. It really seemed to be a trip about asking questions, but there was so much going on. And we never really talked a lot about getting any answers. I guess I'm finding it difficult committing to asking questions that we never seem to resolve, that we never shut down."*
>
> *"There's a lot going on during an experience like that. And it's early in the programme. The range of students in the group, and the staff as well, come from such diverse learning and cultural backgrounds ..."*
>
> *"Yeah. Some of the other students seemed so comfortable. They've been in a canoe or kayak before so they're not dealing with trying to learn the skills. Others seem more familiar with the style of thinking you seem to be after."*
>
> *"How did you find the paddling side of things on the Spey?"*
>
> *"When I started it felt like I was going to tip in all the time when we were in fast moving water, so I was really tight, just focussing on the water immediately up ahead. Quite scared at times. Then the canoe was loaded with all of our equipment. Things*

> *seemed to be moving so fast at times – decisions being made on the run. When we came to the rapids they just looked like a mess to me. I couldn't see the lines or eddies you were talking about. Then we'd get onto a flat bit and you'd be telling stories about the surrounding landscape, bridges, buildings, the geology or land ownership. To be frank, I felt pretty overwhelmed."*
>
> *Higgins and Wattchow, 2013, p. 25*

Our own work took a similar approach, conveying multiple experiences through composite characters (Clayton & Coates, 2015). Instead of a reflective dialogue, however, we employed a 'real time' story of a 'typical' weekend for traditional (heterosexual, dual) parents who climb. As a more complete story, further techniques of fictional writing were needed, such as a surplus of detail to set the scene and the use of inner dialogue that help to convey meaning and experience in a more fluid way and also help the reader to themselves reside in the scene we depict. We will provide some examples and explanations of how and why we did this later in the chapter.

Why use creative nonfiction?

All of the researchers above used creative nonfiction not because it was in vogue or for any form of qualitative insurgency, but rather because it fit the analytical and dissemination needs of their research agendas. For us, creative nonfiction offered three crucial benefits that could not be found in more traditional approaches. First, given that the British climbing scene is relatively small, we wanted to do all we could to ensure the anonymity of our participants, and the use of fictional methods, such as composite characters and adjustments of time and place, can be useful to help protect the identities of those involved, without the need to lose the rich detail of actual happenings (see also Coffey & Atkinson, 1996).

Second, the lived experiences of our participants were fraught with complexity, full of competing discourses and contradictory feelings and emotions of many layers. This 'truth', we thought, could not be done justice through conventional means of detached description and presentation of interview quotes, and could not be interpreted from a singular standpoint. Frank (2000, p. 483) wrote that there is "a possibility of portraying a complexity of lived experience in fiction that might not always come across in a theoretical explication". She is not alone in this thinking. Across social science disciplines, the use of fiction, and fictional techniques in nonfiction, has been heralded as a possible solution to issues of what Bauman (2000) refers to as the "expertocracy" of social science, interpreting away the plight of individuals. Many exponents of creative nonfiction, including ourselves, regularly cite Richardson and the difference between 'knowing' through narrative and 'telling' through traditional academic writing (Richardson, 1994, 1997) to support their case that stories might better account for complexities, and especially contradictions, that are inherent in lived experiences, and that cannot be contained within any particular scientific frame or located within any one metanarrative. In other words, creative nonfiction can help researchers to face the consequences of the "endemic contingency and uncertainty of human condition" (Bauman, 2000, p. 213).

Third, because our research arguably catered to a limited audience and represented a fairly exclusive group of people, but, we felt, simultaneously generated themes and outlooks important for a wider population, we wanted better access to that wider population. Gutkind (1997) argues that the most basic function of creative nonfiction is to capture a subject in such a way that anyone and everyone will find it interesting and want to read more about it. Traditional academic texts, it is argued, are quite simply boring (Caulley, 2008; Richardson, 2000) and boring texts do not generate the same level of engagement, learning and diffusion as do interesting, absorbing and entertaining ones.

How to do creative nonfiction

As we have already seen from the excerpts of creative nonfictions presented here, there is no single, accepted form that they should take and, equally, there is no clear and accepted method of generating a creative nonfiction. What follows in this section of the chapter, then, is a discussion of *our* vision and the techniques we used and the reasoning behind them. We will begin here with some more general discussion or debate about the place and use of theory in creative nonfiction writing, and then suggest how we wanted our story to be judged as effective, and how we tried to ensure that effectiveness.

Theory in creative nonfiction

The use of theory in creative nonfiction remains a point of contention. Many argue that creative nonfiction can be a useful technique for *generating* theory, where different readings of a story can compete with one another and 'open up' theory production (Frank, 2010; Smith, McGannon & Williams, 2016), but should the writers themselves employ theory frameworks to explain their stories? Conceivably, all works of scholarly writing are at least tinged with theory because one cannot simply detach oneself from one's macro-knowledge during the writing or research processes. But it is what the writer does about this knowledge, how and how much they 'confess' or make explicit to the reader, which is up for debate. We have seen above that central to researchers' rationales for choosing creative nonfiction is the desire to allow 'alternative readings' of their stories. In this vein, some creative nonfiction writers offer no explanation of their stories at all (e.g. Bruce, 2000; Douglas & Carless, 2010) because it is "futile to try to summarize the insights the story provides [since] these insights are best expressed through the story itself" (Douglas & Carless, 2010, p. 347). Others, however, such as Clayton (2010) and Gearity and Mills (2012) more explicitly impose a theory framework on the reader, interrupting their stories with explanatory passages that draw on particular theorists, revealing to the reader the writer's ongoing processes of clarification and explanation.

In our creative nonfiction of parents who are also 'serious' climbers we employed the ideas of Foucault to understand the experiences depicted in our story, but we were deliberately cautious in our approach, noting that while our "analytic lens and written account are coloured, but not saturated by a Foucauldian and late-modern thesis, this is intended to be neither hidden nor obtrusive and [is noted] only as a pre-story and epistemological confession of sorts" (Clayton & Coates, 2015, p. 236). Here we wanted to show *our* workings whilst simultaneously acknowledging and encouraging alternative readings. For us, the very idea of *alternative* readings implies that any one reading may have credence but may not take precedence over another reading.

The debate about how theory should be used in creative nonfiction is almost certainly set to continue, but of importance is the separation of the story and the theoretical explication. All stories contain a theoretical signposting (Jones, 2006) and may show theory or resonate with theory but, significantly, they must allow other theoretical possibilities to emerge (Smith, McGannon & Williams, 2016).

Writing and judging creative nonfiction

Creative nonfiction, while purposely unlike more conventional qualitative writing practices, is not exempt from the requirements of quality, rigour and relevance that are demanded of all academic endeavour. However, it would be unreasonable to subject 'new' and 'traditional' writing practices to the same measures of quality. No such measures can be absolute and, in all qualitative research practices, it is important that accepted characteristics and measures continue to evolve within communities of practice to develop standards (Preissle, 2013). To that end, useful provisional lists of some of the main characteristics of 'good' creative nonfictions are provided by Sparkes (2002) and Smith, McGannon and Williams (2016). For us, however, we wanted our story of parents who climb to be judged on just three intersected and inseparable principles, and we used different techniques to try to ensure effectiveness.

First, we wanted to make a *substantive contribution* to understandings of the impact of children/parenting on serious leisure time and climbing identity, and to theories of gendered parenting. We also wanted to make a contribution to methodology in the outdoors by showing how the use of fictional techniques may – perhaps paradoxically – better represent the complex reality of lived experience. The data that informed our story were generated by detailed interviews with couples and individuals and substantial periods of time spent with these families. The story was constructed to be as true to a 'typical weekend' for our participants as was possible and was written with a level of detail sufficient for the reader to live through the weekend and use their own experiences (their own stories) to determine some explanation – or theory – for the events shown. While we did engage what we thought was a fitting theory framework to give our own explanation, we were careful to keep this separate from the main story – interjected as a clear change of discursive tone and direction – so that it could be ignored or tweaked as the reader wished. Similarly, our story was also set in the middle of a more familiar academic structure of an explanatory abstract and contextualising prologue and epilogue, which allowed us to better pronounce our substantive contribution.

To make a substantive contribution, we also needed to engage our second principle for a good creative nonfiction, which was to create a *plausible reality* for climbers and parents, and parents who climb. That is to say, the creative nonfiction writer can only make their contribution to knowledge, theory, and methodology if the reader is able to "viscerally inhabit" (Rinehart, 1998, p. 204) the world that they present. Plausibility is to be found in empathy, which can be achieved through verisimilitude.

> *Jack turns off the shower tap and emerges through the cloud of steam, vigorously towelling his deep brown hair. He wraps the towel around his waist, lodges a toothbrush into his cheek and walks back to the bedroom. He peeks through the curtains to reveal an unwelcoming dull grey and then riffles through his drawers and wardrobe in search of his thermals, khaki trousers and fleece.*
>
> *"What do you think? Somewhere with bouldering options?" He calls down the stairs. "Maybe Stanage Plantation? Car park at ten?"*
>
> *"Yeah, that's maybe a better idea", comes a delayed response. "It's looking too cold for Sam to be sitting around all day, but the wind is low so we could do some routes first. I'll text Emma".*
>
> *Jack grabs his phone from the bedside table and slumps on the bed. He puffs out his cheeks and with a purposeful, rapid exhale pulls up Mike's number. "Mike will understand", Jack thinks to himself. "Plans change when you have kids. Mike knows that".*
>
> *He hurriedly texts the new plan and throws down his phone on the pillow and trots downstairs. Sam is on the kitchen floor pulling a train around an unsoundly and illogically designed wooden track while Liz slices sandwiches into neat triangles, wraps them and places them methodically into the lunch-bag. Jack fumbles through various drawers and cupboards and lines-up the day's supplies on the table: gear, rope, harnesses, climbing shoes, chalk bags, finger-tape, first-aid kit, baby wipes, spare nappies, toy cars and picture books. Bouldering mats are already in the car.*
>
> *Clayton and Coates, 2015, p. 240*

While descriptions of Sam playing with his trains and Liz making a packed lunch may seem inessential, they work in combination with further rich detail-giving to help the reader to imagine and experience the scene and, moreover, remain infinitesimally faithful to lived experience, built as they were on witnessed events while in the homes of research participants. Similarly, the conversations were 'real'; the words were taken from interview responses or naturally occurring talk and either explicitly relayed or paraphrased, or sometimes presented as inner dialogue, as was necessary to maintain both 'truth' and verisimilitude. Using dialogue between two or more characters can enhance action by giving it immediacy, and show the personalities of the characters, as well as delivering *real* thoughts, feelings and utterances.

Third, we wanted our story to be *aesthetically captivating*. This was not only essential if our previous two quality principles were to be achieved, because the content, sentiment and aesthetics are inseparable, but it is also important that the creative nonfiction writer maintains the values and standing of the genre s/he employs. Story writers want their stories to be heard, to be allowed to breathe, and have some kind of impact on as many people as possible. We were seeking the potential for more public scholarship by creating a 'user-friendly', appealing, and not boring text that might be picked-up by academics, non-academics and students alike. To do this, we employed techniques of literary writing. Some of these, like rich detail, captured conversations, and inner dialogue, have already been discussed, but we also tried to employ scene-by-scene writing, metaphor and observational humour to engage the reader and help them to follow and enjoy the story:

> *Darkness descends over the Nottingham suburb. Jack swings his bike from the road, the beam of his mounted cycle lamp momentarily slicing across the rose bushes and the front of his house before saturating the garage door with an eerie glow. He glides to a near-stop and expertly dismounts and, one hand on the rear of the saddle, pushes his bike through the gate, the back garden and into the shed. The light from the kitchen and upstairs bedroom tumbles across the lawn, providing just enough visibility for Jack to secure the padlock on the shed door and dodge the strewn plastic toys as he makes his way to the house. On entering, he contentedly inhales the roving aroma of tomato, garlic and onion that is escaping the pan on the stove.*
>
> *"Liz? Sam?"*
>
> *"Up here."*
>
> *"Be up in a minute." Jack jangles his keys and drops them to the table where a messy spread of fresh finger-paintings near covers the somewhat congested calendar. Appointments, meetings, work event, Sam's nursery, Sam's play date. Saturday: Sam's friend's birthday party, recently struck through with red pen and "CHICKENPOX" written boldly underneath.*
>
> *"Good!" Thinks Jack to himself with only a momentary sense of remorse. Sunday: Sam's swimming class, only an hour but slap-bang in the middle of the day! "Out for a climb Saturday then."*
>
> *Clayton and Coates, 2015, p. 238*

The above excerpt is the opening scene of our story and it served the purposes of introducing one of our main actors and his persona, introducing the 'problem' of finding time for climbing in family life, and, perhaps most importantly, immediately involving and hopefully hooking the reader with vivid text (see Caulley, 2008). Like in subsequent scenes, we write in a way that is thick with metaphor that creates richness and litany that helps the story move along at pace. These techniques, we hoped, also set a tone that was simply more convivial and equalising, inviting the reader to relax and enjoy the story for the story's sake.

Conclusion

The purpose of this chapter was to introduce the idea of creative nonfiction as a methodological alternative in outdoor studies. While there are currently few examples of the practice within the field, we have argued using examples and advice from our own work and other creative nonfiction scholars, that stories can provide a more inclusive, more engaging and more 'real' account of the complex processes and experiences encountered in the outdoors.

References

Barone, T. (2008). Creative nonfiction and social research. In J.G. Knowles & A. Cole (Eds.) *Handbook of the arts in qualitative research* (pp. 105–116). Thousand Oaks, CA: SAGE.

Bauman, Z. (2000). On writing sociology. *Theory, Culture and Society, 17*(1), 79–90.

Behar, R. (2001). Yellow marigolds for Ochun: An experiment in feminist ethnographic fiction. *International Journal of Qualitative Studies in Education, 14*(2), 107–116.

Bruce, T. (2000). Never let the bastards see you cry. *Sociology of Sport Journal, 17*(1), 69–74.

Carless, D., Sparkes, A.C., Douglas, K. & Cooke, C. (2014). Disability, inclusive adventurous training and adapted sport: Two soldiers' stories of involvement. *Psychology of Sport and Exercise, 15*(1), 124–131.

Caulley, D. (2008). Making qualitative research reports less boring: The techniques of writing creative nonfiction. *Qualitative Inquiry, 14*(3), 424–449.

Cheney, T. (2001). *Writing creative nonfiction: Fiction techniques for crafting great nonfiction.* Berkley, CA: Ten Speed Press.

Clayton, B. (2010). Ten minutes with the boys, the thoroughly academic task and the semi-naked celebrity: Football masculinities in the classroom, or pursuing security in a "liquid" world. *Qualitative Research in Sport, Exercise and Health, 2*(3), 371–384.

Clayton, B. & Coates, E. (2015). Negotiating the climb: A fictional representation of climbing, gendered parenting and the morality of time. *Annals of Leisure Research, 18*(2), 235–251.

Coffey, A. & Atkinson, P. (1996). *Making sense of qualitative data.* London: SAGE.

Douglas, K. & Carless, D. (2010) Restoring connections in physical activity and mental health research and practice: A confessional tale. *Qualitative Research in Sport and Exercise,* 2(3), 336–353.

Frank, A. (2010). *Letting stories breathe: A socio-narratology.* Chicago, IL: University of Chicago Press.

Frank, K. (2000). "The management of hunger": Using fiction in writing anthropology. *Qualitative Inquiry,* 6(4), 474–488.

Gearity, B. & Mills, J. (2012). Discipline and punish in the weight room. *Sports Coaching Review, 1*(2), 124–134.

Gutkind, L. (1997). *The art of creative nonfiction: Writing and selling the literature of reality.* New York: Wiley.

Gutkind, L. (2012). *You can't make this stuff up: The complete guide to writing creative nonfiction – from memoir to literary journalism and everything in between.* Boston, MA: Lifelong Books.

Higgins, P. & Wattchow, B. (2013). The water of life: Creative non-fiction and lived experience on an interdisciplinary canoe journey on Scotland's River Spey. *Journal of Adventure Education and Outdoor Learning, 13*(1), 18–35.

Inckle, K. (2010). Telling tales? Using ethnographic fictions to speak embodied "truth". *Qualitative Research, 10*(1), 27–47.

Jones, R. (2006). Dilemmas, maintaining "face", and paranoia: An average coaching life. *Qualitative inquiry, 12*(5), 1012–1021.

Leavy, P. (2013). *Fiction as research practice: Short stories, novellas, and novels.* Abingdon: Routledge.

Peacock, S., Carless, D., & McKenna, J. (2018). Inclusive adapted sport and adventure training programme in the PTSD recovery of military personnel: A creative non-fiction. *Psychology of Sport and Exercise,* 35, 151–159.

Preissle, J. (2013) Qualitative futures: Where we might go from where we've been. In N.K. Denzin & Y.S. Lincoln (Eds.) *The landscape of qualitative research* (4th ed., pp. 517–543). London: SAGE.

Richardson, L. (1994). Writing: A method of inquiry. In N.K. Denzin & Y.S. Lincoln (Eds.) *The SAGE handbook of qualitative research* (pp. 516–529). Thousand Oaks, CA: SAGE.

Richardson, L. (1997). *Fields of play: Constructing an academic life.* New Brunswick, NJ: Rutgers University Press.

Richardson, L. (2000). New writing practices in qualitative research. *Sociology of Sport Journal, 17*(1), 5–20.

Rinehart, R. (1998). Fictional methods in ethnography: Believability, specks of glass, and Chekhov. *Qualitative Inquiry, 4*(2), 200–224.

Smith, B. (2013). Sporting spinal cord injuries, social relations, and rehabilitation narratives: An ethnographic creative non-fiction of becoming disabled through sport. *Sociology of Sport Journal, 30*(2), 132–152.

Smith, B., McGannon, K. & Williams, T. (2016). Ethnographic creative nonfiction. In G. Molnar and L. Purdy (Eds.) *Ethnographies in sport and exercise research* (pp. 59–73). Abingdon: Routledge.

Sparkes, A. (2002). *Telling tales in sport and physical activity: A qualitative journey.* Leeds: Human Kinetics.

14

SHARED-STORY APPROACHES IN OUTDOOR STUDIES

The HEAR (Hermeneutics, Auto/Ethnography and Action Research) 'listening' methodological model

Tracy Ann Hayes and Heather Prince

Introduction

Ethical, responsible research demands careful expression: we are in a position of power, our words can influence others and be inadvertently harmful. We explore shared-story approaches as ways in which experiences can be considered, analysed critically and conceptualised to give 'testimony', construct meaning and disseminate research findings. We illustrate the HEAR methodological model with reference to research exploring young people's relationship with nature (Hayes, 2017). A 'listening' model, this conceptual approach is a form of praxiography (a method focused on production of knowledge in practice), rooted in the use of fables and stories to convey findings. Using the example of a story with layered meanings, we demonstrate its effectiveness for engaging attention and introducing alternative ways of thinking about research findings.

The use of shared-story reflects perspectives and experiences through narrative and is a powerful tool for illuminating and problematising practices. We critique its position as a transdisciplinary method within creative qualitative methodologies to optimise and enable inclusivity for participants, practitioners and researchers. We provide guidance on ways of using this approach, whilst exploring the reasons underpinning this method, aiming to encourage further practice.

Exploring shared-story approaches

Our use of the term 'shared-story' emphasises that the story is shared for a reason: we found meaning and felt the need to *share*, to communicate it to others in a way that encourages them to respond and perhaps provide a reciprocal story. The terms story, fable and anecdote are often used interchangeably, so we will pause to provide working definitions so that if you want to adopt this method you will be able to identify the most effective form to use. A story can be fictional, nonfictional or a blend; can be unimaginatively told (for example: an account or report) or imaginatively told, using creative writing techniques (for example: a tale or amusing anecdote). Frank (2012, p. 2) highlights that as well as a means of providing information, stories "... give form – temporal and spatial orientation, coherence, meaning intention, and especially boundaries – to lives that inherently lack form". A tale is a form of story or narrative, particularly one imaginatively told. Like Van Maanen (2011, p. 8) we use the term in deference to the "inherent story-like character of fieldwork accounts". An anecdote is a concise entertaining story about real events or people. A fable is a short story, typically with animal characters, designed to convey a moral. This can be useful within pedagogical situations as exemplified by Carson's seminal text *Silent Spring* (1962). A *Magic Moment Fable*, as defined by Hayes (2017) is a short story, imaginatively told, that interprets a moment subjectively perceived as significant. It conveys a lesson, a moment from which we can learn. Figure 14.1 shows the different forms of stories and relationships between them. Myths, legends and fairytales are included; however, they are not discussed in detail as they represent *extraordinary* stories (our emphasis), viewed as the realm of fantasy and imagination. Whilst we may draw from some of the techniques used to create these fantastical stories, we feel this is a less appropriate method for making sense of carefully elicited research data.

Sharing stories

Stories can engage, captivate, encourage participation and be used to foster comfort, familiarity, make connections and as a hook to gain attention. Frank (2012, p. 3) highlights, "Stories may not actually breathe, but they can animate ... Stories work with people, for people, and always stories work *on* people, affecting what people are able to see as real, as possible, and as worth doing or best avoided". Sharing a story enables others to begin to understand us, to give *testimony* to what we have experienced (Etherington, 2004).

Writing stories

Some researchers use creative/artistic forms to create evocative narrative pieces, others use creative nonfiction often for ethical reasons (see Chapter 13), Stories can be written as a way of illustrating key points and exploring issues in more detail. Here, initially, Tracy, one of the authors, utilised fables written by other people (for example, the traditional Aesop's Fables); subsequently, she developed a collection of her own fables. These highlighted initial key findings from her research, and sharing them evoked a response from the reader/listener, encouraging them to explore their own experiences and values regarding outdoor learning, resulting in mutual response between respondents and researcher. The questions and comments heard, the stories people shared, inveigled their way into her thinking. Thus, her stories came to be included within the body of the data set; writing and sharing them

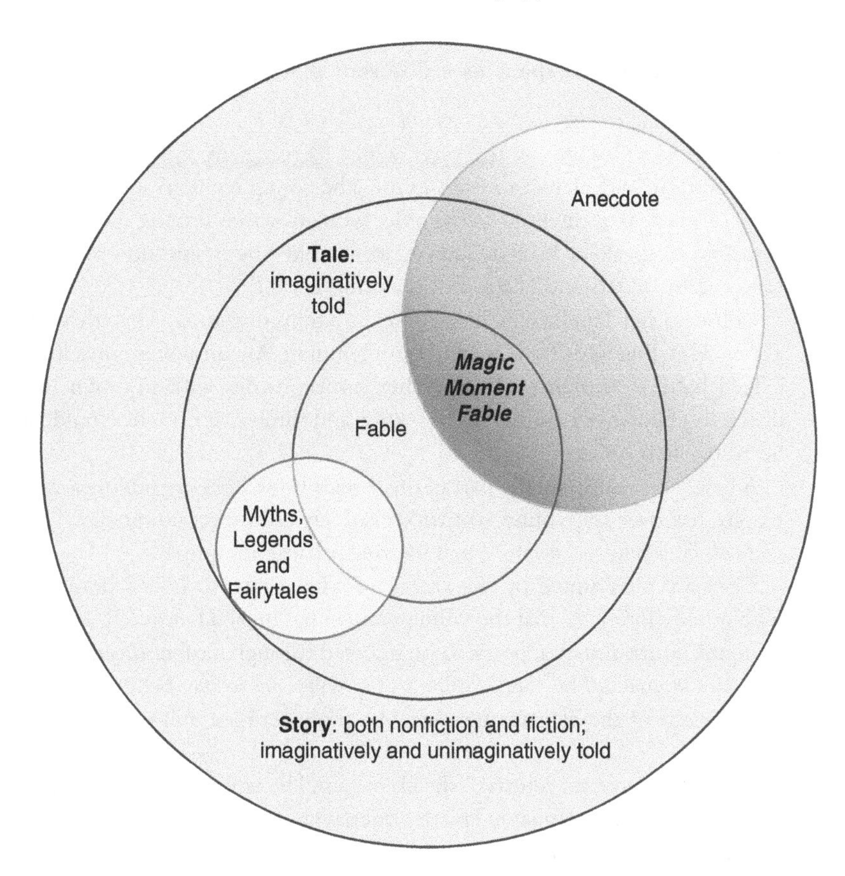

FIGURE 14.1 Diagrammatic illustration of the relationship between the different forms of story

Source: Hayes, 2017

became part of the elicitation process, as a creative approach to engaging with the topic. As you read the next section, we respectfully ask you to consider how could you make use of this approach? What stories do you have to share?

Illustrating shared-story approaches

This next section includes a *Magic Moment Fable* as a way of demonstrating how to utilise a shared-story within research. It is based on research into young people's relationship with nature; however, this approach is applicable elsewhere, for topics that warrant qualitative attention and an exploratory, questioning approach. What is presented here is a joining of voices in a multi-layered textual form (Ellis, 2004) whereby we first invite you to imagine yourself in another's place, embodied within the story, the retelling (Mazzei & Jackson, 2012) of a specific moment, and then provoke you to join in a reflective contemplation of what this may mean in practice.

Exploring a shared space as a different place

"*She loves going outside, and the fresh air is so good for her …*"

I listen to the words of the carer whilst focusing my eyes on the young woman in the wheelchair in front of me. The young woman's eyes are firmly fixed on the top of the head of the older woman who is leaning over her. She is unable to speak, at least in audible words. No-one is sure how much she can hear, or understand of the world around her …

I observe her hands, which the carer is attempting to cover with woolly gloves. Her fingers stiffen, resolutely unbending. An unspoken, unacknowledged battle commences. I remember similar battles with my own (non-disabled) children when they were young, and think that mittens would have been so much easier …

A brief concession on the part of the carer who switches to pulling a warm, woolly hat over the young woman's head, cheerfully pronouncing: "*There, that will keep your ears warm.*" Then the finger battle resumes …

They are interrupted by the arrival of what I take to be a school nurse, who quietly interjects that the young woman has missed her lunch. A syringe of liquid nourishment appears, to be injected through a tube into the young woman's stomach. The carer explains they will have to stay behind while we take the rest of the class to the park. She promises they will join us as soon as they can …

Two hours later, we return to the classroom. The young woman is sat at the front of the class, accompanied by the carer, who cheerfully announces, "*Look, she's having her own nature experience …*"

She is positioned in front of a whiteboard connected to a computer, which is showing whirling, swirling, multi-coloured pictures, accompanied by 'sounds of nature'. My first reaction is frustrated anger. This is not a nature experience. This is someone else's interpretation, an unnatural, synthesised, clinical version in an indoor space. The young woman has been deprived of something vital …

I place myself next to them as the other students go to the cloakroom to remove their outdoor clothing. I rest my hand next to hers, our fingers gently touching. Then I look down. Her fingers are relaxed. Her eyes are focussed on the patterns in front of her. I follow her gaze to the screen and find its effects hypnotic …

On wakening from my trance, I reflexively question "How can I include this experience in a research project that purposefully focuses on young people's voices?"

The ellipses at the end of eight sentences within this story suggest points at which to pause and focus the conversation. We offer our perspectives within an interpretation that takes account of both "context and circumstance" (Mazzei & Jackson, 2012, p. 745) and encourage you to consider your own interpretation.

She loves going outside, and the fresh air is so good for her …

Simple words, casually spoken, however they reveal so much about the carer's approach to caring for her charge. There is an air of assumed positivity, an easy assertion of authority that perhaps aims to convey a commitment to inclusive practice: a belief that we should have equality of opportunity to participate in the activities on offer. Yet there is the first niggle of discomfort that we are forcing this young woman to experience the world on our terms, not hers. This normalising language is from the perspective of the provider; the shared space cannot be contested, but ownership of the decision to be there can, through giving more status to knowledge constructed in everyday life (Fenwick, 2003). The dominant discourses regarding learning outdoors argue that children and young people are disconnected from nature, they are experiencing nature-deficit disorder (Louv, 2005) and, arguably more importantly, that we need to reconnect them. There is further admonishment that activities should be accessible for all, with specific adaptations where required. Has this negative discourse become too dominant (domineering)? How can we know what is important to her?

No-one is sure how much she can hear, or understand of the world around her …

Many questions arise from this simple statement, including how does she make meaning of her world? How do we then make meaning of this? How do we capture 'silent/silenced' voices? What do we mean by 'voice'? Is this different from speech? Can you have a voice without speech? Yes, according to Mazzei and Jackson (2012), if we see it as noiseless rather than silent. There is not sufficient room to address all the questions here, especially those with a biological/neurological aetiology. We focus on meaning making and voice, and on how we, as researchers bearing witness to the encounter, avoid treating her 'voice' in a simplistic and mechanistic manner (Mazzei and Jackson, 2012; Koro-Ljungberg and Mazzei, 2012). We must develop a relationship that is a relational exchange (Gilligan, 1993) which forms part of the process whereby we make sense of our world, and our experiences within it. Cope and Kalantzis (2000) identify six elements of the process of meaning making, only one of which is linguistic; the other five are textual, visual, audio, gestural and spatial modes of meaning. In their later work Cope and Kalantzis (2009) propose the use of open-ended questions about meaning, enabling us to interpret situations and transfer multi-literal meaning to different and possibly unfamiliar settings. There are multiple ways to 'read' the encounter described in the story (Kuntz and Presnall, 2012); the focus on capturing and presenting the "finger battle" within a story avoids "… reification of the transcript as the primary artifact (sic) of the interview" (p. 733), highlighting the embodied and emplaced nature of the interaction. If this was recorded as an interview/focus group, without an oral-voice she would be absent, although perhaps there may be some displaced noises that may be attributed to her. Through shifting perspective, moving from the invisible/inaudible to the visible/audible, embodying the central role in the story, she becomes visible, her presence is heard and felt.

We identify purposeful behaviour on the part of the carer, who demonstrates concern and care, suggesting a shared belief that it is good for her to go outside, a sign of common purpose with us/the project. By interpreting the resultant tensing of the fingers as a deliberate action, rather than an uncontrollable muscle spasm, we are also attributing purpose to this act – a political statement through body language rather than words: a silent rebellion.

Mittens would have been so much easier ...

If mittens rather than gloves had been used, this moment would have been masked, it would have gone unnoticed. How much else are we missing through adherence to more traditional methods? Is there time and space for non-verbal communication and do we have the necessary skills and awareness to adopt this approach with young people, particularly those who have some form of disability? Childhood and adolescence are not phases to be outgrown, young people's experiences in social, cultural and political contexts (Valentine, 2003) are part of a life-long process that shapes and defines a person. Whilst mittens may be appropriate for a young child, young people of a similar age to this young woman are more likely to opt for gloves. Gloves allow for more individual movement of fingers, more freedom for expression.

Then the finger battle resumes ...

There are no answers, merely more questions, particularly with regards to what, if anything, the young woman was trying to convey. Was she hungry? Cold? Unhappy at the thought of going outside? The limitations of being a participant observer in this space are all too clear. The researcher does not have a well-established relationship with either the young woman or her carer; she is there as an accompaniment to the project worker who is facilitating the outdoor experience. She recalls an earlier session when the young woman in the story had joined in, in her own way. She had been wheeled outside by her carer and placed near the centre of the playground. Tracy recalls observing her turn her face towards the sun, like a flower, absorbing the warm rays through her skin. She had appeared calm and relaxed: so very different from today.

She promises they will join us as soon as they can ...

They do not join us in the neighbouring park. If they had, would it have been accessible? Or relevant? We have become adept at considering those who have become excluded, and, in the name of inclusion, developing methods of reaching them, to include them in what we offer. However, we also need to listen in a multi-modal manner. It is not enough to simply provide resources and materials to enable others to participate in what we offer, what we like doing; we need to explore other ways of providing relevant experiences.

Look, she's having her own nature experience ...

Kellert (2012) categorises experience of nature as taking several forms including direct experience (unstructured play and contact with wild places, self-sustaining nature), which is seen as integral to healthy growth and development, referred to as the "naturalistic necessity" (Kellert, 2012), and "Vitamin N" (Louv, 2013). Other forms include indirect experiences (structured/facilitated contact with 'managed' nature, for example, a garden or a pet), and representational experiences of nature, for example through stories, toys, computer or images (Kellert, 2012). In the story shared here we see an example of a representational experience of nature; is this a lesser experience for this young woman? It is certainly a safer one.

The young woman has been deprived of something vital ...

Vital to whom? Her or us? Left to the unmitigated vagaries of nature, this young woman would not be alive; in many ways she lives in an artificially constructed world. Yet even the action of typing these words, attempting to describe her life feels unkind, as if we are denying the value of her existence in the world; we feel we are falling into the "slippery stuff" of conflicting values and controversy (Milligan & Wood, 2009); the things that are "... hard to quantify, measure, produce outcomes for and ultimately rely more on beliefs and opinions than facts" (Wood, 2007, pp. 44–5). The questions continue: what are the benefits to her of spending time outside? How does it impact on her wellbeing? Who is defining the 'criteria'? Placing ourselves next to her, we attempt to enter her space, as defined on the young woman's terms. Within this shared space we consider is it important for her to spend actual (not just virtual) time in nature?

I follow her gaze to the screen and find its effects hypnotic ...

The impact of technology, media, mediated experiences is a highly contested, emergent area of research. However, there appears to be a paucity of relevant literature, with more being published about children and technology, including play (Skår and Krogh, 2009), education/pedagogy (Palmárová and Lovászová, 2012; Heinonen, 2015) and the use of mobile devices. Within a wider discussion on how an over-reliance on virtual, electronic connections may be eroding our connection to actual physical places, Kupfer (2007, p. 39) claims that "the more electronically mediated activity replaces place, the more we become dis-placed". We find no answers from looking at the screen, merely more questions, particularly as to what this troubling encounter may mean in terms of this research project.

How can this experience be included in a research project that purposefully focuses on young people's voices?

The challenge is to find a place for this story within the more mainstream/dominant narratives on (dis)connection to nature. The direct, formalised relationship is with the project, not with the school, staff, young people or their parents. There is a need to stay aware, sensitive to needs and situation, yet also realistic and pragmatic as the research study was not about SEN/D (special education needs and/or disabilities), this was just one facet; although a very interesting one, it should not dominate the bigger picture.

Critiquing shared-story approaches as a transdisciplinary method

Like other researchers exploring the social world (for example, Macartney, 2007; Mazzei and Jackson, 2012) in this study we felt constrained by traditional forms of both qualitative and quantitative methodologies and were determined to find a way of working in a transdisciplinary manner to explore the research question. We agreed the most effective way to do this was to combine three methodologies: **h**ermeneutics (questioning); auto/**e**thnography and **a**ction **r**esearch, and the acronym HEAR implies listening. Taking each methodology in turn, firstly what we mean by applying the term hermeneutics is best elucidated by Fairfield (2011, p. 3): "The logic of hermeneutics is non-linear, non-formal and non-foundational; it is relational, contextual and dialogical. Interpretation does not begin at the beginning and it is without end...". Through the creation of stories, we pass forward this questioning approach, encapsulated in textual form to provoke and stimulate further questioning. The second methodology, and arguably the strongest within the blend, is auto/ethnography. A concept that has been in use for several decades yet still has multiple meanings and interpretations, it involves "a rewriting of the self and the social" (Reed-Danahay, 1997, p. 4; see Chapter 10). Events are remembered and analysed; the moments chosen as a focus tend to be perceived as epiphanies, a turning point (Ellis and Bochner, 2000), those defining moments that make us stop and wonder, to question in an attempt to extract meaning (Denzin, 2014). The experience must be critically analysed, otherwise it becomes merely 'a nice story'. This means recognising that our interpretation and presentation have been filtered through our own experiences. We agree with Macartney (2007, p. 29) that re/presenting "... perspectives and experiences through narrative is a powerful tool for illuminating and problematising practices and approaches based on deficit discourses ..." such as those applied to young people with disabilities, and/or nature-deficit disorder.

The third methodology is action research: the impetus for innovation is a typical feature of action research (McNiff and Whitehead, 2011) and is central to the research. Another distinctive characteristic is that the researcher is typically a practitioner within a workplace setting: "... it is research from inside that setting"

(Somekh and Lewin, 2011, p. 94). This contrasts with other research strategies, which insist on the researcher being objective and external to the practice/setting. It is unashamedly subjective and situated. This story demanded attention due to the jarring of researcher values, the questioning and challenging of beliefs, which proved to be a transformational (Custer, 2014) moment. For many young people their attendance at a non-mainstream school already places them in the socially constructed position of being "ontologically other" (Farrugia, 2009, p. 1013), inhabiting a space that is defined by processes that may serve to stigmatise the young people, and their families; their voices are mediated and interpreted by the teaching staff. How do we know it is their 'voice', how can we even begin to gain an understanding of their perspective, is it possible for us to view the world through their lenses? This questioning highlights the importance of working in a moral and ethical manner that goes beyond ethics panels, and necessitates highly developed self-awareness, empathic skills and a creative imagination (see Chapter 2).

This is not unemotional research, it embraces a view of research as being emotional (of the heart), as well as cognitive (the head) and practical (hand). There is no attempt to hide behind a curtain of academic objectivity. The researcher is included, not as a form of narcissism or self-therapy but "… on behalf of others, a body that invites identification and empathic connection, a body that takes as its charge to be fully human" (Pelias, 2004, p. 1). We agree with this methodological approach of research with people beyond assumed authority, critical argument and establishing the correct criteria. We want to capture the silent, minority voices, to be compassionate, passionate and emotional. However, we also want our research to be credible, to be recognised as contributing to knowledge, to do justice to the people who volunteer to participate and to respect those who read our work.

Tracy became increasingly aware that she was the only one to witness the finger battle, the only one able to bear testimony, and of the privileged position that we are in as interpreters of social interactions. As identified by Landsman (2003) and discussed by Farrugia (2009, p. 1013), parents of children with disabilities "come to locate, interpret, and often to advocate for the personhood of one they would previously have known only as 'the other'". The same may be said of researchers, who also have to find an appropriate methodology that will withstand the demands (vagaries?) of academia. The finger battle becomes a "…methodologically, embodied metaphor [that] works against the logic of abstraction, the oversimplification of processes of human meaning-making […] a means of presenting a depth to human experience" (Kuntz and Presnall, 2012, p. 738).

Bringing this to a close

This experience, her story, impacts on us in that it makes us question the different ways that we access the natural world; there is no universal 'best way', only temporally, spatially and socially constructed ways. Ultimately it is a matter of subjective differences (Ellis, 2004) based on personal values and morals. Understanding may be gained by participation in social situations, through dialogue between researcher and researched; however, there are ethical and moral questions/dilemmas raised through entering the lives of others, participating alongside them, sharing stories and thoughts.

Transdisciplinary approaches need to cross disciplinary divides, reflect a wide range of interests and resist being categorised within one discipline. This necessitates developing methodologies that are imaginative, creative and practical, and most importantly, they need to be ethical. In developing HEAR, we found a way to critically reflect on experiences in a way that made *understanding* the focus of the reflections and then to put this learning into practice – so that it was of practical, as well as methodological, use.

References

Carson, R. (1962, reprint 2000). *Silent spring*. London: Penguin.

Cope, B. & Kalantzis, M. (2009). Multiliteracies: New literacies, new learning. *Pedagogies: An International Journal, 4*(3), 164–195.

Cope, B. & Kalantzis, M. (Eds.). (2000). *Multiliteracies: Literacy learning and the design of social futures*. London: Routledge.

Custer, D. (2014). Autoethnography as a transformative research method. *The Qualitative Report, 19*(37), 1–13.

Denzin, N.K. (2014.) *Interpretive autoethnography*. London: SAGE.

Ellis, C. (2004). *The ethnographic I: A methodological novel about autoethnography*. Walnut Creek, CA: Altamira Press.

Ellis, C. & Bochner, A.P. (2000). Autoethnography, personal narrative, reflexivity. In N.K. Denzin & Y.S. Lincoln (Eds.), *The SAGE handbook of qualitative research* (2nd ed., pp. 763–768). Thousand Oaks, CA: SAGE.

Etherington, K. (2004). *Becoming a reflexive researcher: Using ourselves in research*. London: Jessica Kingsley Publishers.

Fairfield, P. (2011). *Philosophical hermeneutics in relation: Dialogues with existentialism, pragmatism, critical theory, and postmodernism*. London: Continuum International Publishing.

Farrugia, D. (2009). Exploring stigma: medical knowledge and the stigmatisation of parents of children diagnosed with autism spectrum disorder. *Sociology of Health & Illness, 31*(7), 1011–1027.

Fenwick, T. (2003). *Learning through experience – troubling orthodoxies and intersecting questions*. Adelaide: National Centre for Vocational Educational Research.

Frank, A.W. (2012). *Letting stories breathe*. London: University of Chicago Press.

Gilligan, C. (1993). *In a different voice*. Cambridge, MA: Harvard University Press.

Hayes, T.A. (2017). *Making sense of nature: A creative exploration of young people's relationship with the natural environment* (Unpublished doctoral dissertation). University of Cumbria, UK.

Heinonen, P. (2015). *Modern portable technology in environmental education as part of formal curriculum teaching* (Unpublished Master's thesis). University of Cumbria, UK.

Kellert, S. (2012). The naturalistic necessity. In J. Dunlap & S.R. Kellert (Eds.) *Companions in wonder: Children and adults exploring nature together*. London: MIT Press.

Koro-Ljungberg, M. & Mazzei, L.A. (2012). Problematizing methodological simplicity in qualitative research: Editors' introduction. *Qualitative Inquiry, 18*(9), 728–731.

Kuntz, A.M. & Presnall, M.M. (2012). Wandering the tactical: From interview to intraview. *Qualitative Inquiry, 18*(9), 732–744.

Kupfer, J.H. (2007). Mobility, portability, and placelessness. *The Journal of Aesthetic Education, 41*(1), 38–50.

Landsman, G. (2003). Emplotting children's lives: developmental delay vs. disability. *Social Science and Medicine, 56*(9), 1947–1960.

Louv, R. (2005). *Last Child in the Woods*. Chapel Hill, NC: Algonquin Books.

Louv, R. (2013). *Richard Louv*. Retrieved from: http://richardlouv.com/

Macartney, B. (2007). What is normal and why does it matter? Disabling discourses in education and society. *Critical Literacy: Theories and Practices, 1*(2), 29–41.

McNiff, J. & Whitehead, J. (2011). *All you need to know about action research*. London: SAGE.

Mazzei, L.A. & Jackson, A.Y. (2012). Complicating voice in a refusal to "Let participants speak for themselves". *Qualitative Inquiry, 18*(9) 745–751.

Milligan, A. & Wood, B.E. (2010). Conceptual understandings as transition points: Making sense of a complex social world. *Journal of Curriculum Studies, 42*(4), 487–501.

Palmárová, V. & Lovászová, G. (2012). Mobile technology used in an adventurous outdoor learning activity: a case study. *Problems of Education in the 21st Century. 44*, 64–71.

Pelias, R.J. (2004). *A methodology of the heart: Evoking academic and daily life*. Walnut Creek, CA: Altamira Press.

Reed-Danahay, D. (1997). *Auto/Ethnography: Rewriting the self and the social*. Oxford: Berg.

Skår, M. and Krogh, E. (2009). Changes in children's nature based experiences near home: From spontaneous play to adult-controlled, planned and organized activities. *Children's Geographies,* 7(3), 339–354.

Somekh, B. & Lewin, C. (2011). *Theory and methods in social research*. London: SAGE.

Valentine, G. (2003). Boundary crossings: Transitions from childhood to adulthood. *Children's Geographies, 1*(1) 37–52.

Van Maanen, J. (2011). *Tales of the field: On writing ethnography* (2nd ed.). London: The University of Chicago Press.

Wood, B. (2007). Conflict, controversy, and complexity: Avoiding the "slippery stuff" in social studies. *Critical Literacy: Theories and Practices, 1*(2), 42–49.

15

DIGITAL NARRATIVE METHODOLOGY AND MULTISENSORY OUTDOOR ETHNOGRAPHY

Kirsti Pedersen Gurholt

Introduction

This chapter aims to explore the introduction of digital technology and digital narrative methodology as innovative tools to investigate, represent and communicate outdoor fieldwork-pedagogy, landscape perceptions and adventure practices. Thus, the approach compares the legitimisation of outdoor studies as a scholarly field that accentuates the epistemological values of 'being-in-and-exploring-nature' and 'being-on-the-move' to kindle direct embodied experiences, in-depth knowledge and awareness of the environmental surroundings. Research indicates that increasing use of technology in outdoor fieldwork-pedagogy and adventure act as a double-edged sword by transforming its multisensorial human/nature relationships (Cuthbertson, Socha & Potter, 2004). In today's higher education sector, as in kindergartens and schools, general expertise in digital technology is regarded as one of the basic competencies of students' learning, engagement and advancement (Fossland, 2015). While digital narrative methodology was introduced to disparate professional studies to improve (Master) students' professional expertise and engagement in research, its potential in outdoor studies and settings has not been investigated.

This chapter draws on a case study introducing digital narrative methodology to investigate whether the use of everyday digital technology can contribute to kindling human/nature connectedness, and may proceed as tools (young) outdoor scholars can use to investigate outdoor practices and develop their professional and scholarly expertise (Gurholt, 2016a). The case study was part of the Erasmus Mundus international joint Master's programme, Transcultural European Outdoor Studies (TEOS, 2011—2017), deliberately designed around small-scale fieldwork-research for two reasons (Loynes & Gurholt, 2017): To investigate transcultural and comparative perspectives of European outdoor practices *and* establish a common experiential base among students and staff to investigate relationships of technology, landscape perception and outdoor practices. A model developed by the Center for Digital Storytelling (CDS-model, Berkeley, CA) and adapted to Norwegian higher education contexts (Haug, Jamissen & Ohlmann, 2012, p. 15) inspires the methodological framework. Multisensory ethnography paradigm (Pink, 2009), visual methods (Kara, 2015; Mannay, 2016), and narrative theory (Bruner, 2002; van Manen, 1990) add vital perspectives to the understanding and evaluation of introducing everyday digital technology as tools for researching and representing sensorial, material and social human/nature relationships.

The design epitomises an evolving innovative and participatory researching process, in which master-students and myself explored digital narrative methodology to investigate multisensorial outdoor field-experiences. Our intention was to elicit meaningful research questions and research our own and others' experiences of nature, while acknowledging the importance of the body, the senses and landscape as socio-material manifestations. Next, the approach aimed to establish a situation for student researchers to gain knowledge about the qualitative research process, emphasising narrative and visual components. Of interest in this chapter is the process of converting embodied multisensorial explorations, learning, and knowing that was digitally documented into personally told and analytically condensed (three minutes) digital narratives. The fieldwork was conducted in a rather extreme winter mountain landscape.

Digital narrative methodology: Theoretical implications

Across continents and time, academics have agreed: fieldwork-pedagogy is key in subjects such as geography, biology, anthropology, and subsequently, outdoor studies for several reasons. The strongest of these is the hands-on experiences of everyday life, which enable students to gain comprehensive real world understanding. Both academics and students find fieldwork practices to be enjoyable, meaningful, and effective methods for studying complex issues, covering both teaching/supervision, learning and knowledge production (Fuller, Edmondson, France, Higgit & Ratinen, 2011). In a recent review of literature on the use of digital technology in outdoor fieldwork-pedagogy in the 21st century, Thomas and Munge (2017) conclude that the warnings are numerous. Moreover, scholars provide few theoretical frameworks to support judgement about the benefits of technology. Most research concerns the undergraduate level, whilst the potentials of using digital technology at Master's level and in research remain largely unquestioned (Loynes & Gurholt, 2017).

Several authors acknowledge tensions regarding the introduction of advanced technology and intelligent-equipment in outdoor fieldwork, expressing concerns that technology produces distractions that deepen the nature/culture divide and increase the detachment of human/nature (Öhman, Öhman & Sandell, 2016). Consequently, the quality of embodied human/nature interactions, sensitivity and knowledge is seen to diminish to such a degree that humans' ability to act in, know about, appreciate and care for nature are lowered. Despite the safety that (urban) people may feel from technology, digital technology may make practitioners become vulnerable; in case of breakdown, replacements or repairs cannot happen in situ. Concerns also connect with the ecological devastation that results from the production and waste management of modern technology. These negative effects may speed up with new generations, who are exposed to digital technologies from birth alongside few hands-on experiences of nature (Frost, 2010). Consequently, outdoor educators have a tendency to ban (modern) technology and establish outdoor studies as a low technology zone (Beames, 2017).

'Old- and new-time' technology generate different epistemological situations. With knowhow and literacy in journeying and mastering whatever situations occur, an 'old-timer' can only gain from long-term, *lived experiences* of nature. In winter landscapes, for example, observations and sensations of snow, ice, wind, skies, temperature and topography carry essential information for safety and survival. In contrast, desktop computers and digital devices are thought to separate the mind from the body – cognitive from practical ways of learning and knowing – thus, representing alluring power of engagement in largely disembodied, physically passive, individuated and distant forms of interaction with the world (Hall, 2012). For example, computers can never capture the feeling of strong wind or the taste of a snowflake. Contemporary higher education policy and pedagogy regard the use of digital technology as a 'must' to deepen students' engagement, enhance their learning outcomes and prepare for future professions (Fossland, 1995; Haug, Jamissen & Ohlmann, 2012), and young people regard smartphones, smart clothing, GPS, GoPro cameras and iPads as 'must-have-tools' for wayfinding, information seeking, documentation and communication. As such, they reflect the embracing of technology as part of a leisure consumption, which (in Norway) is growing at a faster rate than everyday consumption (Aall, Klepp, Engeset, Skuland & Støa, 2011).

The gap between technology pessimism and optimism calls for diverse actions, either to ban use of digital devices or commence innovative studies of how technology can enhance and transform knowledge acquisition. Although it may be of the highest quality, digital technology alone can hardly replace the literacy and judgement gained from wide-ranging lived experiences of nature. Obviously, the use of digital technology represents different ways of experiencing, knowing and navigating. Being literate in old *and* new technology, such as in digital narrative methodology, may open innovative ways of knowledge production by integrating embodied and digitally acquired information of the sensorial, material and social engagements that outdoor fieldwork comprises.

Phenomenology of the body inspires sensory ethnography and "takes as its starting point the multisensoriality of experience, perception, knowing and practice" (Pink, 2009, p. 1). The approach understands the senses, body and mind – doing, feeling and knowing – as interconnected and inseparable. Thus, acting is regarded a way of perceiving and knowing. Furthermore, visual methods and digital technologies are seen to provide access to "privileged insights into human relationships to their material environments" (Pink, 2009, p. 97), which conventional ethnography cannot apprehend (Kara, 2015; Mannay, 2016). Combining sensory ethnography, visual methods and digital narratives gives space for multisensorial field-explorations and knowledge acquisition, recognising humans as embodied and emplaced organisms who act, think and feel at the same time.

Following Bruner (2002), storytelling is the cultural medium humans have for thinking and making sense of their lived experiences – of their adventures, surprises, and oddities in life. The Norwegian concept *eventyr,* which translates as adventure and fairytale, covers this ancient cultural connection of journeying and storytelling (Gurholt, 2014, 2016b). However, story making and storytelling do far more than entertain. Stories inform, educate, move, inspire, motivate, persuade and represent powerful tools making embodied experiences explicit and accessible for reflection. Storytelling requires engagement from those creating and telling them along with proficiency in listening among the audience. Over the last decades, digital storytelling has developed as an innovative method to enhance professional expertise, in (physical education) teacher education, nursery schools and engineering (Fossland, 2015; Hall, 2012; Haug, Jamissen & Ohlmann, 2012; Robin, 2008), showing that creative computing "can promote discovery, delight, curiosity, creativity, self-expression, and pleasure in learning" and research (Plowman & Stephen, 2003, p. 160). Oslo Metropolitan University, who made digital storytelling mandatory in professional training, rhetorically asserts, "The digital age is over us, and the classic storytelling gradually moves from camp-fires to the PC-screens",[1] thus diminishing the ancient cultural relationship of adventure and storytelling, which still is alive in outdoor experiences and fieldwork-pedagogy.

The digital narrative methodology presented here keeps the ancient oral tradition of storytelling around a campfire in an outdoor field – in this case study, from a snowy mountain around candle-lights in a snow-cave. Deliberately, it incorporates two additional loops of storytelling and listening, what the CDS-model identifies as story circles. One story circle comprises a process, in which the co-researchers retell, re-live and re-explore an array of embodied and digital field-data as part of transformation into digital narratives employing multimodal expressions. The other story circle emerges when the participatory group of co-researchers gather to celebrate and review all digital narratives accomplished.

Undertaking embodied and digitalising field-explorations

Four cohorts of international master-students of the TEOS-programme (2014–2017) participated in the digital narrative case study, intending to develop a student-active and inclusive co-researching approach on how everyday digital technology may fertilise investigations in landscape perceptions and experiences. Five experienced Norwegian undergraduate outdoor students and two senior colleagues facilitated the practical four-day winter mountain fieldwork. To assure that the purpose of the digital narrative project and the TEOS-master-students' ethnographic investigations were fulfilled, I designed their tasks and assisted the colleagues in the planning and evaluation of the fieldwork.

The practical induction endorsed personal explorations, experiences and conceptualisation, including gaining new skills and insights through journeying and dwelling – skiing, walking and snowshoeing in and across the southernmost mountain landscape in Europe to have an Arctic climate. In addition, the student researchers used all senses to explore the qualities of the surroundings – the various surfaces of snow and ice, and the wind, temperature and topography while playing, preparing food and making snow shelters for the nights. Embodied experiences, feelings and reflections were documented as directly as possible by taking diary-notes and using the digital devices at hand. Data acquisition embraced written notes, collected texts, recorded soundtracks, visual images (still photos, video-clips, and sketches), collages, maps and artefacts.

The approach exemplifies the rich sensorial, material and social nuances that outdoor fieldwork exploration and investigation provokes. Located in natural landscapes is a rich range of socio-cultural reminiscences and contemporary configurations, inhabitants and visitors, and other-than-human elements: topography, surfaces, forms, colours, tastes, smells, sounds but also snowdrifts, skies, shivering trees, animals, and so on. Obviously, digital devices can document visual and aural appearances of the present manifestations more completely than merely written notes, despite limitations; no technology can capture the feelings of fear or ice-cold wind creeping under your jacket. The digital narrative methodology gave space for comprehensive multisensorial, aesthetical investigation of the role and relationships of digital and conventional approaches in outdoor fieldwork-pedagogy and research. It also gave a means of apprehending analogue and digital ways of data documentation and representation. Included was a questioning of the contemporary dominance of visual expression and representation, and the limitations inherent in spoken and written language in analysing and representing the richness of oral, visual, aural, tactile and textual modalities.

Personal stories, innovative scholarly writing and multimodal expressions

The CDS-model builds on two principles: the powerful process of storytelling using personal voice, and 'the magic' of story circles for scaffolding and expanding the student researchers' self-representation, perspectives and knowledge acquisition (Haug, Jamissen & Ohlman, 2012). Hence, the digital editing process can be organised differently, by means of assorted digital editing tools that can be downloaded free of charge, such as iMovie and Movie Maker (which was used in this case study). After return from the field, the digital narrative processing continued, encircling single outstanding key-experiences and analytical ideas arising from the field-experiences and data generated. Through an intensive two-day guided workshop, each researcher highlighted one single key-insight and represented it in the format of a multimodal digital narrative.

Using a free-flow-writing technique, 'without lifting pen from paper' for about twenty minutes, a 'stream of experiences' was captured as notes, mind-mappings and/or sketches on paper. Next, in an atmosphere of mutual trust and fairness, the co-researchers were given the same time, attention and scaffolding in a story circle to read his and her free-flow-text and receive constructive comments and associations intending to elaborate the key-themes presented. Hence, the winter mountain explorations were retold, re-lived, re-reflected, and revised individually and collectively, several times.

The digital narrative methodology provides a context for innovative narrative style of scholarly writing, in which a phase of idea elaboration turns into a phase of idea-condensation and analytical refinement; a process involving repeated individual rewriting, rereading and reediting, intending to crystallise valid ideas. In this case study, each co-researcher detached one single analytical idea and once again presented his/her idea in a story circle before completion. Thus, they were expressing its essence by giving it a final narrative structure of 300 words maximum – including an opening, the evolving idea, a turning point and closing. Finally, authors tape-record their narrative, using their own voice.

The process continues by synthesizing disparate visual and aural elements with the tape-recorded narrative into a condensed choreographed multimodal digital narrative of 3–4 minutes. The narrative may include video-clips of moving bodies and natural elements, accompanied by soundtracks, for example, of the wind, dialogues recorded in situ or just stillness, and adding music or any other effects the researchers may include, recalling field-experiences. The research approach opens for strengthening the presence of the researcher and for creative ways of expressing lived experiences, reflections, and ways of viewing; using one's voice and personality by singing, whispering, breathing, or even accompanying by playing a piano or guitar. Finally, by the use of the digital editing tool the final digital narratives appear as nonfictional, though choreographed, comprehensive multimodal representations of key experience, communicating complex meanings that cannot be expressed solely by words.

Multimodal expressions and multiple tales

A final story circle was organised collectively to celebrate the digital narratives accomplished. The variations of key-themes, narrative structure and dramaturgical effects originating within common field-experiences were overwhelming. Even though particular visual elements and ideas were used several times among the co-researchers, different and unusual ways of choreographing released novel perspectives and ways of viewing that added insights and levels of meaning. While watching and listening everyone involved became deeply touched; people laughed, dried up tears, became quiet, and felt astonished. In particular, hearing one's personal voice was ambiguously demanding, though the student researchers were gratified with approval and pride, having a personally edifying experience. This illustrates the ways in which the collective sharing of multimodal expressions represents partially evoked feelings, unspoken reflections and unexpected interpretation and re-interpretation.

An overall insight grew: *experiencing* is individual; learning is social and cultural, as articulated by a student-researcher who emphasised shared feelings of mutual interdependency in the communal process of knowledge production and when encountering and embodying the powers of winter mountain nature. To stay safe, warm and healthy a sensitivity towards personal and human vulnerability matured. Literally, everyone had to trust each other to make life and research in the snow viable. The embodied insight was underscored by the tangible and demanding excavating of snow shelters – bivouacs, igloos, a Quincy and 'snow-crypts' – all of which were connected by tunnels to become home for the nights. Some of these 'buildings' were copies of typical ancient shelters used by native people and even animals of the northern hemisphere; others were place-adapted constructions. In a metaphorical and literary sense, a paradoxical, indeed astonishing insight began to grow during nights spent sleeping in a cluster of connected snow-caves – that the student-field-researchers named "an underground snow-sculpted hotel"; "snow can take your life, but also save your life!" Human life and nature are inseparable!

The digital narrative methodology has much in common with experiential and inquiry-based approaches central to conventional outdoor fieldwork-pedagogy. Knowledge becomes personally meaningful but is created in scholarly dialogues connecting theory and practice. Likewise, the diversity of ideas mirrors the ample potential of knowledge acquisition in fieldwork-research, which in this case embraces themes such as the metamorphoses of the 'life' of snowflakes becoming a digitally narrated metaphor of the educational journey of the transcultural group of master-students, studying in three countries. Another narrative explored the "digging of tons of snow" as a metaphor for fieldwork-pedagogy as place-making and familiarising processes in unknown environments. The role of the researcher's biography became visible in a digital narrative comparing the multisensorial feelings of moving in sandy versus snowy landscapes.

In *Tales of the fields: On writing ethnography*, Van Maanen (1988) critically recognised three conventional narrative genres in ethnographic writing – the realist, confessional and impressionist tales. Confessional tales, which focus the fieldworker's experiences, were easily recognisable among the digital narratives given the emphasis this case study put on the (becoming) researchers' voices and points of views. Besides, the confessional style was present as 'private letters' and self-therapeutic tales. Accordingly, one student-researcher spontaneously responded, "If we had not known, we would have recognized who was the storyteller", an insight exemplifying the complexity of issues of ethnographic and any scholarly writing that brings reflection to the forefront. Other examples, critically addressing the concepts of inter-subjectivity, include use of first-person tales, relevance of researchers' biography and positioning in the field, including interpretation of visual and conventional data in qualitative research knowledge production.

Realist tales, characterised by a dispassionate, third-person voice providing rather direct, matter-of-fact representations of what happened, were also present. These were exemplified in digital narratives encompassing the cultural history and contemporary (re)making of the mountain landscape as a tourist site and railway station. Such tales investigated the presence of the railway, the hiker's cabin, the old telephone poles, marked trails, snow-covered private cabins, along with a visit to the local museum and reflections on the community house which served as a safety 'back-up' during the 4-day field-explorations.

Several of the digital narratives may be said to represent impressionist tales, which according to Van Maanen (1988, p. 7), are "personalised accounts of fleeting moments of fieldwork cast in dramatic form", portraying emblematic elements of both realist and confessional 'nature'. Consciously, some researchers introduced irony and humour to provoke critical reflection, for example on tourism developments and sustainability issues, or by resisting including photos in their digital narratives to oppose the dominance of visual images conventionally portraying outdoor field-experiences in public spaces. Minimalism was also expressed by reducing the amount of multimodal effects to refine, for example, the embodied feeling of backcountry skiing. Connecting the body's rhythmical breathing and the 'swishing' sounds of gliding skis towards disparate surfaces of snow created an astonishing emotional effect.

Customary romantic nature tales, frequently presented in commercial announcements of adventure tourism and much outdoor education literature, favour narratives depicting the adventure landscapes according to romantic clichés. Additionally, the romantically inspired digital narratives tend to depict the winter mountain landscape unreflectively by reproducing illusionary images of the winter mountains as untouched, uninhabited, and white, infinite, and sublime panoramic beauty, accompanied by well-known classical romantic composers. Thus, the romantically inspired digital narratives represented a counter-discourse to the realist and confessional tales that were told. The diversity of tales and themes exemplifies the concept of nature as culturally discursive. Conversely, there is a need always to reflect critically on the researchers' positioning and choice of narrative genre, which the digital narrative methodology 'spell[s] out'. However, a discursive perspective implies, and reveals, an understanding that different genres or discourses frequently overlap and intersect (Ryall, Schimanski & Howlid Wærp, 2010).

Conclusion and implications

Introducing digital narrative methodology in outdoor fieldwork-pedagogy and research, was never intended to replace what Wattchow (2001, p. 25, as cited by Thomas & Munge, 2017, p. 9) terms "our organic links with the world". Nor was the intent to disfavour low technology outdoor practices. Rather, the aim was to explore innovative and high-quality use of everyday digital technology as research tools to expand ways of gaining new knowledge and representation of the interconnections of sensorial, material and social manifestations of human/nature (landscape) relationships. The design combined multisensorial ethnography, visual methods, and digital narratives – embodied and digitalised outdoor field-experiences – while also offering a theoretical framework for outdoor fieldwork-pedagogy and research. The approach has much in common with what Kara (2015) and Mannay (2016) characterise as creative research methods, using visual methods as a tool of inquiry and drawing on methodologies reflecting the positioning of the researcher in real world situations.

The role of technology continues to be controversial in outdoor studies in the higher education sector; uncertainties embrace ethical issues, environmental and epistemological significances and practical implications. The presented approach draws attention to digital technology as a tool for explorations, sense-making, critical reflection and analysis of outdoor experiences, never as a goal in its own terms. Certainly, discussing potential distractions, disadvantages and tensions are significant. The digital narrative methodology may inspire reproduction of the predictable and publicly dominant genre of nature writing, misuse of copyright material, or even pre-making of narratives 'from the field' to fulfil public self-promotion (cf. Gray, Norton, Breault-Hood, Christie & Taylor, 2018). Some student researchers may have felt too vulnerable to share their personal story and withdraw from the process. All of which are concerns germane to the presented case study.

However, the methodology inspired master-students to use digital narrative methodology to explore its knowledge-producing potential in investigating other topics relevant to outdoor research. Conversely, they submitted conventional academic essays and master theses, but integrated digital narratives to communicate insights gained from combining conventional and digital processing. In master-student research, the digital narrative methodology is used in co-researching with children and young adults, investigating ways in which young people conceptualise nature and make sense of outdoor field-experiences.

The interplay of the 'old ways' and multiple uses of digital devices release a richness of data acquisition and multimodality in combining visual, aural and textual expressions. The digital narrative approach encompasses participatory approaches, allowing for connecting, contrasting and comparing various ways of experiencing, learning and knowing in dialogue with the views and reviews, scholarly reflection and theory, of co-researchers. The approach allows the (becoming) researchers to work in ways that promote curiosity, engagement, discovery, creativity, delight, self-expression and pleasure in research. The multimodal digital narratives illustrate the values of combining analogue and digitalised data acquisitions to give a complete representation of the embodied emplacement that outdoor field-experience implies.

The overall conclusion is that digital narrative methodology stimulates strong engagement among (young) researchers, making field-studies become meaningful and creative, whilst carrying an openness triggering unpredictability and excitement. The approach stimulates critical awareness, academic knowledge production, and questioning relevant for future research in ways that give space for self-representation and analysis anchored in social and scholarly contexts. Besides, it allowed the (becoming) researchers to explore experientially, within a given structure, the qualitative research process, emphasising its hermeneutic phenomenological 'dance' of repeated participatory observations, seeing/hearing with writing/reading/listening/peer-reviewing and multimodal digital representation.

The approach offers potential for multisensorial explorations and knowledge production that allow for more authentic and comprehensive multimodal expressions, representation, and communication of 'real world' outdoor field-experiences.

Note

1 Retrieved 3 March 2014 from: www.jbi.hio.no/FOU/FOU/digital-historiefortelling.html.

Acknowledgements

I am grateful to colleagues Jannicke Høyem, who introduced me to digital storytelling, and Trond Augestad and Thomas Vold, who conducted the snowy mountain fieldwork as part of the digital narrative project. The contributions of social youth worker and photographer Joe Sowerby, who conducted the practical digital workshops, and the digital narratives created by TEOS-students (2014–2017), allowing me to use their stories for research purposes, are sincerely appreciated.

References

Aall, C., Klepp, I.G., Engeset, A.G., Skuland, S.E. & Støa, E. (2011). Leisure and sustainable development in Norway: Part of the solution and the problem. *Leisure Studies, 30*(4), 453–476.

Beames, S. (2017). Innovation and outdoor education. *Journal of Outdoor and Environmental Education. 20*(1), 2–6.

Bruner, J. (2002). *Making stories: Law, literature, life*. Cambridge, MA: Harvard University Press.

Cuthbertson, B., Socha, T.L. & Potter, T.G. (2004). The double-edged sword: Critical reflections on traditional and modern technology in outdoor education. *Journal of Adventure Education and Outdoor Learning, 4*(2), 133–144.

Fossland, T. (2015). *Digitale læringsformer i høyere utdanning*. Oslo: Universitetsforlaget.

Frost, J.L. (2010). *A history of children's play and play environment*. New York, NY, London: Routledge.

Fuller, I., Edmondson, S., France, D., Higgit, D. & Ratinen, I. (2011). International perspective on the effectiveness of geography fieldwork for learning. *Journal of Geography in Higher Education, 30*(1), 89–101.

Gray, T., Norton, C., Breault-Hood, J., Christie, B. & Taylor, N. (2018). Curating a public self: Exploring social media images of women in the outdoors. *Journal of Outdoor Recreation, Education, and Leadership, 10*(2), 153–170.

Gurholt, K.P. (2014). Joy of nature, *friluftsliv* education and self: combining narrative and cultural–ecological approaches to environmental sustainability. *Journal of Adventure Education and Outdoor Learning, 14*(3), 233–246.

Gurholt, K.P. (2016a). Digital narratives of snowy mountain experiences: An aesthetic pedagogical design to stimulate students' academic reflection. Paper presented at the Seventh International Outdoor Education Research Conference, Cape Breton, Canada.

Gurholt, K.P. (2016b). *Friluftsliv*: Nature-friendly adventures for all? In Humberstone, B., Prince, H. & Henderson, K.A. (Eds.) *International handbook of outdoor studies* (pp. 288–296). Oxford, New York, NY: Routledge.

Hall, T. (2012). Emplotment, embodiment, engagement: Narrative technology in support of physical education, sport and physical activity. *Quest, 64,* 105–115.

Haug, K.H., Jamissen, G. & Ohlmann, C. (Eds.) (2012). *Digitalt fortalte historier: refleksjon for læring.* Oslo: Cappelen Damm.

Kara, H. (2015). *Creative research methods in the social sciences: A practical guide.* Bristol: Policy Press.

Loynes, C. & Gurholt, K.P. (2017). The journey as a transcultural experience for international students. *Journal of Geography in Higher Education, 41*(4), 532–548.

Mannay, D. (2016). *Visual, narrative and creative research: Application, reflection and ethics.* Oxford, New York, NY: Routledge.

Pink, S. (2009). *Doing sensory ethnography.* Los Angeles, London: SAGE.

Plowman, L. & Stephen, C. (2003). A "benign addition"? A review of research on ICT and pre-school children. *Journal of Computer-Assisted Learning, 19*(2), 149–164.

Robin, B.R. (2008). Digital storytelling: A powerful technology tool for the 21st century classroom. *Theory into Practice, 47*(3), 220–228.

Ryall, A., Schimanski, J. & Howlid Wærp, H. (2010). Arctic discourses: An introduction. In Ryall, A., Schimanski, J. & Howlid Wærp, H. (Eds.), *Arctic discourses* (pp. ix–xxii). Cambridge: Cambridge Scholars.

Thomas, G. & Munge, B. (2017). Innovative outdoor fieldwork pedagogies in the higher education sector: Optimising the use of technology. *Journal of Outdoor and Environmental Education, 20*(1), 7–3.

Van Maanen, J. (1988). *Tales of the field: On writing ethnography.* Chicago, IL: University of Chicago Press.

van Manen, M. (1990). *Researching lived experience.* London, ON: University of Western Ontario.

Öhman, J., Öhman, M. & Sandell, K. (2016). Outdoor recreation in exergames: a new step in the detachment from nature? *Journal of Adventure Education and Outdoor Learning.* doi:10.1080/14729679.2016.1147965

16

PRACTISING FEMINIST REFLEXIVITY

Collaborative letter writing as method

Pip Lynch, Martha Bell, Marg Cosgriff and Robyn Zink

Introduction

In the Troubling Terrains study (Bell, Cosgriff, Lynch & Zink, 2018), we explored the ways in which four Pākehā[1] women in the outdoors reacted to contradictory constructions of identity and gendered physicality. The purpose of that study was to examine aspects of our professional experiences as women outdoor leaders that revealed how gendered situations had influenced our professional outdoor lives. We had each worked in outdoor education over at least three decades in varied roles, including as instructors, outdoor guides, teachers, researchers, programme administrators and, for one, policy advisor. The study caused us to return to moments in time that still held meaning for our individual understandings of lived relations of gender. Reflecting on our experiences unearthed memories of ways in which working and being outdoors had nourished each of us. It also exposed deeply troubling[2] situations and practices and demanded that we each provide an account of, and collectively make meaning from, those situations and experiences. We made no attempt to connect the memories and treated them only as a basis for a period in time which gave "a conception of gender as a constituted *social temporality*" (Butler, 1990, p. 141; italics in original). We paid particular attention to phenomena that appeared to cut into such temporal enactments of gender.

In order to consider the particular and the general in our analysis, we conceptualised our subjective experiences in light of the objective conditions of the remembered time. To do so, we drew on a feminist methodology of analysing lived experience of gender (Gluck & Patai, 1991; Stanley & Wise, 1990). The methods we employed included writing letters to each other and developing a framework to guide analysis of the letters. We sorted the content of the letters into thematic "terrain[s] of signification" (Butler, 1990, p. 148). Through collective, abductive analysis (Bryman, 2012, p. 401), our experiences in these terrains were then troubled, that is, unsettled and critically re-examined in the context of the understandings we hold now of this profession.

We had been inspired by the Letters Project. This enquiry invited early career academics to write narratives of their work experiences and pose questions for participants at senior levels in the academy. One member of our research group had heard its Australian researchers present their method of composite letter writing at a conference and she invited us to collaborate. At the time, the Letters Project had yet to be published. However, as professionals in the field of outdoor and experiential education, we were familiar with writing practices of journal keeping and other forms of reflection. In the Letters Project, three co-researchers analysed 30 solicited narratives which they had collected asking about the joys and challenges of "being and becoming academics" (Alfrey, Enright & Rynne, 2017; Enright, Rynne & Alfrey, 2017; Rynne, Enright & Alfrey, 2017). The co-researchers then created three composite letters based on narrative analyses of these texts. Eleven participants in professorial positions read these "letters as narrative" (Alfrey et al., 2017, p. 8) and responded by writing letters back to the generalised set of junior colleagues. This chapter examines our collaborative narrative approach, focusing on why and how we went about exploring our professional outdoor experiences and how our collective analysis produced an interpretive framework of four terrains of contestation related to pedagogies, work, skills and bodies.

Narrative inquiry as methodology

Narrative research methodologies emerged in sociological and anthropological studies (Richardson, 1990), as well as more recent critical approaches to pedagogical inquiry. The latter focused on the desire of teacher educators to hear teachers' own experiences of their practice (Ciuffetelli Parker, 2011; Clandinin, 2013; Clandinin & Connelly, 2000; Knowles & Cole, 1994). The narrative approach allows practitioners to give accounts informed by their own lived knowledge about the dynamics of classroom, school and wider socio-cultural power relations through stories that show the logic of an event in retrospect. A narrative approach is often selected to reveal or retell stories to a wider audience for the purposes of "dialogue about complex moral matters and about the need for social change" (Chase, 2011, p. 429) in the public sphere. The audience for such research plays an important part in both listening to narrated experience and responding to the content of those experiences, often in light of shared storylines.

The Personal Narratives Group (1989) is one interdisciplinary collective of feminist writers who came together to examine histories and biographies structured by gendered experience. Members of the group contributed fieldwork to a research-based anthology of narratives. Another example is the Emotion and Gender writing collective (Crawford, Kippax, Onyx, Gault & Benton, 1992) formed by a group of feminist psychologists. This group challenged members' professional practice by examining the social psychological construction of gendered emotion based on their personal memories as girls and women. Both projects employed reflexivity;

that is, not reflection on but reconstruction and representation of lives lived – historical, empirical and personal – in order to depict conditions shaping those lives (Crawford et al., 1992, p. 39; Personal Narratives Group, 1989, pp. 4–6). In fact, 'reflexivity' is a social practice in which a constant reconstruction of 'self' identity is required by 'modern' individuals to make sense of the structural conditions of which they become aware (Adams, 2006, p. 513).

Narrative inquiry can include interview settings in which participants are encouraged to remember and tell stories (Bryman, 2012; Crossley, 2000; Riessman, 2008). The subjects of the inquiry reflect on experience as a narrative whole unfolding over a period of time. Crawford and colleagues (1992) found that writing down memories rather than events helped to elicit stories. They employed the method of "memory-work" (Haug, 1987) in which the group decided on a specific memory "trigger" and each woman wrote about her own experience in the third person. The collective met to read the memories, discuss the experiences that writers had recalled and analyse the emotions that arose in interactive situations, particularly when the unexpected occurred. Memories often relate to an inherent conflict or contradiction at the time. Such a conflict can often reside between the personal explanation for behaviour and the social world "which renders the events problematic" (Crawford et al., 1992, p. 154). At the time, there may not have even been an explanation available as to how an event fitted in a social sequence. As a memory, therefore, it could be retrieved and reopened to analytical interpretation that would resolve the knot of conflict and confusion (Crawford et al., 1992, p. 163). Memory studies emphasise that memories enable "the meaning of the past [to be considered] in relation to the present" (Keightley, 2010, p. 57).

Memory writing is similar to the creativity involved in letter writing (Crawford et al., 1992, p. 37). Both data elicitation methods are "generative" (Rynne et al., 2017, p. 141) by creating data and analytical points for interpreting the data. Since the letters emerge from within the field of inquiry between collaborators, the letters are a type of "field text" (Clandinin & Connelly, 2000, p. 107). In letters, we, as writers, "give an account of ourselves" and "make meaning of our experiences", while we also "attempt to establish and maintain relationships among ourselves, our experience, and the experience of others" (Clandinin & Connelly, 2000, p. 106). "One of the merits of letters is the equality established, the give-and-take of conversation" (Clandinin & Connelly, 2000, p. 106), enhancing collaboration.

'Epistolary' writing yields a relationship that opens an expectant space for dialogue over aspects that the writer explores (Alfrey et al., 2017; Enright et al., 2017; Rynne et al., 2017). Dialogue is particularly structured through the interchangeable relation between letter writer and letter reader. The reader shifts to writer and the writer then waits to become reader; neither is passive in the relationship, ensuring reciprocity without hierarchy (Jolly & Stanley, 2005, pp. 94–95). The act of letter writing is "referential" to real world content, "purposeful" following a specific intent and socially shaped in format while personal in content (Jolly & Stanley,

2005, pp. 94–95). For this reason, letter writing and reading can generate intense emotional affect which is socially grounded yet personal. Letter writing as method creates the means of interpreting data. Following the Australian project, we began by asking each other to write narratives of nourishing and troubling aspects of our professional work in the outdoors.

Collaborative data generation

In our first letters in the Troubling Terrains study we each wrote approximately 500 words about "subjectively significant" (Haug, 1987, p. 40) experiences from our professional lives in the outdoors. Each researcher responded individually to the letters received from the others, summarising the main points, raising any questions or contradictions and suggesting points for further development. Thus, the iterative aspect of data generation and thematic analysis began at that point. The probing by our co-researchers gave us direction as letter writers, but also encouraged a wider 'epistolary' engagement with the unknown potential of the space of dialogue. Our revised, second letters gave more contextual depth and, where possible, referred to relevant published literature. The four letters ranged in length from 1300 to 2400 words and formed the data set for the Troubling Terrains project.

We chose to work collaboratively with self-generated texts in order to investigate lived experiences while maximising the coherence, plausibility and credibility of the research findings, meeting three key tests of the trustworthiness of such research (Riessman, 2008). The advantages of our approach were that it involved two phases of narrative construction, the second being in response to critical commentary, thus enhancing coherence; it required narratives to respond to published theory, thus addressing plausibility; and it drew on existing interpersonal relationships to enhance credibility.

Credibility, in particular, is often a sticking point for qualitative researchers when faced with critical appraisal of research. We had known each other, as researchers and teachers in the outdoor education field, for some 30 years. We each knew something of the others' career trajectories, life histories, personalities and professional profiles and were similar in generation. We had all lived in Aotearoa New Zealand during adulthood and had all worked in the outdoor field there and overseas. Our experiences thus arose from a socio-historic-geographical point in time. We had each adopted critical perspectives in our writing (see Bell, 1997; Cosgriff et al., 2012; Lynch, 2000; Zink, 2010) and confronted dominant group privilege in our professional activities. While we had not previously communicated in detail about the experiences described in the letters, we recognised the consistencies between the content of the letters and our long-term knowledge of one another. The four-way cross-referencing reinforced the credibility obtained through our approach. We also wrote with the intent of reading each other's writing (Personal Narratives Group, 1989). Knowing, and participating in, the audience gave our writing some intensity. To maintain the integrity of the process, we were firm about finishing and sending our own letter before reading the letters delivered to us. We were not so well known to each other that we could anticipate the content of one another's letters. Arguably, this aspect of our relationships freed the analysis from the undue inferences that close familiarity can produce.

Generating the texts

One pragmatic rationale for biographical methods in social science research is that "the stories and ideas that one creates should be useful for solving further problems in one's professional life" (Smith, 1994, p. 302). Our collaboration shed light on how our histories had impacted on, for example, our teaching and research. Particularly thought provoking were experiences of an inability to change troubling gender relations in outdoor settings so that our students and colleagues could avoid the difficulties we had experienced. Other research methods, such as autoethnographical and interview-based methods, can be employed to similar ends, however a collective process such as ours can facilitate multiple, contextually informed, examinations of each biographical text, producing a deeply nuanced data set.

To illustrate, we take the topic mentioned above – impacts on our teaching and research – and trace our analysis from one researcher's letter. In her first letter, Writer 1 wrote:

> I encouraged men and women to consider critically their own practices and those of others and to be active in making the outdoors an open learning space for diverse people. However, I remain troubled by what I observe in field teaching and read in student journals: women feeling trapped into stereotypically female roles such as cooking … when they prefer to also chop wood and make fires. In particular, when the terrain is difficult or weather adverse, gender stereotypes become more prevalent.
>
> *W1, lines 31–37*

One co-researcher challenged this writer to examine:

> how you now reconcile the determination you felt to absolutely overcome that situation of discrimination with the ineffectualness of the role of outdoor leader to incite such determination in your women students … I can see from your writing how troubling that would be.
>
> *Response from W4 to W1, p. 4*

Another encouraged the writer to reflect on:

> what sort of feminist framework/practice in outdoor leadership work might have supported [you] to voice [your] critiques of leadership, challenge colleagues and support other women and men towards alternative gender practices in their outdoor participation.
>
> *Response from W3 to W1, p. 4*

And the third co-researcher commented:

> I wonder if [leadership and gender] can be understood through the neo-liberal rubric of individual choice which elides the ways in which social structures, including gender, shape behaviour and judgement. Do you see this linking into how you have not been more successful at creating nourishing learning environments for women?
>
> *Response from W2 to W1 p. 2*

Writer 1 was also encouraged to consider feminist theory with regard to whether her own practice of feminism was useful to a younger generation.

These challenges led Writer 1 to interrogate her experiences of engagement with feminist pedagogy at multiple levels, thereby enhancing the narrative's coherence. In her second letter, she referred to her "low-visibility response" (Emerson, 2009, p. 540) to the troubling situation as a deliberately "low-profile engagement with feminist research … in response to my personal-professional situation … Taking on the gender battle (feminist praxis) at household level left me with little desire to take it on at work, either institutionally or in my research" (W1, lines 57–64). She referred to theorising on careership by writing that decisions such as hers "are not purely technical decisions, but are partly emotional, and are made in relation to the range of opportunities that a person perceives as available and possible, their 'horizon for action' (Hodkinson & Sparkes, 1997, p. 34)" (W1, lines 66–69). She wrote that her own "'horizon for action' was influenced by 'structures of opportunity' (Allin & Humberstone, 2006, p. 138) and [her] individual habitus". Through these references, Writer 1 added rigorous and credible interpretation from relevant theory to her narrative.

Similarly, in further reflection on leadership and pedagogy, Writer 1 stated:

> Why did my teaching fail [students] in this regard? It failed them because my colleagues and I did not address gender directly in the curriculum. Our approach ran counter to Henderson's (1996) argument that for feminist leadership to be transformative it is necessary to focus on the social (gender) dynamics within the group. Similarly, Loeffler (1996, p. 98) included "educat[ing] staff and participants about gender issues" among her strategies for increasing women's participation in outdoor leadership. However, we regularly had more women than men in the class … Most of the troubling incidents occurred while on trips without staff. The students concerned did not bring up the gender stereotyping during post-trip reflection in class … It can be very challenging to bring up uncomfortable topics, especially if a student is not sure how much support will be gained from others. Such social dynamic issues as these are not considered by Breunig (2005), who reports having deep critical discussions in her classrooms.
>
> *W1, lines 142–162*

The deeper reflections in the second letters, such as in the example above, comprised the data we analysed.

Collective analysis

Our data analysis began with the construction of a conceptual framework (Ritchie & Lewis, 2003) to facilitate disciplined interpretation of issues raised. Preliminary categories for the framework were identified by two researchers reading the first round of letters. These categories become further "terrains" to extend theorisation of the "gendered terrain" of outdoor leadership (see Bell, 1996; Newbery, 2002). They were utilised and refined during reading, rereading and coding of the subsequent round of letters. The completed framework was then shared with the whole research team for agreement. Four terrains could be cross-referenced across all the letters: outdoor pedagogies, outdoor work, outdoor bodies and outdoor skills. Following on from the extract discussed above, we take two sections from our first terrain of 'outdoor pedagogies' to illustrate how the framework provided a means of excavating meaningful content from the field texts.

The framework consisted of a six-column table. In practice, we had one long, multi-page table for each separate terrain. An extract of two short sections (see Table 16.1) shows how we sorted data chunks from the letters into the framework, noting their original source. No longer did data belong to one writer; all writers had contributed to the development of each of the second letters. In this way, the framework encapsulated collective concerns, giving it plausibility and strengthening its credibility. We developed linkages between sections of meaningful content in each terrain and across each terrain by commenting on its relevance to the larger project aims. The Comment column generated analytical concepts through which to interpret the framework as a whole.

Meaning making

The framework provided a disciplined basis for interpretive "memos" (see Emerson, Fretz & Shaw, 1995), which were then recorded in the Memo column; in the extract above, Memo 9 was recorded.

> Although expressed variously, permeating the letters of women 1, 2 and 3 was a deep sense of questioning themselves and their apparent (self-declared) 'failure' to more actively draw on and explicate feminist theory and feminist pedagogy explicitly in their work over time.
>
> *Memo 9, lines 3–5*

This memo, titled *The F Word (Failure),* drew out the meanings of 'failure' in the data set: the writers' engagement in feminist outdoor initiatives yet individual frustrations with the tensions of enacting feminist pedagogy and leadership;

TABLE 16.1 Extract from analytic framework for sorting data

What's troubling about this terrain?	*How did this shape a gendered identity?*	*How did we trouble or challenge?*	*Comment re: relevance to this project*	*Review of literature sources*	*Memo?*
Section: Our lack of success at implementing feminist pedagogy/	The troubling nature of pedagogy *"I'm troubled by how best to 'navigate' and foreground gender (and sexuality, class, ethnicity, and ability) theoretically and pedagogically…"* (W3, L2, lines 154–155)	By deep questioning (blaming?) of self and [own] expression of concerns	*"Why is 'doing gender' women's work? This is the 'cul de sac'"* (W4, Response to W3; W3, L2, lines 166–167).	Henderson, Loeffler, Warren, Roberts, Breunig & Alvarez, Zink, Zink & Leberman	Memo #9
Section: Our failure to teach and/or research gender	*"One question I ask myself as I reflect on the outdoor education I have engaged in is 'why have I not more explicitly asked how outdoor education and gender intersect, or drawn on feminist theory?'" (W2, L2, lines 2–4)* *"I have reflected a great deal on why we women have failed to create a nourishing feminist outdoor theorising practice" (W2, L2, lines 42–43)*	By asking questions of practice, [being] reflexive in practice (why only of ourselves?)	How to talk and 'do' gender productively?		
	"It troubles me that, as a leader, I have not been more successful in creating more nourishing learning environments for these women in their outdoor learning" (W1, L2, lines 137–139) *"I'm bothered by my slowness to more tenaciously pursue gender in professional outdoor forums and the overwhelming feelings of how much is at stake when I do"* (W3, L2, lines 147–149)	By struggling with (and living with) tensions	What is possible re: nourishing feminist outdoor practice within [an] outdoor labour market?		

questioning whether these perceived failures were of themselves or of all outdoor leaders (of whatever gender); failure to recognise one's own feminist leadership in practice; perceived failure of courage to engage in the gendered politics of one's profession; the failure of feminist politics; and failure "to account for the hostility in the labour market at the time ([linking to] Memo #6)" (Memo 9, lines 41–42). It was one of nine memos constructing our analysis to tease out overarching narrative concerns.

Hence, the discussion of various failures was further condensed in the process of analysis. The findings of that project were presented through analytical concepts. For the purposes of this illustration, 'failure' helped to explain the concerns of the fourth concept, "How Does Gender Become Women's Problem?" in which we concluded that:

> Gender can become girls' or women's 'problem' through the re-embodiment of restraint on their labour to the extent that they do not want to raise gender as an issue; this was apparent in all of our letters. The risk of speaking out and becoming 'one of them', or the feminist voice, is of taking on the problem that is not ours. In the end, the failure of feminism to create a nourishing practice as each of the writers moved into tertiary sites of pedagogy and research was part of an ongoing struggle.
>
> *Bell, Cosgriff, Lynch and Zink, 2018, p. 211*

Conclusion: Letter writing in narrative research

As we exemplified in the Troubling Terrains study, letter writing as method creates textual data and invites textual interpretation. Moreover, letters are written to be read and can generate intense emotional affect experienced with socially meaningful feelings. We began by writing narratives of nourishing and troubling aspects of our professional work in the outdoors addressed to each other. As co-researchers, we read and annotated the first letters we received from each other and wrote to each writer in response. Thus the iterative aspect of data generation and thematic analysis began at that point. The probing by our co-researchers gave us direction as letter writers, but also encouraged a wider epistolary engagement with the unknown potential of the space of dialogue. Collaborative writing is effective when researching interactional experience, because it offers an interactional process itself.

This method of reflexive narrative inquiry opens up fruitful opportunities for outdoor studies research. Within such a kinaesthetic professional practice, outdoor educators and students often encounter 'uncomfortable' feelings and dynamics which remain as silences and tensions over time. To excavate social temporalities in practices and pedagogies requires a reader willing to become a writer who engages in a space of non-hierarchical and reciprocal dialogue. Embodied socio-cultural politics are not resolved through the norm of introspective journal writing. The method of writing to a reader leads to the holistic dialogue needed in a people-centred outdoor studies research paradigm.

Acknowledgements

The authors acknowledge Vickie Zhang's helpful comments.

Notes

1 Pākehā is a New Zealand Māori word for people of non-Māori descent.
2 'Troubling' experiences are those in which interpersonal difficulty or conflict arises with some ambiguity as to who is responsible for reparation, and thus a problematic relationship or situation remains unresolved (see Emerson, 2009).

References

Adams, M. (2006). Hybridizing habitus and reflexivity: Towards an understanding of contemporary identity? *Sociology, 40*(3), 511–528.

Alfrey, L., Enright, E. & Rynne, S. (2017). Letters from early career academics: The physical education and sport pedagogy field of play. *Sport, Education and Society, 22*(1), 5–21.

Bell, M. (1996). Feminists challenging assumptions about outdoor leadership. In K. Warren (Ed.) *Women's voices in experiential education* (pp. 141–156). Dubuque, IA: Kendall/Hunt.

Bell, M. (1997). Gendered experience: Social theory and experiential practice. *Journal of Experiential Education, 20*(3), 143–151.

Bell, M., Cosgriff, M., Lynch, P. & Zink, R. (2018). Nourishing terrains? Troubling terrains? Women's outdoor work in Aotearoa New Zealand. In T. Gray & D. Mitten (Eds.), *The Palgrave international handbook of women and outdoor learning* (pp. 199–215). London: Palgrave Macmillan.

Bryman, A. (2012). *Social research methods* (4th ed.). Oxford: Oxford University Press.

Butler, J. (1990). *Gender trouble: Feminism and the subversion of identity.* New York, NY: Routledge.

Chase, S.E. (2011). Narrative inquiry: Still a field in the making. In N.K. Denzin & Y.S. Lincoln (Eds.) *The SAGE handbook of qualitative research* (4th ed., pp. 421–434). Los Angeles, CA: SAGE.

Ciuffetelli Parker, D. (2011). Related literacy narratives: Letters as a narrative inquiry method in teacher education. In J. Kitchen, D. Ciuffetelli Parker & D. Pushor (Eds.) *Narrative inquiries into curriculum making in teacher education* (pp. 131–149). Bingley: Emerald.

Clandinin, D.J. (2013). *Engaging in narrative inquiry.* Walnut Creek, CA: Left Coast Press.

Clandinin, D.J. & Connelly, F.M. (2000). *Narrative inquiry: Experience and story in qualitative research.* San Francisco, CA: Jossey-Bass.

Cosgriff, M., Legge, M., Brown, M., Boyes, M., Zink, R. & Irwin, D. (2012). Outdoor learning in Aotearoa New Zealand: Voices past, present and future. *Journal of Adventure Education and Outdoor Learning, 12*(3), 221–235.

Crawford, J., Kippax, S., Onyx, J., Gault, U. & Benton, P. (1992). *Emotion and gender: Constructing meaning from memory.* London: Sage.

Crossley, M.L. (2000). *Introducing narrative psychology.* Buckingham: Open University Press.

Emerson, R.M. (2009). Ethnography, interaction and ordinary trouble. *Ethnography, 10*(4), 535–548.

Emerson, R.M., Fretz, R.I. & Shaw, L.L. (1995). Processing fieldnotes: Coding and memoing. In R. M. Emerson, R. I. Fretz & L. L. Shaw (Eds.) *Writing ethnographic fieldnotes* (pp. 142–168). London, Chicago, IL: University of Chicago Press.

Enright, E., Rynne, S.B. & Alfrey, L. (2017). "Letters to an early career academic": Learning from advice of the physical education and sport pedagogy professoriate. *Sport, Education and Society, 22*(1), 22–39.

Gluck, S.B. & Patai, D. (Eds.). (1991). *Women's words: The feminist practice of oral history.* London, New York, NY: Routledge.

Haug, F. (Ed.). (1987). *Female sexualization: A collective work of memory.* London: Verso.

Jolly, M. & Stanley, L. (2005). Letters as/not a genre. *LifeWriting, 2*(2), 91–118.

Keightley, E. (2010). Remembering research: Memory and methodology in the social sciences. *International Journal of Social Research Methodology, 13*(1), 55–70.

Knowles, J.G. & Cole, A.L. (1994). We're just like the beginning teachers we study: Letters and reflections on our first year as beginning professors. *Curriculum Inquiry, 24*(1), 27–52.

Lynch, P. (2000). Fitting in and getting on: Learning in the school of the outdoors. *Childrenz Issues, 4*(2), 32–35.

Newbery, L. (2002). "Mirror, mirror on the wall, who's the fairest one of all?": Troubling gendered identities. *Resources for Feminist Research/Documentation sur la recherche féministe (RFR/DRF), 29*(3/4), 19–38.

Personal Narratives Group (Ed.). (1989). *Interpreting women's lives: Feminist theory and personal narratives.* Bloomington, IN: Indiana University Press.

Richardson, L. (1990). Narrative and sociology. *Journal of Contemporary Ethnography, 19*(1), 116–135.

Riessman, C.K. (Ed.). (2008). *Narrative methods for the human sciences.* Thousand Oaks, CA: SAGE.

Ritchie, J. & Lewis, J. (Eds.). (2003). *Qualitative research practice: A guide for social science students and researchers.* London: SAGE.

Rynne, S.B., Enright, E. & Alfrey, L. (2017). Researching up and across in physical education and sport pedagogy: Methodological lessons learned from an intergenerational narrative inquiry. *Sport, Education and Society, 22*(1), 140–156.

Smith, L.M. (1994). Biographical method. In N.K. Denzin and Y.S. Lincoln (Eds.) *The SAGE handbook of qualitative research* (1st ed., pp. 286–305). Thousand Oaks, CA: SAGE.

Stanley, L. & Wise, S. (1990). Method, methodology and epistemology in feminist research processes. In L. Stanley (Ed.) *Feminist praxis: Research, theory and epistemology in feminist sociology* (pp. 20–60). London, New York, NY: Routledge.

Zink, R. (2010). Coming to know oneself through experiential education. *Discourse: Studies in the Cultural Politics of Education, 31*(2), 209–219.

17

POST-QUALITATIVE INQUIRY IN OUTDOOR STUDIES

A radical (non-)methodology

Jamie Mcphie and David A.G. Clarke

Introduction

We feel Elizabeth St Pierre introduces post-qualitative inquiry best when she reminisces:

> Looking back now, I know that I read Deleuze so early in my doctoral program that the ontology of humanist qualitative methodology could never make sense. For me and others like me, that methodology was ruined from the start, though we didn't quite know it at the time.
>
> *St Pierre, 2014, p. 3*

With this statement, St Pierre demonstrates the manner in which institutionalised methodology has neglected what theory can do. A recent and controversial addition to academic methodological practice, post-qualitative inquiry diffracts dominant qualitative methodologies to produce different paths to the habitually trodden ones in academia (St Pierre, Jackson & Mazzei, 2016). In this type of research, each researcher will undoubtedly create their own "*remix, mash-up, assemblage*, a *becoming* of inquiry that is not *a priori*, inevitable, necessary, stable, or repeatable but is, rather, created spontaneously in the middle of the task at hand" (St Pierre, 2011, p. 620, original emphasis). St Pierre (2011) believes that

> this has always been the case but that researchers have been trained to believe in and thus are constrained by the pre-given concepts/categories of the invented but normalised structure of 'qualitative methodology', its 'designs' and 'methods', that are as positivist as they are interpretive, often more so
>
> *p. 620*

Post-qualitative inquiry seeks to destabilise this representational trend of knowledge *re*-production.

Emerging novel post-qualitative (non-)methodologies[1] challenge the researcher to produce knowledge differently by "refusing a closed system for fixed meaning" in order to "keep meaning on the move" (Jackson & Mazzei, 2012, p. i). These fixed meanings could involve "mechanistic coding", which St Pierre (2011, p. 622) infers "is a positivist social science of the 1920's and 1930's", or "reducing data to themes", which Jackson and Mazzei (2012) suggest "do little to critique the complexities of social life" as "such simplistic approaches preclude dense and multi-layered treatment of data" (p. i). Put another way, "to convert what we owe to the world into 'data' that we have extracted from it is to expunge knowing from being" (Ingold, 2013, p. 5).

In the translator's foreword to Deleuze and Guattari (2004), Brian Massumi pointed out:

> The question is not: is it true? But: does it work? What new thoughts does it make possible to think? What new emotions does it make it possible to feel? What new sensations and perceptions does it open in the body?
>
> *xv–xvi*

What does it 'do'?

In this chapter we provide examples from our own inquiries, demonstrating how thinking with a post-qualitative itinerary attempts to tease research – ethics, ontologies, epistemologies and methodologies – out of the Enlightenment agenda and profoundly transform them to create new opportunities for learning. Specifically, we describe our post-qualitative inquiries in outdoor environmental education and therapeutic landscapes (Clarke and Mcphie, 2016; Mcphie, 2017). We recount how our (non-)methodological approaches to these areas changed the very nature of the realities we thought we were inquiring into, and what this, in turn, made (im)possible[2] for us as practitioner-researchers. We describe thinking with post-qualitative insights for performing creative research and illustrate how our research data were presented and analysed towards new conceptions of research in outdoor studies "after method" (Law, 2004).

We urge researchers in outdoor studies not to fall "paradigms behind" (Patton, 2008, p.269) by avoiding the temptation, common in mainstream qualitative research methods, to separate *thinking and theory* from *research practices* (Guttorm, Hohti & Paakkari, 2015; Jackson and Mazzei, 2012). We recommend a post-qualitative agenda as a positive discrimination to challenge the (un)comfortable binaries of research. Research itself becomes another theory to deconstruct and think *with*, hopefully to create new epistemological pathways to further social and environmental equity.

The following discussion is deliberately conversational in style to disrupt the subject-object distancing that a more formal academic writing style adopts. We do this to witness its performance on the page and see how this might transfer to your own interpretation of the text. Simultaneously, we deliberately alienate you from a positivistic representational academic text to remind you that this is not *the truth*. We talk with each other, but also draw the work of other authors into the conversation, taking their original quotes 'out of context' to create yet another context – another (non-)methodology.

A discussion

Jamie: Dave, this is how I came to post-qualitative inquiry … created spontaneously in the middle of my co-operative action research (see Heron & Reason, 2001). I hadn't 'planned' it beforehand; I hadn't prepared it. I simply couldn't find a suitable method of 'analysis' – that didn't adhere to some kind of mythical epistemological 'truth' – until 'it' came to me as I came to it simultaneously. Vanessa, what was *your* intention?

Vanessa: There is no use asking me what I intended with this text: this text wrote itself into being, so my relationship with it is the same as that of a reader – what it did to me will be different from what it does to you (de Oliveira Andreotti, 2016, p. 80).

Jamie: Well, I can tell you that I was unsatisfied with suggestions of coding and theming in qualitative methodologies – they created too many boundaries and hierarchies. It seemed too superficial. I was also beginning to wonder about what makes *primary* empirical evidence supposedly more 'reliable' or 'trustworthy' than *secondary* empirical evidence. It seemed to be 'made-up' to fit a particular patriarchal/hierarchical agenda. I found a quote – or perhaps, the quote found me, I can't quite remember. Elizabeth, would you remind us please?

Elizabeth: [T]here is no primary empirical depth we must defer to in post analyses as there is in the ontology and empiricism of conventional humanist qualitative methodology. That is, in post ontologies it makes no sense to privilege language spoken and heard 'face-to-face' as if it has some primary empirical purity or value, as if it's the origin of science (St Pierre, 2014, p. 12).

Jamie: Thanks. Dave, I can't tell you how relieved I was by this quote. St Pierre (2011) suggested qualitative inquiry was born out of a positivistic paradigm, and as such had failed to escape it at the ontological level. Yes, of course! I had always felt uncomfortable with prescriptive methodologies. Justifying my discomfort with quantitative research designs was easy (see Parsons, 2003). I found a little solace in qualitative designs, especially more creative approaches, yet there was still a lingering knot preventing a free-flow of movement between myself and traditional academic research practices.

Both *validity* and *credibility* are judged against a set of rules and voices that came into use from middle French *validité* (Harper, 2016a) and from the medieval Latin *credibilitas* (Harper, 2016b). By assuming that research is credible, we are also assuming that the ideals of the institutional paradigm are set or fixed and true. Patti Lather

(1993) chose to problematise validity "in order to both circulate and break with the signs that code it" as well as wrestle with "all the baggage that it carries plus, in a doubled-movement, what it means to rupture validity as a regime of truth" (Lather, 1993, p. 674). In a similar vein, Maggie MacLure (2015) problematised "critique in qualitative inquiry". Maggie, could you elucidate?

Maggie: [It] assumes that the world is demarcated or divided into asymmetrically valued categories: authentic and inauthentic, true and false, good and bad, and aspires to negate one side in the interests of a greater moral authority, or a smarter take on what's really going on (p. 5).

Jamie: The invented concept 'rigour' – used to judge the merits, worth and trustworthiness of modern research – is always embedded in the historicisation of hierarchical knowledge production. It presupposes a strict disciplinary "adherence to the truth" (Allende, 2012), a way of perceiving the world that became deeply entrenched from the Italian Renaissance through to the Enlightenment to legislate an ethical, ontological and epistemological stranglehold on the Western world.

Jorge: Rigour is also being methodical commitment [sic] to experimental procedure, to the need of controlling all parameters that can affect the results of our tests [...] it is to disrobe ourselves of our prejudices and enthusiasm when we interpret our results (Allende, 2012, paras. 5–6).

Jamie: But Jorge, procedures, inaccuracies, controls, parameters and preciseness are always already prejudiced due to their Occidental framing that subjugates other ways of knowing and being. So, one of the co-emerging purposes of my PhD thesis became an attempt to "produce different knowledge and produce knowledge differently" (Lather, 2013, p. 653; St Pierre, 1997, p. 175). This process allowed me to deconstruct prevailing conceptualisations of outdoor therapies (as 'natural' landscapes influencing peoples' mental health and wellbeing) and instead create new concepts to describe/explore the implications of an immanent ontology on conceptualisations of mental health and environments – as metaphysically and ethically inseparable. For example, I found no reason why a city centre couldn't be identified as being driven insane (Mcphie, 2018).

Dave: Another example for outdoor studies is Reinertsen (2014), who takes up a post-qualitative approach to explore an outdoor education project being used by a high school in Norway to draw assessment and learning closer together. Reinertsen (2014) develops a playful (non-)methodology where words are seen as bodies with material effects and in which she performs "research as deep mappings and/or diffractive readings as spatial or topographical fractural analysis of other objects emerging" (Reinertsen, 2014, p.1023). This research produces a novel and affecting language to help complexify the topics of study: teachers' experience, assessment, outdoor education, etc. Here, research creates interference patterns through thinking *with* data and the theories of Dewey/Derrida/Deleuze. Ultimately this research invents new concepts of researcher/outdoor education/assessment and keeps the research problems on the move in a novel manner.

Jamie: I used *thinking with* as well: my reading of theory happened alongside my reading of data until a gradual multidirectional co-production emerged. For example, I recorded some of my co-participants'/co-researchers' comments (just as they did with other data) that were always already informed by literature, embodied memory, etc. In turn, the comments inspired an expedition of inquiry *that took me* along a particular path of investigation, constantly informed by myriad influences, such as phrases from websites that I simply couldn't let go of due to their shocking inequitable impact.[3] I was on a ride that I was not in control of – and I liked it that way. The inquiry took me for a walk.

Maggie: [We] are obliged to acknowledge that data have their ways of making themselves intelligible to us [...] On those occasions, agency feels distributed and undecidable, as if we have chosen something that has chosen us [...] In a previous article, I described that kind of encounter in terms of the data beginning to '**glow**' (MacLure, 2013, pp. 660–661, **glow** added).

Jamie: Maggie, the initial comments I heard in Liverpool glowed a little too, a sort of *blush*, enough for me to feel the need to record them. The focus group meetings post visit (re)enforced the glowing of particular data, encouraging them to *bloom*. Looking back at the photos, the videos, the journals, my notes, also (re)enforced the *blossoming* of certain paths/events. Discussions and readings all merged to inform what I initially thought were 'my' choices. When I think back to how I could possibly justify what *type* of inquiry this was in terms of what influenced what (theory⇄practice) or how 'I' might have 'chosen' a particular route to take the study, the closest I can get to an answer is that it was like participating in/with a murmuration of starlings. Of course, neither came *first* as they were never transcendently bounded in *clock time* in the first place. This is how the assemblages of my thesis were written. Each assemblage was co-produced by multiple 'things' – events, processes, materials – coming together from multiple directions, rhythms and temporalities. Intention didn't seem to 'begin' *in* me. It was always relational, multi-agential, topological and 'intra-active' (Barad, 2007). I really don't think I had much of a choice in the matter. I came to conceptualise 'environ(mental) health' (as co-produced) and this was a major outcome of my inquiry – transforming the manner in which those in outdoor studies can think about therapeutic practices.

One of the most striking things that happened during my inquiry was what the inquiry itself did – how it performed. I always knew that it was futile to attempt to take the 'I' out of the research, but I never really thought about taking the research out of the research. For example, after a year of post-qualitative co-operative action research on mental health and wellbeing, all of the co-participants/co-researchers reported becoming 'healthier'. The research process became a therapeutic tool. *Doing* the research itself – the regular outings, the 'data' collection, the social interactions, the focus group meetings, the group analyses, the debates, the writing, the reading, the thinking, etc. – seemed to co-produce contextualised effects that I could not separate from what it was I intended for the inquiry to 'find out'. I realised that I could never again simply 'do' research without the research itself doing something back. Anyway, I passed!

Dave: Well done! Post-qualitative research is sometimes described as thinking how to reach the new, and how to reach the new can't be described, as it hasn't yet arrived (Massumi, 2010). This is obviously problematic for someone wondering 'how to do it'! So, rather than discussing procedure, I'll start with the 'why', or the ethics, of post-qualitative research and how this is linked with thinking. Firstly, methodology is philosophy at its foundation (excuse this turn to depths!). That is why when students are asked to write their methodology sections they are often asked to talk about ontology and epistemology. The emergence of post-qualitative research is a consequence of how these philosophical concepts are understood in contemporary thought, but it also drives these debates. Beginning to understand post-qualitative research therefore involves lots of reading. When we were students of Adventure Education in the early 2000s, Jamie, I remember us both reading Allison and Pomeroy (2000). The authors critique the dominant positivist approach to research in experiential education in the 1990s. They argued that researchers needed to shift their philosophical assumptions from positivism to an approach based within a constructivist paradigm to allow access to the processes of experiential learning. That paper changed entirely the way I *think* about research and importantly it taught me that *my thinking about methodology could change radically*. Of course, since then my thinking has again changed, but it is papers like that one that affect you at certain points in your life, that are hugely ethically important, because altering the way you think the world *is* also creates what methodology is possible for you. Reading is one of the processes of this learning, *but so is enacting research*. Methodology is pedagogy; it teaches as we perform it. I am not concerned here with the obvious fact that it teaches you about the research question you are attempting to answer or your subject of study, rather enacting research teaches the researcher about the very possibilities of being; how could focusing on ontology and epistemology so directly (and the manner in which they seep into every consideration of our practice) *not* do this? What happens is a sort of research↔learning, where the learning is metaphysical. Perhaps this is why your participants became 'healthier', Jamie? Research creates worlds in its process. All methods do this, yet not to the same ends. For instance, in post-qualitative research this research↔learning is *posthuman* – that is, it realises that the enlightenment human subject might be nothing more than an idea, and that this idea has had (some catastrophic) world-changing effects. This understanding is important for ethical living/research in the face of injustice, climate collapse and mass extinction. It sanctions attempts to articulate other ways of thinking/creating the world. It wants to create research that implies other worlds. Consider, for example, this understanding of ethics from Deleuze:

Gilles: In an ethics, it is completely different [to morality], you do not judge. [...] *you relate the thing or the statement to the mode of existence that it implies, that it envelops in itself. How must it be in order to say that? Which manner of Being does this imply?* (Deleuze, 1980, np, my emphasis)

Dave: Now, when I read or write, I try to ask myself what mode of existence the writing implies. I try to background what a paper is *saying*, and instead focus on what a paper *does*: what subjectivities does it create? What worlds are implied? What is the research↔learning for me, and the reader? To spot what modes of Being a piece of research implies it is useful to have some concepts to *think with*. According to Deleuze and Guattari, concepts are tools to put to work in the world. For example, when we used their concepts of the *rhizome* and *becoming* in our examination of the Scottish education policy *Learning for Sustainability* (Clarke and Mcphie, 2016) it allowed us to connect our thinking in ways that humanistic qualitative inquiry obstructed. We thought *with* theory and the policy to dislodge the stable notions of indoors and outdoors, learning, place, and the learner as inferred by the policy. We asked what mode of existence the policy implied and, thinking with the rhizome and becoming, we saw that the policy could imply other modes of existence. And so, we interpreted (read 'created') it differently: communication in learning events became expressive rather than seen as the transfer of information; the learners/teachers became co-constitutive events or processes (haecceities) rather than individual enlightenment subjects; and we were able to destabilise the prevailing distinction of places as locatable, delineated, geographical sites, instead to envision places becoming, as paths of learning. It is important to recognise that this interrogation was not a critique as is generally understood in academic terms, as a putting down or judgement of the *Learning for Sustainability* policy. Rather, it was what MacLure (2015) calls an immanent critique; designed not to shut thought down, but to be productive. This, I think, is more seismic than it might sound. The entire nature of academic critique changes in post-qualitative research, and I want to talk a little about that now.

Gerrad, Rudolph and Sriprakash (2017) raise several points of (traditional) critique against post-qualitative inquiry. They suggest that, in what they see as its complexity and difficulty, it can fail to acknowledge the exclusionary boundaries it creates. They are concerned about the potential 'mystification' of the research process and suggest the focus on the 'new' can reinforce settler colonialism in research practices. In thinking about these points, I could follow a rational logic to agree with or critique these critiques, and this is tempting, but, in recognition of an immanent critique as the mode of thought of post-qualitative research, I instead wonder at the potential for research to open up the concerns that Gerrad, Rudolph and Sriprakash (2017) describe; to riff off them. Or to acknowledge that even they, critiques, are immanently affective within post-qualitative research; they imply modes of existence, they do things. Post-qualitative research doesn't attempt to operate from a perspective of critical objectivity, but rather acknowledges the situated, partial, ethical, relational, posthuman and responsive ways of knowing that have been developed in feminist studies. It is non-oppositional. In this way post-qualitative research might not best be described as an approach, but as a series of understandings linking with other understandings, even critiques, in the pursuit of ethical research. Personally, I feel I am learning about post-qualitative research all the time. It always feels like an attempt, or something that informs my thinking about research. For now, I think that philosophical concepts garnered from reading *are* methods, but, at the same time, are tools that allow possibilities for living (Lenz Taguchi & St Pierre, 2017). In post-qualitative inquiry philosophy is the coal face of practice; it expands the realm of the possible, and acknowledges that research creates worlds (Law, 2004).

Conclusion

Post-qualitative inquiry is now gaining ground in qualitative research handbooks and journals (Honan & Bright, 2016; Jackson & Mazzei, 2012; Lather & St Pierre, 2013; St Pierre, 2011; St Pierre, Jackson & Mazzei, 2016). However, Greene "expresses concerns: first, about whether post-qualitative research can still be considered research; second, where it is going; and third, what is being lost in the new inquiry" (Lather & St Pierre, 2013, p. 632). Greene (2013) imagines post-qualitative inquiry "as a kind of retreat *into the mind*" (p. 753, emphasis added). We think the *Cartesian ghost* still haunts Greene's (2013) onto-epistemological position as she perceives post-qualitative inquiries as challenging her *mind*, but not engaging her *body* (p. 754). But *thinking with* post-qualitative inquiry, for example, upends this understanding, removing the mind-body dualism and highlighting Spinoza's point of the mind as an idea of the body (Dolphijn & van der Tuin, 2012). Therefore, we would inflect Greene's stance with an affirmation that post-qualitative inquiries are more like transgressive and ethico-political advances out of *the non-physical mind* and into a physical world other than merely human.

> If we cease to privilege knowing over being; if we refuse positivist and phenomenological assumptions about the nature of lived experience and the world; if we give up representational and binary logics; if we see language, the human, and the material not as separate entities mixed together but as completely imbricated 'on the surface'– if we do all that and the 'more' it will open up – will qualitative inquiry as we know it be possible? Perhaps not.
>
> *Lather & St Pierre, 2013, pp. 629–630*

The dominant paradigms that have forced their hand in the world of academia need an overhaul to find better stories than the current one being traced repeatedly, as we attempt to "produce different knowledge and produce knowledge differently" (St Pierre, 1997, p. 175). This becomes an ethical imperative of what we *can do*.

Notes

1 Ken Gale (2018) refers to *methodogenesis* in which "conceptualisation and inventive research process is given precedence over the fixities of set methodological representation and signification" (p. 44). So, (non-) methodological approaches attempt to co-produce a more fluid "enactive understanding" (Massumi, 2015, p. 94) of the world – knowledge 'making'. This is not an anti-methodology, as we see the importance of inquiry and analysis of inquiry methods. So, the 'non' is bracketed to denote a particular problematising

of many methodological approaches and understandings, often born out of patriarchal, logocentric and Euclidean onto-epistemologies.

2 The bracketing (im) denotes that either or both statements are possible.

3 Liverpool ONE website: The company who built Liverpool ONE, Grosvenor, proposed *eliminating* "anti-social elements such as vagrants and beggars" from their privately owned public space.

References

Allende, J. (2012). Rigor – The essence of scientific work. *Electronic Journal of Biotechnology*, 7(1). Retrieved from: www.ejbiotechnology.info/index.php/ejbiotechnology/article/view/1112/1494

Allison, P. & Pomeroy, E. (2000). How shall we "know?" Epistemological concerns in research in experiential education. *Journal of Experiential Education*, *23*(2), 91–98.

Barad, K. (2007). *Meeting the universe halfway: Quantum physics and the entanglement of matter and meaning.* Durham, NC: Duke University Press.

Clarke, D.A. & Mcphie, J. (2016). From places to paths: Learning for sustainability, teacher education and a philosophy of becoming. *Environmental Education Research*, *22*(7), 1002–1024.

de Oliveira Andreotti, V. (2016). (re)imagining education as an un-coercive re-arrangement of desires. *Other Education: The Journal of Educational Alternatives*, *5*(1), 79–88.

Deleuze, G. (1980). *Cours Vincennes: Ontologies-Ethique.* Les cours de Gilles Deleuze. Retrieved from: www.webdeleuze.com/textes/190andgroupe=Spinozaandlangue=2

Deleuze, G. & Guattari, F. (2004). *A thousand plateaus: Capitalism and schizophrenia.* (Trans. B. Massumi). London: Continuum.

Dolphijn, R. & van der Tuin, I. (2012). *New materialism: Interviews & cartographies.* University of Michigan Library, Ann Arbor: Open Humanities Press.

Gale, K. (2018) *Madness as methodology: Bringing concepts to life in contemporary theorising and inquiry.* Oxford: Routledge.

Gerrard, J., Rudolph, S. & Sriprakash, A. (2017). The politics of post-qualitative inquiry: History and power. *Qualitative Inquiry*, *23*(5), 384–394.

Greene, J.C. (2013). On rhizomes, lines of flight, mangles, and other assemblages. *International Journal of Qualitative Studies in Education*, *26*(6), 749–758.

Guttorm, H., Hohti, R. & Paakkari, A. (2015). "Do the next thing": an interview with Elizabeth Adams St Pierre on post-qualitative methodology. *Reconceptualizing Educational Research Methodology*, *6*(1), 15–22.

Harper, D. (2016a). *Validity.* Retrieved from: www.etymonline.com/index.php?term=validity

Harper, D. (2016b). *Credibility.* Retrieved from: www.etymonline.com/index.php?term=credibility

Heron, J. & Reason, P. (2001). The practice of co-operative inquiry: Research "with" rather than "on" people. In P. Reason and H. Bradbury (eds.) *Handbook of action research: Participative inquiry and practice.* London: SAGE.

Honan, E. & Bright, D. (2016). Writing a thesis differently. *International Journal of Qualitative Studies in Education*, *29*(5), 731–743.

Ingold, T. (2013). *Making: Anthropology, archaeology, art and architecture.* Oxford: Routledge.

Jackson, A.Y. & Mazzei, L.A. (2012). *Thinking with theory in qualitative research: Viewing data across multiple perspectives.* London, New York, NY: Routledge.

Lather, P. (1993). Fertile obsession: Validity after poststructuralism. *Sociological Quarterly*, *34*(4), 673–693.

Lather, P. (2013). Methodology-21: What do we do in the afterward? *International Journal of Qualitative Studies in Education, 26*, 634–645.

Lather, P. & St Pierre, E.A. (2013). Post-qualitative research. *International Journal of Qualitative Studies in Education, 26*(6), 629–633.

Law, J. (2004). *After method: Mess in social science research*. London, New York, NY: Routledge.

Lenz Taguchi, H. & St Pierre, E.A. (2017). Using concept as method in educational and social science inquiry. *Qualitative Inquiry, 23*(9), 643–648.

MacLure, M. (2013). Researching without representation? Language and materiality in post-qualitative methodology. *International Journal of Qualitative Studies in Education, 26*(6), 658–667.

MacLure, M. (2015). The "new materialisms": A thorn in the flesh of critical qualitative inquiry? In G. Cannella, M.S. Perez & P. Pasque (eds.) *Critical qualitative inquiry: Foundations and futures*. Walnut Creek, CA: Left Coast Press. Retrieved from: www.e-space.mmu.ac.uk/e-space/bitstream/2173/596897/3/post-criticalLATEST.pdf

Massumi, B. (2010). What concepts do: Preface to the Chinese translation of A Thousand Plateaus. *Deleuze Studies, 4*, 1–15.

Massumi, B. (2015) *The politics of affect*. Cambridge: Polity Press.

Mcphie, J. (2017). *The accidental death of Mr. Happy: A post-qualitative rhizoanalysis of mental health and wellbeing*. (Unpublished doctoral dissertation). Lancaster University, UK.

Mcphie, J. (2018). I knock at the stone's front door: Performative pedagogies beyond the human story. *Parallax. 24*(3), 306–323.

Parsons, K. (Ed.). (2003). *The science wars: Debating scientific knowledge and technology*. Amherst, NY: Prometheus Books.

Patton, C. (2008). Finding "fields" in the field. *International Review of Qualitative Research, 1*(2), 255–274.

Reinertsen, A.B. (2014). Outdoor becoming thinking with data Dewey Derrida Deleuze: Assessment for health and happiness. *Qualitative Inquiry, 20*(8), 1022–1032.

St Pierre, E.A. (1997). Methodology in the fold and the irruption of transgressive data. *International Journal of Qualitative Studies in Education, 10*(2), 175–189.

St Pierre, E.A. (2011). Post qualitative research: The critique and the coming after. In N.K. Denzin and Y.S. Lincoln (Eds.) *The SAGE handbook of qualitative research* (4th ed., pp. 611–626). London: SAGE.

St Pierre, E.A. (2014). A brief and personal history of post qualitative research: Toward "post inquiry". *Journal of Curriculum Theorizing, 30*(2), 2–19.

St Pierre, E.A., Jackson, A.Y. & Mazzei, L.A. (2016). New empiricisms and new materialisms: conditions for new inquiry. *Cultural Studies ⇔ Critical Methodologies, 16*(2), 99–110.

18

TOGETHER ALONG THE WAY

Applying mobilities through praxis in outdoor studies field research

Philip M. Mullins

In this chapter I introduce and show the relevance of the mobilities paradigm and praxis as an approach to its operationalisation in outdoors studies research. I explain the "commonplace journey methodology" (Mullins, 2014b) to show one possible way of integrating mobilities and praxis into field research. My intent is to identify the concepts and explain the process so they can be adapted to different contexts and research questions.

On epistemology/ontology

Feminist and ecofeminist research has long critiqued that Western epistemological/ontological dualisms and their associated values (e.g. mind over body, rational over intuitive, theory over practice ...) obscure other possibilities, gloss over complexities and reinforce existing power, privilege and approaches to the more-than-human world (Humberstone, 1998). These dualisms and values shape research methodologies, methods and reports, and therefore the ways in which the world is understood. The theory/practice dualism can be seen in the normalisation of post-positivist methodologies concerned with universal/generalised knowledge, and phenomenological methodologies conceived as purely subjective 'lived experiences', a position problematically privileged in phenomenological research (Barnacle, 2004; Hanley, 1998). Within this dualism, praxis brings together abstracted thought and subjective experiences; beyond the dualism, phenomenology and praxis outdoors hold possibilities for critique, new practice and insight, and societal change towards socio-ecological understandings (Mullins, 2014a).

Mobilities paradigm

Urry (2000) and others (Büscher, Urry & Witchger, 2010; Sheller & Urry, 2006) have pushed Western sociological research to examine the functions and impacts of flows and movements (and lack thereof) of people, capital, goods and materials globally, but also, importantly, to include the more-than-human world. Urry rebuffed the presumed "chasm between nature and society" (Urry, 2000, p. 10) and suggested that research needs to acknowledge and trace their interconnections. "What we will find", Ingold described, "is not so much an interplay between two kinds of history, human and non-human, as a history comprised by the interplay of diverse human and non-human agents in their mutual relationships" (Ingold, 2005, p. 506). Such mutual relationships spread across environments and settings. They continually form people, places and landscapes over time including ourselves as researchers and our participants, as well as the places and landscapes of outdoor studies (Büscher & Urry, 2009; Ingold, 2005; Mullins, 2009, 2018; Sheller & Urry, 2004, 2006). Therefore, as scholars in outdoor studies, we need to think carefully about participants' practices as well as our own.

Within the mobilities paradigm, to research outdoor activities, people, places, landscapes, flows, and/or wildlife is to study their interconnected movements, actions, lives and stories as well as the impacts of these on one another and their shared environment. Ingold explained a connection between a journey, story and knowledge:

> Telling the story of the journey ..., I weave a narrative thread that wanders from topic to topic, just as in my walk I wandered from place to place. This story recounts just one chapter in the never-ending journey that is life itself, and it is through this journey – with all its twists and turns – that we grow into a knowledge of the world about us.
>
> *Ingold, 2007, p. 87*

Moreover, people and their stories move through and across landscapes, as well as digital and non-digital realms (Mullins, 2009; Smith & Kirby, 2015; Stinson, 2017). Storying is as true for researchers as it is for participants; we encounter, critique, use and write particular stories/theories about the world and inhabitants. Therefore, as scholars, we need to think carefully about participants' as well as our own theories, stories and representations.

The mobilities paradigm is relevant to generating critical and creative understandings and approaches to research and practice in outdoor studies. We shape spaces, places, bodies, experiences and identities through activities, behaviours, products and stories that have far-reaching impacts through mobilities of various type, degree and range. Mobilities are found in the performance and experience of outdoor recreation, education and tourism but also – importantly – their production and delivery, which further engage larger social, ecological and economic forces and realities. Mobile methods respond to the limited ability of typical research methods to deal with phenomena that are fleeting, distributed, malleable, complex, emotive, spiritual and kinaesthetic (Büscher et al., 2010). Mobile methods in 'the outdoors' have begun responding to traditional research methods positioning researchers as passive observers and landscapes as static backdrops, rather than both being active within the production of experiences and knowledge (Coates, Hockley,

Humberstone & Stan, 2015; Kennedy, MacPhail & Varley, 2018; Mullins, 2014b). Basic metaphors of outdoor mobility require considerable critical reflection and careful use by scholars and practitioners alike because they come embedded with assumptions and shape our practices, stories and understandings of phenomena.

Researchers can find ways to examine movement from outside the activity, they can also move or participate with the activity as part of the research. Each approach provides different insight and relative value to a project's question and contribution to knowledge; both approaches influence the type and qualities of data collected. Importantly, they also influence researchers' representation of phenomena and findings. Pragmatic considerations such as cost, access and timelines may enable or constrain methodological choices. Nevertheless, the researcher's mobility and its influence on the project, findings and writing is an important consideration.

Praxis and mobility for outdoor studies

Praxis reaches back to ancient Greece, through permutations of phenomenology offered by Husserl and Heidegger among others, and is explicated and further re-interpreted in educational and political thinking and research. Seeking social change and liberation through education, Freire (1970) described students coming to understand their world as "a reality in process, in transformation" (p. 71) and their word as their own interpretation and voice resulting from praxis, described as "*reflection* and *action* directed at the structures to be transformed" (1970, p. 120). Praxis is undertaken by people with their leaders/teachers working together in dialogue reflecting "simultaneously on themselves and the world" and "without dichotomizing this reflection from action" (1970, p. 71) because it responds to participants' lived world. In political theory, Arendt (1998) described praxis as acting together with purpose; for her, praxis depends on a public realm performed by a plurality of people in the real world, in which collective narratives are created, shared, and questioned—in which participatory democracy occurs—and from which deliberative action can flow (Melaney, 2006). Humberstone (1998) argued that praxis may enable outdoor research to engage diverse relationships with lived, grounded, contexts that are social but also, crucially, ecological.

From these authors, we can understand praxis as the combination of thought and practice with the intent of creating changed ways of understanding and acting in the world. Both Arendt and Freire framed praxis as people coming together within their lived world and based on their experience to participate in change while engaging in reflection and dialogue that shares, creates and interrogates narratives/stories and actions/practices. Such an approach respects and enriches both participatory democracy and the fundamental humanity and ecology of participants.

Mobilities and praxis fit well professionally and in a scholarly way within outdoor studies for reasons including the nature of the activities, the methods used to teach outdoors, and commitments to social and ecological justice. Outdoor studies has strong practical ties, origins and applications with rich and expanding theoretical traditions, and with a self-reflexive streak that values thought, ethics and experiential learning. Methodologies incorporating praxis are therefore congruent with outdoor studies, as well as valuable because they can resist reduction to, and division along, the theory/practice dualism (Breunig, 2005). Incorporating mobilities and praxis also opens possibilities to act within and understand the outdoors as a nuanced and informative realm that can offer deep epistemological and ontological critique and insight, as well as societal change within and beyond the field. As Law and Urry (2004) noted, social inquiry through mobile methods not only describe the world, they enact and make worlds. Research, then, provides an opportunity – with responsibility – to engage participants and contribute to shaping the world.

A mobile methodology for praxis in field research

The commonplace journey methodology was inspired by a process, ethic and tradition of experiential learning in outdoor expeditions, and uses the rhythms and patterns of outdoor travel to facilitate inquiry. The methodology facilitates people participating together while learning from one another, from events and their surroundings, and from different theoretical or practical approaches. The methodology was informed by Ingold's (2000) notion that an "education of attention" (p. 146) can occur while travelling with others, and by Sumara's (2002) use of commonplace techniques to facilitate his own and his students' self-reflection and insight.

The commonplace journey methodology was first developed for research during Paddling the Big Sky, an extended canoe trip that engaged the group of fellow travellers-practitioners-thinkers in creative dialogue (Mullins, 2014b). The methodology facilitates working with participants as quasi co-researchers to pursue inquiry through praxis.

I draw on two projects in addition to Big Sky. The first involved university students sailing through and learning about the North and Central Coast of British Columbia, Canada (a.k.a. the Great Bear Rainforest or GBR) while debate raged over the proposed Northern Gateway Pipeline Project. The second project is *Living and Learning Social and Environmental Relations* (LLSER)[1] which worked to understand how fly anglers relate to one another and their surroundings through fishing. I will describe the process as developed for Big Sky and use GBR and LLSER to note adaptations and limitations.

Setting the commonplace journey

A research project occurs within a pragmatic setting, but it also occurs within an academic setting of movements and priorities in the field, entrenched ideas, typical and alternative conceptions, new critiques and diverse theories. Having these two contexts inform and complement one another is important to praxis and the knowledge created through a research project.

As much as possible, data collection for the commonplace journey occurs in the course and context of practice, which helps inform understandings. Landscapes and settings can be used as a common world shared by participants, and are evocative and active in the unfolding phenomenon. Moreover, being in situ provides the researcher access to unexpected aspects of the activity. Car rides, down time, and other ancillary facets of an activity can be very informative, and I encourage researchers to conceive of the activity broadly across phases and settings (Mullins, 2014a). Moreover, participant behaviour, skill and language in the course of practice express understandings and knowledge that participants might not be aware of or able to explain. Conducting research in situ makes this type of knowledge available and represented in data. Fieldwork might be continuous during a longer journey (as during Big Sky, or sailing trip in GBR), or require separate visits with groups of participants (as in LLSER's day trips to one field location), or happen periodically depending on whether and how data will be analysed for change.

The researcher should be purposeful about the settings in relation to the research objectives, and reflexive about the possible impacts on data analysis and findings. The route and/or setting can help facilitate inquiry: landscapes, communities and places are engaged in an order, and the itinerary shapes social and environmental engagement and meanings. The researcher might consider, for example, how the route is related to particular issues, and participants' degree of familiarity with a place.

A single typical setting might be used, as in LLSER because normalizing the setting (as much as possible) helped us see and compare participants' relative angling skill and specialisation. Alternatively, places and/or routes that help explore particular phenomena or create opportunity for insight might be used (as we did with Big Sky, which crossed multiple ecoregions and landscapes, Mullins, 2009). Researchers should recognise in their analysis and discussion that the data and interpretation come from specific settings and activities, and are not necessarily applicable more widely.

Establishing commonplaces

Commonplace is not intended to mean 'usual' or 'mundane', rather it refers to a technique developed by Sumara (2002) of establishing pedagogical or methodological meeting points that provide touchstones for participants' inquiry. Sumara described commonplaces as activities that support interpretation of human experience through data collected in response to instances of indeterminacy. One's journal could be used to record notes, passages and insights over time (the historic practice described by Blair, 1992); or a commonplace could be a shared text or annotated work of literature revisited periodically, showing one's change over time (Sumara, 2002), or it could be a shared route or activity (Mullins, 2009, and Mullins & Maher, 2007). The basic structure and innovation of the commonplace journey methodology is to use prompts to link together one or more commonplace activities into cycles that facilitate praxis. This uses participant engagement with phenomena coupled with individual and group reflection and discussion to provide both insight and data (see Figure 18.1).

Depending on the research objectives, prompts can be written, and commonplace activities can be developed and structured to help describe phenomena, investigate

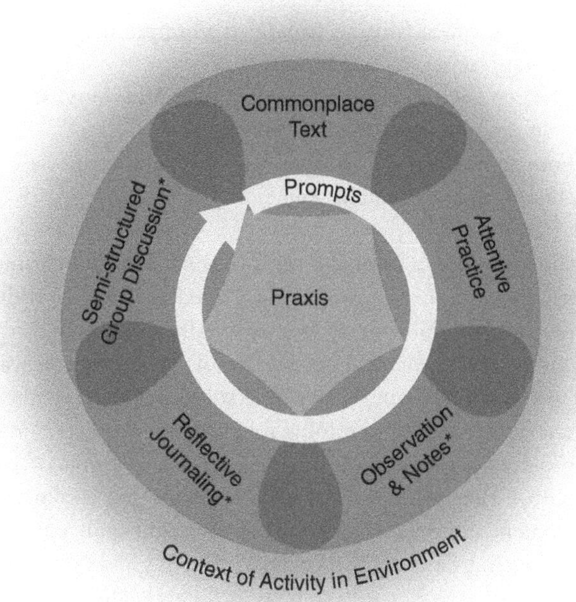

FIGURE 18.1 Model of the commonplace cycle from Big Sky. Commonplace activities indicated with asterisks generated data for post-trip analysis

Source: Adapted from Mullins, 2014b, courtesy of SAGE Publishing

existing theories and practices, and/or explore innovative practices, critiques and theories. Crucially, researchers must be clear at the outset about such objectives, design the study accordingly, and acknowledge and respect these limits when interpreting and discussing the findings of the research.

Prompting participants

Prompts are used within the commonplace cycles to guide the content and focus of praxis vis-à-vis the research question(s). Prompts also serve as a way to integrate theory explicitly or implicitly. The prompts on Big Sky concerned central themes of skill, place, interrelationships and stories because we were inquiring about human-environment relations in the practice of canoe travel and, specifically, the implications and insights offered by diverse critiques and Ingold's (2000) dwelling perspective. We introduced between one and five prompts per cycle. For GBR, by contrast, we kept the same prompts for the duration of the trip in order to explore insight and change over the course and events of the journey. For GBR, the prompts were:

1. Tell me what these words mean to you: consumption, responsibility, wilderness, stewardship, conservation, place.
2. Where and when have you noticed community? Who or what does it include?
3. What are you disconnecting from/connecting to?
4. Tell me about an experience that you had with another species?
5. What has life on the boat taught you?

Prompts must be written at a level and with content that participants can engage. Prompts might ask questions, suggest trying altered practice, or address language. Prompts can serve different purposes. They may address aspects of the phenomenon in question based on the theoretical perspective of the research, as was the case on LLSER where we inquired about the personal, social, ecological and environmental relations enacted by anglers. Or, prompts might centre on topics and respond to realities in the place/locale that are of interest for the group of participants, as was the case with GBR where we learned about the sociopolitical ecology and resource development along the coast. Or, prompts might apply concepts familiar to participants from readings, as was the case during Big Sky and GBR. Finally, prompts might seek to explore critiques and/or new or different theoretical approaches in practice, as on Big Sky.

Moving through commonplace cycles

The activities comprising a commonplace cycle need to cohere with the daily routines, practices, rhythms and phases of the activity and journey, with repeated cycles and changing settings or conditions providing opportunity to extend praxis, data collection and insight. On Big Sky we used prompts to integrate five commonplace activities into a cycle that built on traditions of outdoor travel, learning and exploration, and which generated data of various kinds (see Figure 18.1):

1. The use of a commonplace text built on a philosophical and literary tradition in the outdoors and, methodologically, provided an alternative theoretical approach integrated through prompts. Ingold (2000) was used on Big Sky, enabling us to question and rethink our practices, behaviours and assumptions, as well as glimpse alternative possibilities we would not otherwise have been able to see.
2. Engaging in attentive practice within the setting and activity built on a tradition of experiential education outdoors. Participants actively tried to became critically aware, curious and perceptive of their practices, environment and encounters rather than simply taking them, and their meanings, for granted.
3. Using notebooks to record observations and thoughts while participating built on traditions of scientific expeditions, and they informed reflections and conversations.[2]

4. Keeping journals to individually explore thoughts and reflections built on familiar outdoor educational and professional practices. Journals enabled participants to work through and record their ideas and observations, and to bring these into group discussions.
5. Meeting for semi-structured group discussions to create dialogue and further insight fit well into the tradition of coming together around a campfire or meal. Group discussions happened periodically and concluded each commonplace cycle. Conversations were audio recorded, later transcribed into text, and used as data.

These practices worked together to form a cyclic process: prompt, engage, observe, reflect, discuss ... repeat. On Big Sky, this cycle recurred ten times over the 100 days of the trip. In LLSER we had four groups of anglers arranged by level of specialisation; each group participated in their own daylong cycle comprising personal video recording and narration, prompted reflection during free practice, and a semi-structured group discussion after fishing. In GBR the cycles lasted roughly two or three days and included class readings, site visits and activities, ongoing creation of a personal digital story of the trip, and engaging in periodic semi-structured group discussions.

Elements of commonplace cycles should be logically structured to facilitate inquiry and praxis, and because some of the activities double as data collection methods they need to be sequenced to result in valid data that has not been confounded by another commonplace activity. Given the objectives of LLSER, for example, anglers recorded and narrated video of their own skilled practice in real time as the first activity of the day/cycle so that prompts and a heuristic model discussed later would not influence these data.

Keeping a research journal and observing participants

Throughout the journey and alongside commonplace cycles, I kept a field research journal in which to record thoughts, reflections and interpretations related to the prompts and research questions as well as the methodology, progress, changes in, or limitations of the research. Doing so helped track and make sense of how praxis was progressing, and aided reflexivity in the moment as well as post-trip during analysis and writing.

I also engaged in participant observation guided by a set of pre-established themes or cues, while remaining open for emergent events or realities. Such observation was crucially important. It recorded significant events, behaviours, practices, language and conversations in the moment, unfiltered through the self-reflection and self-representation present in journal entries and group discussions. Without participant observation, such data and anecdotes might not have become part of the data set. Observations further informed prompts and discussions, reflecting the group back to itself, and aiding praxis. Observation and note taking helped me perceive the flow of activity. Reviewing notes at the end of the day helped reflexivity regarding the activity and its re-interpretation, allowed critical questions and new possibilities to emerge, and ultimately helped me make sense of what was going on. Finally, observations and anecdotes helped represent and explain research findings and implications.

Summary of research practice

Participants received (and/or contributed to writing) prompts that they explored in practice, observing and recording salient events, experiences, thoughts and conversations. Participants reflected upon their experience and the prompts in personal journals, and then, finally, collectively discussed thoughts and observations on each prompt. These activities comprised commonplace cycles. The semi-structured group discussions punctuated each cycle, after which a new one began. In this way, the journey engaged in praxis out of which emerged new experiences and insights recorded in practice-rich data of various types, as well as the researcher's field journal. After the trip, or outside the context of practice, these data were transcribed and analysed using established methods such as thematic coding and narrative analysis to derive findings.

The methodology is valuable because it allows for the investigation of phenomena and collection of data in situ, informed by both practice and theory. Furthermore, it positions participants and the researcher as being and acting within dynamic surroundings that are also involved and represented in the research and knowledge produced. Nevertheless, these qualities present limitations. The realities of travel and practice mean that the research is subject to the elements, surprises and vagaries of the human and non-human world, which can delay, prevent or require changes to the planned activity or research.

Relative to the research question, the research design needs to balance the theory used, integrated and represented, with the nature of the practice and setting, and the selection and characteristics of participants. The methodology asks a lot from participants, and so their tolerance and ability to engage, cope with and complete the research while having a rewarding experience is important. The researcher should therefore carefully consider the frequency, intensity and diversity of tasks asked of participants.

The methodology also presents challenges for the researcher as practitioner and trip participant. The researcher may need technical and interpersonal skills, abilities and experience to conduct the journey and engage in the setting, and these will be relative to those of other participants. Participants may come with diverse social and cultural norms or expectations, and these can therefore influence the research process, may need to be negotiated by the researcher as both insider and outsider (Kennedy et al., 2018), and can present ethical dilemmas.

The methodology produces large amounts of data in diverse forms. Although developed using qualitative data, quantitative data collection could be integrated. The data set produced may also be complex given that participant experiences, thoughts and ideas shift along the journey. Data management and tracking is therefore important, as the researcher needs to be able to handle, link and integrate multiple forms of data into analysis and interpretation. This limitation and the nature of outdoor travel mean the methodology tends to work with relatively small groups of participants. Therefore, analysis and discussion in reports focus on dominant themes, insights, changes and re-interpretations that resulted for participants, the group and the researcher based on the data.

Conclusion

Having a dynamic world represented in study data and writing is consistent with the epistemological and ontological foundations of the commonplace journey methodology, and it reflects the socio-ecological realities and complexities of our field of study and practice, for participants and researchers. Mobilities and praxis embrace the situated nature of recreational, educational and touristic activities and research outdoors.

Notes

1 Living and Learning Social and Environmental Relations: Exploring Skill and Embodied Knowledge for Outdoor Education has been generously funded by the Social Sciences and Humanities Research Council of Canada.
2 Importantly, some traditions of expeditionary exploration are tied up with conquest and colonialism (Loynes, 2010; Tuhiwai Smith, 1999).

References

Arendt, H. (1998). *The human condition*. Chicago, IL: University of Chicago Press.

Barnacle, R. (2004). Reflection on lived experience in educational research. *Educational Philosophy and Theory, 36*(1), 57–67.

Blair, A. (1992). Humanist methods in natural philosophy: The commonplace book. *Journal of the History of Ideas, 53*(4), 541–551.

Breunig, M. (2005). Turning experiential education and critical pedagogy theory into praxis. *Journal of Experiential Education, 28*(2), 106–122.

Büscher, M. & Urry, J. (2009). Mobile methods and the empirical. *European Journal of Social Theory, 12*(1), 99–116.

Büscher, M., Urry, J. & Witchger, K. (Eds.). (2010). *Mobile methods*. New York, NY: Routledge.

Coates, E., Hockley, A., Humberstone, B. & Stan, I. (2015). Shifting perspectives on research in the outdoors. In B. Humberstone, H. Prince & K.A. Henderson (Eds.) *International handbook of outdoor studies* (pp. 69–77). Oxford: Routledge.

Freire, P. (1970). *Pedagogy of the oppressed*. New York, NY: The Seabury Press.

Hanley, C. (1998). Theory and praxis in Aristotle and Heidegger. Presented at the Twentieth World Congress of Philosophy, Boston, MA, USA. Retrieved from: www.bu.edu/wcp/Papers/Acti/ActiHanl.htm

Humberstone, B. (1998). Re-creation and connections in and with nature: Synthesizing ecological and feminist discourses and praxis? *International Review for the Sociology of Sport, 33*(4), 381–392.

Ingold, T. (2000). *The perception of the environment: Essays on livelihood, dwelling and skill*. New York, NY: Routledge.

Ingold, T. (2005). Epilogue: Towards a politics of dwelling. *Conservation and Society, 3*(2), 501–508.

Ingold, T. (2007). *Lines: a brief history*. London; New York: Routledge.

Kennedy, S., MacPhail, A. & Varley, P.J. (2018). Expedition (auto)ethnography: an adventurer-researcher's journey. *Journal of Adventure Education and Outdoor Learning, 18*(1), 1–15.

Law, J. & Urry, J. (2004). Enacting the social. *Economy and Society, 33*(3), 390–410.

Loynes, C. (2010). The British youth expedition: Cultural and historical perspectives. In S. K. Beames (Ed.) *Understanding educational expeditions* (pp. 1–15). Rotterdam: Sense.

Melaney, W.D. (2006). Arendt's revision of praxis: On plurality and narrative experience. In A.-T. Tymieniecka (Ed.) *Logos of phenomenology and phenomenology of the logos* (Book 3, Vol. 90, pp. 465–479). Berlin/Heidelberg: Springer-Verlag.

Mullins, P.M. (2009). Living stories of the landscape: Perception of place through canoeing in Canada's north. *Tourism Geographies, 11*(2), 233–255.

Mullins, P.M. (2014a). Conceptualizing skill within a participatory ecological approach to outdoor adventure. *Journal of Experiential Education, 37*(4), 320–334.

Mullins, P.M. (2014b). The commonplace journey methodology: Exploring outdoor recreation activities through theoretically-informed reflective practice. *Qualitative Research, 14*(5), 567–585.

Mullins, P.M. (2018). Toward a participatory ecological ethic for outdoor activities: Reconsidering traces. In B.S.R. Grimwood, K. Caton & L. Cooke (Eds.) *New moral natures in tourism* (pp. 149–164). Abingdon, New York, NY: Routledge.

Mullins, P.M. & Maher, P.T. (2007). Paddling the Big Sky: Reflections on place-based education and experience. *Science and Stewardship to Protect and Sustain Wilderness Values*. Retrieved from: www.fs.fed.us/rm/pubs/rmrs_p049/rmrs_p049_402_410.pdf

Sheller, M. & Urry, J. (Eds.). (2004). *Tourism mobilities: Places to play, places in play*. London, New York, NY: Routledge.

Sheller, M. & Urry, J. (2006). The new mobilities paradigm. *Environment and Planning A, 38*(2), 207–226.

Smith, K. & Kirby, M. (2015). Wilderness 2.0: what does wilderness mean to the Millennials? *Journal of Environmental Studies and Sciences, 5*(3), 262–271.

Stinson, J. (2017). Re-creating Wilderness 2.0: Or getting back to work in a virtual nature. *Geoforum, 79*, 174–187.

Sumara, D. (2002). Creating commonplaces for interpretation: Literary anthropology and literacy education research. *Journal of Literacy Research, 34*(2), 237–260.

Tuhiwai Smith, L. (1999). *Decolonizing methodologies: Research and indigenous peoples*. London: Zed Books.

Urry, J. (2000). *Sociology beyond societies: Mobilities for the twenty-first century*. London, New York, NY: Routledge.

19

MOBILE METHODS IN OUTDOOR STUDIES

Walking interviews with educators

Jonathan Lynch

Introduction

> New material feminisms, post-humanism, actor network theory, complexity theory, science and technology studies, material culture studies and Deleuzeian philosophy name just some of the main strands that call us to reappraise what counts as knowledge and to re-examine the purpose of education. Together these strands shift the focus away from individualised acts of cognition and encourage us to view education in terms of change, flows, mobilities, multiplicities, assemblages, materialities and processes.
>
> *Taylor & Ivinson, 2013, p. 665*

This quote supports the purpose of this chapter well because it calls on us to see education in more mobile terms. Shifting our attention towards the mobile nature of our lives is particularly useful to educators in outdoor settings as much of our practice concerns movement. For example, deliberate movement such as on expedition or deliberate non-movement such as undertaking a solo experience. Whilst taking a research approach that involves mobilities seems an obvious thing to do, it is underutilised in education outdoors. This chapter is one explanation of how mobilities might inform research in education in outdoor settings.

I will show how mobilities research and mobile methods offer nuanced ways to research practice in education in outdoor settings, and how the turn towards mobilities in social science is part of a contemporary approach to qualitative research. Mobilities research can be understood as part of a wider project that challenges humanism in the social sciences. Notable work can be found in cultural geography (Thrift, 2004; Urry, 2007; Whatmore, 2002), anthropology (Ingold, 2011; 2013), and post-qualitative research (Lather, 2001; Lather & St Pierre, 2013). Theorists in these domains of knowledge are concerned with expanding our understanding of the social world in ways not limited to humanist registers such as language.

At the core of these ideas is a view of the ontological situation that rejects dualist thinking and any notion of the human as separated from the world. Instead, our existence can be understood in more relational terms; that we exist in relations and come to know our world through these relations in their many forms. Thrift (2004, p. 87) argues for this view as part of his non-representational project in human geographies when he notes that "Behaviour is not localised in 'individuals' but is understood as a relational structure that constitutes what might be termed an 'extended organism'". This view of the ontological situation is gaining traction in education in outdoor settings (Gough, 2016; Mcphie & Clarke, 2015; Rautio, 2013; Somerville, 2016) and is one that rejects humanism for flatter ontologies such as in new materialism (Coole & Frost, 2010). These views call on us to understand outdoor places where we cannot be other than intertwined with the social, material and discursive. Mobilities research is concerned with relations, not discrete bounded subjects nor dualisms of indoor/outdoor.

This chapter is structured in the following ways: First I explain how mobilities and mobile methods can usefully inform research in education in outdoor settings; then I show the inspiration to use mobilities theory and mobile methods in my research. I explain my research decisions through three provoking questions: (1) How are we to understand the ontological situation, especially the outdoors? (2) How do we collect data in ways that fit into our view of the ontological situation? (3) How might we analyse these data? Arguably, these are questions that every researcher faces, but here I show how using mobilities theory and mobile methods were particularly appropriate for researching education in outdoor settings. I conclude the chapter by suggesting how other researchers in outdoor-related fields might choose to work with mobile methods and mobility inspired research practices.

This research is based on primary data from in-service teachers in the UK, who used outdoor learning on a regular basis in areas beyond the school grounds. The research was focused on how the more-than-human[1] (Whatmore, 2002, 2006) aspects of place were harnessed into the curriculum planning and enactment of outdoor learning. The methods I used to collect data needed to be nuanced to the non-anthropocentric dimensions of place. As a result, I devised methods that were sensitive to the more-than-human features of place in the research setting. I understood the outdoor places in a posthumanist[2] inspired way; as a field of relations (Ingold, 2004).

Mobilities and place

In this section, I use provoking question (1) to show how, and why, I drew upon mobilities inspired theories of place: How are we to understand the ontological situation, especially the outdoors?

In my research, I was interested in how I could understand place to include the more-than-human as I was keen to resist humanising the outdoors through constructs such as science or adventure. Instead I wanted to conceptualise place in ways that did not just privilege the human. To do this I drew on theorising in mobilities literature, and posthumanist concepts of place.

The term mobilities is well defined by Urry who argues that in the past, social science has not sufficiently attended to the mobile nature of our world (Urry, 2007). In response, Urry and others seek to establish "a movement-driven social science" (2007, p. 18). Mobilities informed research is more than paying attention to that in our lives that is mobile/immobile; the mobilities paradigm is part of a project to transform theoretical and methodological landscapes. For example, Büscher, Urry and Witchger (2011) write,

> The term 'mobilities' refers not just to movement but to this broader project of establishing a 'movement-driven' social science in which movement, potential movement and blocked movement, as well as voluntary/temporary immobilities, practices of dwelling and 'nomadic' place-making are all viewed as constitutive of economic, social and political relations
>
> *2011, p. 4*

In mobilities research, place, space and learning can be understood as interrelated (Leander, Phillips & Taylor, 2010). Writing in educational research, Fenwick, Edwards and Shawchuck (2011) explain the effects of a "spatial turn" in social science in the 1990s (2011, p. 11) and, drawing on postmodernism and critical theory, they show how space is not just a static background or container where our lives are acted out. For these authors, space is dynamic and relational with sustained and constant change. As a result, spaces can impinge on learning in how they influence certain interactions or relationships. Writing in the mobilities literature in cultural geography, Urry notes "such spatial structuring makes a significant difference to social relations" (2007, p. 34). As a result, space and relations can be understood as not discrete or separate. Going further, some posthumanist theorists in outdoor-related fields such as environmental education (Sonu & Snaza, 2015) and early years education outdoors (Pacini-Ketchabaw, 2013; Rautio, 2013) see these relations as more than interrelated, they argue that humans and the more-than-human are co-extensive.

In my research, I came to see place in ways inspired by these ideas because it gave me a way to understand how the human and the more-than-human features could be accounted for in teachers' curriculum planning and enactment. I was also inspired by the anthropological writing of Ingold (2004, 2011) who argues for a relational view of place. In a relational view, place can be understood as a meshwork of relations that are ongoing, in formation and becoming. In terms of education, places can be pedagogical through the relations we might form and maintain with the more-than-human.

Place as relations

A relational view of place is found in the literature in cultural geography (Whatmore, 2006) and anthropology (Ingold, 2011). Cultural geographers such as Wylie (2007) articulate a relational view of place that includes the material and co-constitutive nature of place and landscapes. Similarly, Ingold's (2011) concept of place was particularly useful for my research because, like some posthumanist theorists (Braidotti,

2013; Whatmore, 2002), he argues for the ontological situation as being non-hierarchical. In other words, humans do not have a privileged, objective perspective. He explains this view of the ontological situation through our perception of the environment. Ingold argues that as humans we are not separated from the world, but instead are points of emergence in a "relational field" (2004, p. 304),

> In short, organisms [for example humans, bacteria, bees, trees] no more interact with the environment than do individuals with society. Rather, ecological relations – like social relations – are the lines along which organisms-persons through their processes of growth, are mutually implicated in each other's coming into being.
>
> *2004, p. 306*

In later work, Ingold argues that this strongly influences how we *perceive* our environment,

> In short, to perceive the environment is not to look back on the things to be found in it, or to discern their congealed shapes and layouts, but to join with them in the material flows and movements contributing to their – and our – ongoing formation.
>
> *2011, p. 88*

Ingold's argument is that we are not separate from place, we are intertwined with place in material and co-extensive ways. He seeks to challenge the Western cultural notion of the 'individual', detached from the world around it. This view accommodates the non-human components of place as not separate, but that with which we are relationally intertwined. This was my answer to question (1) How are we to understand the ontological situation, especially the outdoors?

Methodological choices

The methodological choices in my research were driven by the relational view of place and the acknowledgment that to collect useful data I needed to devise place-sensitive methods. Working within a multicase study methodology (Stake, 2006), I devised methods that were particularly attuned to place, mobilities and the material. Together, these methods were used to collect data on the mobile worlds of outdoor learning with the teachers in the study.

Mobile methods

In this section, I express how question (2) was considered, using mobilities theory and mobile methods: How do we collect data in ways that fit into our view of the ontological situation?

For my research, I wanted to gather data on the harnessing of the more-than-human in outdoor learning curriculum planning and enactment. To do this,

I devised a walking interview that was sensitive to an understanding of place as a field of material and discursive relations. I was inspired by Mazzei's (2013) posthumanist views, which identify the importance of the relational and situated nature of the interview as method. She notes that,

> Perhaps more attention needs to be given to the *where* of the interview, and the *when* of the interview, and the *if* of the interview. If we are to make sense of these material and discursive material constructions and joining of forces, perhaps we must think practices that disavow an over-reliance on words as the primary source of meaning.
>
> *2013, p. 739*

For my research, these ideas around interviewing encouraged me to rethink the interview as a method that could be more nuanced to relational place data than an indoor or static one.

The walking interview was devised as a mobile method to collect data on the emplaced experiences of the teachers' planning and enactment of outdoor learning in their chosen outdoor learning sites. It is covered in more detail in Lynch and Mannion (2016) but in summary, these interviews were seen as material and discursive; where the social fabric of life is not static (Lorimer, 2008) and neither are the practices within it (Pink, 2007, 2012).

The data collected during the walking interviews included voice recordings, photographs from the places of learning, and any field notes that I generated. This choice of method was vital to my research because it allowed me (the researcher) to be in the unified relational field (Ingold, 2004, 2011) and collect data on the material and discursive unfolding world. Ingold views our knowledge-building in the world not just with place but as *movement* through places; he calls this "wayfaring" (2011, p. 148). For Ingold, place itself is less important than the movement through/along places: "The path, and not the place, is the primary condition of being, or rather of becoming" (2011, p. 12). Ingold believes that it is in moving through the world (wayfaring) that knowledge is created, "that scientific knowledge, as much as the knowledge of inhabitants, is generated within the practices of wayfaring" (2011, p. 155).

With a walking interview method in a unified relational field the world is experienced through wayfaring. One way of being in this wayfaring is to walk with the practitioner in their place of outdoor learning. As Ingold writes, it is these practitioners [in place] who are key: "Practitioners, I contend are wanderers, wayfarers, whose skill lies in their ability to find the gain of the world's becoming and to follow its course whilst bending it to their evolving purpose" (2011, p. 211). This quote describes well what I was seeking to do in the walking interviews; following the wayfarers in their places of outdoor learning. Wayfaring with educators was a mobile method sensitive to the relational view of place and the mobile nature of knowledge bound up in the sites of practice.

The second method I used was a memory-box interview. This interview was focused around materials that teachers collected in a cardboard shoe box that were related to any harnessing of the more-than-human in outdoor learning. Like the walking interview, this method was attuned to the role of materials in the co-production of meaning and is also congruent with the new materialist theoretical framework of the research (Barad, 2003; Braidotti, 2013; Coole & Frost, 2010). For example, Taylor and Ivinson (2013) note:

> In 'new' materialism, matter is not inert, neither does it form an empty stage for, or background space to, human activity. Instead, matter is conceptualised as agentic and all sorts of bodies, not just human bodies, are recognised as having agency.
>
> *2013, p. 666*

As a result, the relations between the material and discursive components of these interviews both played a role in the production of meaning.

Rhizoanalysis

A mobilities inspired walking interview method required a form of analysis that did not return to a dualistic view of the world. Methodologically, to be congruent with the relational view of place in my research, I chose to use a post-qualitative inspired approach to analysis. I see the post-qualitative approach as also congruent with the project of a more mobile social science argued for by Büscher et al. (2011). This explains my answer to question (3): How might we analyse these data?

The analysis of the data was post-qualitative and I chose a format called rhizoanalysis. Post-qualitative research has evolved in certain directions of late that have accounted for shifts in critical and cultural theory. These are developments that have been labelled post-qualitative (Lather, 2013; Lather & St Pierre, 2013). The post-qualitative field is not a finished affair and could be conceptualised as many authors working with common aims in different ways. Some of these common aims are concerned with acknowledging the materiality in the structure of, and ordering of, our social worlds. For example, in ways not limited to a purely linguistic ordering (Lenz Taguchi, 2012) and that resist the representation of knowledge though language alone (MacLure, 2013). In post-qualitative research, ontologically it is understood that we are *always part* of ongoing relations, never separated (Lather & St Pierre, 2013).

The term 'rhizoanalysis' has its genesis in the geophilosophy of Deleuze and Guattari (2004), the focus of which is their use of the term "rhizomatic" (p. 7). Deleuze and Guattari (2004) understand thought as rhizomatic; random, multiple, and propagating. Rhizoanalysis is useful to post-qualitative research because it is concerned with relations over a discrete autonomous subject (for a deeper exploration of rhizoanalysis in education see the work of Masny (2013a, 2013b)). In my research, the rhizoanalysis sought not to represent the data but to see what new thinking and conceptualising about outdoor learning can be produced (Fox & Alldred, 2015).

In my research, data were analysed through the production of vignettes and four rhizoanalysis tools (Masny, 2014). The vignettes were produced through the coming together of the different data sets of photographs, transcripts and field notes. I used Masny's (2014) four rhizoanalysis tools to read the data and identify affective changes. One of these tools was "palpating" (p. 358). Palpating data is an ambiguous, difficult and challenging task. In my research, I saw palpating the data as a process of noticing changes in a body (mine, a teacher, or the more-than-human), and how affects and sensations are involved in these changes. In the vignettes, I see the reader is *also* part of the ongoing performance of the research. As data are 'read', any reader will also have thoughts and ideas that are produced by the vignette.

Next, I present a vignette from the research, which was about curriculum planning. I provide an overview of the case this vignette was from, and the events on which I collected data. I then follow by describing in more detail how meaning was produced with the vignette through the process of rhizoanalysis and the palpation of the data.

Vignette example on curriculum planning: "Teacher's noticing of missing biodiversity"

This vignette includes a photograph taken during the walking interview of the forest and another of a curriculum planning document the teacher showed to me. The excerpt of text is from the field notes I took after the walking interview. This vignette provides a starting point for portraying how this teacher is drawn to noticing the more-than-human in outdoor places in her curriculum planning.

During the walking interview the teacher expressed how she noticed a lack of biodiversity in the site, especially a lack of flowering plants (Figure 19.1a). To her, this was a problem that influenced the more-than-human she felt able to harness in her outdoor learning. She stated how this lack of biodiversity meant she found it hard to do the games and activities that she had planned beforehand (Figure 19.1b). This is important to note because it depicts how the reality of the place, and the more-than-human capacities there can impinge on the curriculum planning process.

"TEACHER'S NOTICING OF MISSING BIODIVERSITY"

From field notes ...

Listening to her ... she couldn't explain how she planned her session. She would get an idea that might come to her when she was doing something else, perhaps at home, that would build into a session plan in her head that would develop into a session when she was there [in the wood]. *One big thing I hear is that her outdoor learning journey from the beginning has taken her to the place where she now uses a child-led approach. She has the confidence to do that now ... there was a bit of a place-led approach coming across in the* [walking] *interview but it was not articulated, as it is in my head when I think about the features of a place-led approach.* ***One of the problems she notes of the site is that it lacks biodiversity (it's a shelter belt plantation – closely planted non-indigenous species and no light to the canopy floor). She notes there are no small plants or flowers growing and so the games or activities she might like to do where the children go to find stuff are not as possible.***

FIGURE 19.1A The spruce plantation with limited biodiversity, which in turn limits the activities and games that might rely on the features and presence of flowering plants

Source: 'Story Circle', Lynch, 2013

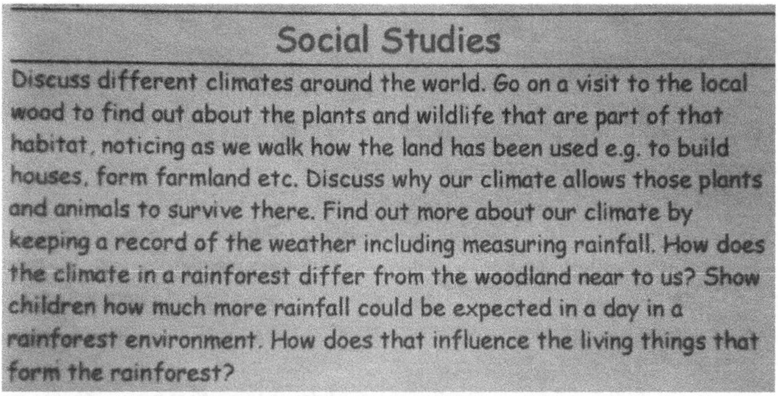

Social Studies

Discuss different climates around the world. Go on a visit to the local wood to find out about the plants and wildlife that are part of that habitat, noticing as we walk how the land has been used e.g. to build houses, form farmland etc. Discuss why our climate allows those plants and animals to survive there. Find out more about our climate by keeping a record of the weather including measuring rainfall. How does the climate in a rainforest differ from the woodland near to us? Show children how much more rainfall could be expected in a day in a rainforest environment. How does that influence the living things that form the rainforest?

FIGURE 19.1B Part of a curriculum planning document the teacher shared

Source: 'Lesson Plan', Lynch, 2013

Meaning making

This vignette portrays how, ontologically, this teacher is unfolding in the world. Her pre-planned activities and topics become problematic when she is in this place. As I palpated the data, I got a sense of the changes that she underwent around her noticing the lack of biodiversity in the forest and the problems this posed to her planning of outdoor learning. What this vignette may be starting to disclose are the ways in which the more-than-human needs to be *noticed* first so it can be harnessed in any planning of outdoor learning.

This vignette allows me to look productively with signposts for what other educators need to think about if they are going to harness the more-than-human into outdoor learning in purposeful ways. As I palpated the data in this vignette, I could sense the teacher is comparing rainforests with the local forest. In my reading of this vignette, I see the agencies of the more-than-human are being noticed for curriculum planning in ways that are dominantly about science, such as climate and levels of rainfall. This vignette produces questions for me about how we harness the more-than-human in the planning of outdoor learning. How can we develop teachers' abilities to notice, and harness, the more-than-human in ways not limited to science? I suggest developing educators' abilities to pay attention to the more-than-human as something we are co-implicated with is one way forward to do this. This vignette is one example of a post-qualitative approach to research and data analysis that sought to work *with* the relational and mobile nature of the world.

Conclusion

In this chapter I have explained how mobilities informed research, and mobile methods are nuanced approaches to further understand education in outdoor settings. I have sought to do this through an example from my research and through an explanation of the methodological choices I made via three guiding questions. By accepting the mobile nature of place and drawing on posthumanist theoretical resources, I was able to research education in outdoor settings in ways that could accommodate the more-than-human. The mobile method of a walking interview allowed me to collect data that was 'on the move' and that was located in the site of outdoor learning practice. However, the mobile method was expensive in terms of time and resources (i.e. travelling to the outdoor site and spending time to walk there) and not all data are on the move (Spinney, 2015).

Finally, by using post-qualitative inspired analysis I was able to work with the data in ways that did not enforce interpretation onto it. Instead, meaning was produced in ways that left room for the acknowledgement of the more-than-human features of place and the relations that are constituted in places. I hope that these three provoking questions might usefully guide other researchers who seek to embrace the mobile nature of the outdoors in their research in mobilities and education in outdoor settings.

Notes

1 The term more-than-human signifies a way of understanding 'nature' that does not reduce it to something less important than the human. Secondly, the term denotes all that we encounter in 'nature' but rejects a priori meaning.
2 Posthumanism rejects the primacy of humanism in cultural theory and philosophy. It is a nascent field with many definitions and expressions. In this chapter it is used to denote a worldview where the human is decentred, and the ontological situation is understood as being less reliant on human experience for all meaning. In other words, non-humans, materiality and objects can impinge on our lives in ways we have not yet imagined or fully understood.

References

Barad, K. (2003). Posthumanist performativity: Toward an understanding of how matter comes to matter. *Signs: Journal of Women in Culture and Society, 28*(3), 801–831.

Braidotti, R. (2013). *The posthuman.* Cambridge: Polity Books.

Büscher, M., Urry, J. & Witchger, K. (2011). *Mobile methods.* Abingdon: Routledge.

Coole, D.H. & Frost, S. (2010). *New materialisms: Ontology, agency, and politics.* Durham, NC: Duke University Press.

Deleuze, G. & Guattari, F. (2004). *A thousand plateaus: Capitalism and schizophrenia.* London: Continuum.

Fenwick, T., Edwards, R. & Sawchuck, P. (2011). *Emerging approaches to educational research: Tracing the sociomaterial.* London: Routledge.

Fox, N.J. & Alldred, P. (2015). New Materialist social inquiry: Designs, methods and the research-assemblage. *International Journal of Social Research Methodology, 18*(4), 399–414.

Gough, N. (2016). Postparadigmatic materialisms: A "new movement of thought" for outdoor environmental education research? *Journal of Outdoor and Environmental Education, 19*(2), 51–65.

Ingold, T. (2004). Two reflections on ecological knowledge. In G. Sanga & G. Ortalli. (Eds.) *Nature knowledge: Ethnoscience, cognition, and utility* (pp. 301–311). Oxford: Berghahn Books.

Ingold, T. (2011). *Being alive: Essays on movement knowledge and description.* London: Routledge.

Ingold, T. (2013). *Making: Anthopology, archaeology, art and architecture.* London: Routledge.

Lather, P. (2001). Postbook: Working the ruins of feminist ethnography. *Signs: Journal of Women in Culture and Society, 27*(1), 199–227.

Lather, P. (2013). Methodology-21: What do we do in the afterward? *International Journal of Qualitative Studies in Education, 26*(January), 634–645.

Lather, P. & St Pierre, E. (2013). Post-qualitative research. *International Journal of Qualitative Studies in Education, 26*(6), 629–633.

Leander, K.M., Phillips, N.C. & Taylor, K.H. (2010). The changing social spaces of learning: Mapping new mobilities. *Review of Research in Education, 34*(1), 329–394.

Lenz Taguchi, H. (2012). A diffractive and Deleuzian approach to analysing interview data. *Feminist Theory, 13*(3), 265–281.

Lorimer, H. (2008). Cultural geography: non-representational conditions and concerns. *Progress in Human Geography, 32*(4), 551–559.

Lynch, J. & Mannion, G. (2016). Enacting a place-responsive research methodology: Walking interviews with educators. *Journal of Adventure Education and Outdoor Learning, 16*(4), 330–345.

MacLure, M. (2013). Researching without representation? Language and materiality in post-qualitative methodology. *International Journal of Qualitative Studies in Education, 26*(6), 658–667.

Masny, D. (2013a). Rhizoanalytic pathways in qualitative research. *Qualitative Inquiry, 19*(5), 339–348.

Masny, D. (2013b). *Cartographies of becoming in education: A Deleuze-Guattari perspective.* Rotterdam: Sense Publishers.

Masny, D. (2014). Disrupting ethnography through rhizoanalysis. *Qualitative Research in Education, 3*(3), 345–363.

Mazzei, L.A. (2013). A voice without organs: Interviewing in posthumanist research. *International Journal of Qualitative Studies in Education, 26*(6), 732–740.

Mcphie, J. & Clarke, D.A.G. (2015). A walk in the park: Considering practice for outdoor environmental education through an immanent take on the material turn. *Journal of Environmental Education, 46*(4), 230–250.

Pacini-Ketchabaw, V. (2013). Frictions in forest pedagogies: Common worlds in settler colonial spaces. *Global Studies of Childhood, 3*(4), 355–365.

Pink, S. (2007). Walking with video. *Visual Studies, 22*(3), 240–252.

Pink, S. (2012). *Situating everyday life: Practices and places.* London: SAGE.

Rautio, P. (2013). Children who carry stones in their pockets: On autotelic material practices in everyday life, *Children's Geographies,* (July), 37–41.

Somerville, M. (2016). The post-human I: Encountering 'data' in new materialism. *International Journal of Qualitative Studies in Education, 29*(9), 1162–1172.

Sonu, D. & Snaza, N. (2015). The fragility of ecological pedagogy: Elementary social studies standards and possibilities of new materialism. *Journal of Curriculum and Pedagogy, 12*(3), 258–277.

Spinney, J. (2015). Close encounters? Mobile methods, (post)phenomenology and affect. *Cultural Geographies, 22*(2), 231–246.

Stake, R. (2006). *Multiple case study analysis.* New York, NY: The Guildford Press.

Thrift, N. (2004). Summoning life. In P. Cloake, P. Crang & M. Goodwin (Eds.) *Envisioning Human Geographies* (pp. 81–104). London: Arnold.

Taylor, C.A. & Ivinson, G. (2013). Material feminisms: new directions for education. *Gender and Education, 25*(6), 665–670.

Urry, J. (2007). *Mobilities.* Cambridge: Polity Press.

Whatmore, S. (2002). *Hybrid geographies: Natures, cultures, spaces.* London: SAGE.

Whatmore, S. (2006). Materialist returns: Practising cultural geography in and for a more-than-human world. *Cultural Geographies, 13*(4), 600–609.

Wylie, J. (2007). *Landscape.* London: Routledge.

20

SENSING THE OUTDOORS THROUGH RESEARCH

Multisensory, multimedia, multimodal and multiliteracy possibilities

lisahunter

Introduction

Humans engage with, understand, navigate, exploit, nurture and shape their worlds, whether seascapes, landscapes or other classifications of spacetimes, through their multiple senses – although with an arguably ocularcentric bias in contemporary Western times (Allen-Collinson & Hockey, 2010). Sensory and sensual methodologies to explore this engagement are becoming more commonplace in research, although feminists, indigenous, performance and artistic scholars have been using such methodologies for some time. In outdoor studies, the study of relationships between humans in nature and with nature is also not new, but gaps remain, with the research field dominated by binaries, onto-epistemological blind spots, disembodied representations and unimodal text-based sensorality. Answering calls to the under-theorisation of materiality, senses and lived bodily experiences (e.g. Sparkes, 2017a), an exciting burgeoning field is emerging through the use of multimedia and its engagement with the multisensorial, multimodal and multiliteracy possibilities.

Audio-visual records, and explorations, of learning in the outdoors like those I use as examples below, are becoming more available. My own forays into using multimedia as a research methodology that attends to the lived, material body have not been without challenges, but offer new research directions and connect with new audiences. The power of multimedia in engaging with an audience is exemplified by the nearly four million views of Maisey Rika's multimedia work exploring her relationships with Tangaroa, the Maori god of the sea, Tangaroa Whakamautai.[1] In a world where pedagogues, outdoor researchers and capitalists are vying for peoples' attention through 'eyeballs', I argue that research needs to consider such multimedia resources for multimodal research and dissemination (lisahunter, Wubbels, Clandinin & Hamilton, 2014), engaging with the multiple sensorial experiences (lisahunter & emerald, 2016) of and beyond the human, from different perspectives and in the intermingling of (non)human in the outdoors (lisahunter, 2013).

Work by scholars in outdoor studies/education (hereafter outdoor studies) and physical culture (e.g. Allen-Collinson & Hockey, 2010; Humberstone 2011; Sparkes 2017a) has demonstrated the importance of and link between corporeality, embodiment, emplacement and attention to the senses. Yet much of the research is recorded, analysed or at least re-presented as research texts that are linear, unimodal, sensorially flattened and text based. Some researchers address this, as illustrations in this chapter from and beyond the academy suggest. This chapter explores the place of multisensorial research (lisahunter & emerald, 2016) in outdoor studies by considering multimodal or multimedia techniques for the creation of field texts, analytical or interim research texts and research texts used to disseminate the research (Clandinin & Connelly, 2000). As such, the chapter seeks the senses 'to uncover meaning and feelings' (Coates, Hockley, Humberstone & Stan, 2016, pp. 74–75). After some examples that illustrate engagement with multimedia in forms of outdoor research, I turn to some of the theoretical bases for multimedia, multisensorial and multi-perspective research, and explore some opportunities and limitations the 'multi' may offer.

Multimedia illustrations: exploring the senses

1. Eliciting seaspacetimes as field texts: Researching surfing

As part of an ongoing ethnography associated with the sea, and in this case surfing, I have used a variety of practices to create field texts. These texts are evidence for me to understand experiences with and beyond the human: initially to document my emplaced movement, embodied memory, and affects of an aquatic assemblage through the personal example of surfing with a mind to use such methods with other research participants. Using a surfboard-mounted camera facing me, a chest-mounted camera, a head-mounted camera, an underwater camera attached to the board and a beach based camera mounted on a tripod, I was experimenting with ways and perspectives to notice and capture engagements with/in/under/on/through/after the ocean in the process of surfing. The footage enfolded multi-perspective, mobile, multimedia technologies with other field texts such as a post-surf body mapping activity (Sweet & Escalante, 2015) where I drew my body on the sand after exiting the water and reflected on whereon my body I was sensing what. Another 'fold' of re-viewing the footage through audio-visual elicitation (Stockall, 2001) and recording my responses allowed me to notice what I had said and done, triggered memories of sensations and emotions that were not explicitly articulated in the film footage, and also captured what I had missed noticing at the time (such as the amount of time I sat on the board waiting for a wave, underwater movements by me and other creatures, changes in the nature of the sea and air, and unconscious movements such as my fingertips stroking the water). The exercise clarified my (in)ability to separate surfing from researching, the importance of the mundane, the 'just being', moments of transition, and moments of not surfing:

> In this 'being in' and 'looking back' through text and video, in the moment and in the moments after, I was able to take stock of what it is to engage with the sea. In surfing, it is not all in the wave. The wave is not the only articulation of the sea-space-time assemblage. It is the waiting. The paddling. The not surfing.
>
> *lisahunter, 2018, p. 105*

In the process, I discovered/created new realisations unavailable whilst surfing. These different 'seaspacetimes' proved valuable for deeper reflection and exposure of some of the sensory experiences I had, and continued to have, as a result of the surf and research.

2. Digital stories of Warreeny: Connecting Georgia to sea country

Engaging with traditional Aboriginal and contemporary marine science perspectives, a group of year eight students (aged 12–13 years) explored Port Phillip Bay in Victoria as part of a cultural project connecting multimedia skills to learning about sea country. Connecting two worlds, Boonwurrung Elder Arweet Aunty Carolyn Briggs shared her Aboriginal knowledge and Harry Breidhal shared his marine science knowledge with the students through workshops and a sailing expedition. The students' experiences were captured in a series of fifteen short digital stories made available online.[2] The website text notes students' learning experiences (see Box) but for me, with even greater multimodal power, Georgia, one of the students, tells us and shows us the audio-visual capturing of her experience over which a reflective Georgia narrates the final output.

SOME OF THE STUDENT EXPERIENCES

"What I learnt is to respect the land and the baang (water); to think about the barerarengar (country), warreeny (ocean); to look after our bubup (children)". – Maria

"I learnt the importance of sharing stories. If they aren't shared they'll be lost forever!" – Latisha

"The two stories from Harry and Aunty Carolyn were identical, knowing the science was proving that Aunty Carolyn's stories were true, was amazing. It was beautiful seeing the ocean's beauty and feeling the water and knowing how it was cared for years ago by the people of the land and the creatures of the warreeny". –Yemurraki

Georgia's story reveals what she has sensed, learned and experienced and illustrates a form of digital diary used by many researchers, what might be called a field text, but also potentially an interim research text wherein the participatory research process of editing to construct the narration and the final documentary are a preliminary form of analysis of the experience.

3. *Beau's sea kayaking journey*

In his visual autoethnography, Beau Miles interacts with some of the same waters as Georgia: the Southern Ocean. But it is a very different experience for the white-identified, sea kayaking adult as he journeys between two coastlines of Australia, from Victoria to Tasmania. The audiovisually recorded journey acts as field texts for his doctorate (Miles, 2017). The edited six-part documentary *Bass by Kayak*[3] is a multimedia text, moving between field, interim and research texts in partnership with the normative written thesis "the visual artefact itself [a] descriptive, analytical artefact in and for itself" (p. 118). His interaction between the genres of multimedia and monomodal text are:

> (a) in the first degree, a product and process of interpretation and is balanced by (auto)ethnographic text and (b) in the form of phenomenological, existential insights (Abram, 1996; van Manen, 1990). Film and text in combination are therefore put forward as reduced, produced, refined, storied, and theorised content. As findings, both the film series and existentials are labelled and talked of in terms of my personal, phasic, and expeditionary 'episodes'. As a whole (conjoined as a film series) this represents my lifeworld in terms of both the broad (long-term) and day-to-day essences.
>
> *Miles, 2017, p. 119*

His reflection-driven insight into himself, outdoor research and expedition are significantly shaped by multimedia engagements; in their making, their re-viewing, their re-production and sharing with other audiences. These texts are the basis for his bigger project of critical intervention via the pedagogical potential of sea kayak journeying and an ongoing project using a blog.[4]

Reading the thesis and then watching the films, I wondered how many versions of the documentary could be made, as different productions doing different work for different audiences, a thought I often pondered during the (post)production of the documentary film *Reaching Higher Ground* (Angel, 2017) and my ongoing analysis of this experience (lisahunter, 2018). Albeit not outdoor research, rather a critical project in social justice and critical pedagogy, my involvement in the reflexive and learning process of this documentary-making echoed that of other multimedia researchers (Miles, 2017; Pink, 2006, 2015; Wood & Brown, 2010). Documentary, and other forms of multimedia-as-research, with all its ontological, epistemological, theoretical and methodological challenges, has much to offer outdoor studies in exploration of methods to engage with the lived body and senses. Before addressing some of the potential and challenges, what of the theoretical and analytic bases to multimedia research?

Theoretical frameworks and analytic perspectives

Research aims to understand and tell stories. Borrowing from Clandinin and Connelly (2000) if we take three points in the research process, identifying or constructing evidence (field texts), analysing that evidence (interim research texts) and creating a way to communicate the findings and implications (research text), there are many options for, and considerations and implications when engaging with lived bodily experiences of and beyond the human. These might be considered through four epistemes – the sensorial, sensual, emplaced and sensational (Table 20.1).

Narrative and ethnographic research focused on the senses (e.g. Mason & Davies, 2009; Pink, 2006, 2013) has inspired me to experiment at these epistemic intersections with technologies in outdoor research to facilitate communication that is 'multi', that is, engaging with multiple senses, multiple modes of communication and employing mutiperspectival multimedia to engage in/beyond human experiences and multiple timespaces.

Sensory studies set out to explore senses that have been largely ignored, and to understand the cultural understandings beyond the Westernised that have routinely framed the senses (touch, taste, sound, smell and vision). Feminist calls last century for lived body and sensory/sensuous scholarship had anthropologists and ethnographers becoming more attuned to using their own bodies and senses for analyses and writing about their experiences (e.g. Allen-Collinson and Hockey's haptics of scuba diving, 2010, and Humberstone's experience of balance in windsurfing, 2011). Researchers may have employed multimedia to record different senses such as sound, still or moving images, but, with the exception of film studies and an emerging field of documentary research (Ahmed, 2010), using multimedia for analysis or research text construction was still in its infancy in social science through to the 2010s.

With attention to the visual, sensory, narrative, mobility and embodiment the use of new methodologies to address multiple modal research included video diaries, visual diaries, photo-voice, cultural studies of film, and documentary-as-method. These developed relatively quickly in the new century, hastened by the widespread availability of digital and mobile technologies and by concern for participatory ethics. The advances in technology and its accessibility brought opportunities for theoretical developments as well as challenges and dilemmas.

Films made outside the intention of research also act as potential field texts and I would argue, to some extent, enable a form of analysis nuanced by the representational intentions of the makers. The use of arts-based practices with documentary "… has contributed a new filmic affect that can better open up and articulate an aesthetic appreciation of experiences, both inside and outside of work, which cannot be captured easily by conventional text on its own" (Wood & Brown, 2010, p. 536). Both documentary and fictional films have been important in the documenting and production of culture, as exemplified in outdoor pursuits such as surfing (Ormrod, 2009). Surfing history and the shaping of values in surfing is illustrated in surf films:

TABLE 20.1 Intersections of three moments and four epistemes when considering sensory narratives (lisahunter & emerald, 2016)

	Senses	*Sensual*	*Sensory geographies of emplacement*	*Sensational – learning points and turning points*
Field texts considerations of 'evidence' or 'data'	Collecting sights, sounds (and silences) tastes, touches and smells. How do we 'gather' or 'record' movement? How to gather what these evoke in participants?	Recording the quality of senses within a – how does touch, for example, play out as pleasure, pain, nostalgia for example? Recording what evokes stories of happiness, anger, disgust and so on.	Capturing and documenting the entanglements of place/space/time/materiality/biology.	The question of 'what captures your attention'? Sense experiences that change us or turn us. The critical incident.
Interim texts considerations of analysis	How to analyse sights, sounds (and silences) tastes, touches and smells? Is there a grammar? A code?	What do pleasure, pain and so on mean to participants and to research? How do we 'analyse' these qualities?	How to analyse and organise?	The reflective and reflexive turn – Why did *that* capture my attention, how has is turned me, how/what did I learn? Noticing patterns in my own and my participants
Research texts issues of representation	Problems of using language to represent the senses: what are the languages of taste, touch, smell, sound and so on? Challenges representing movement, temperature. Writing evocatively. Issues of representation arts, words, visuals, etc. Trusting the reader and trusting the text. *Using* the senses to create a text and creating a text that can be engaged sensorially: can it be touched, smelt, tasted, can a research text evoke pleasure or pain, where/when is it in place/space/time, how can a text capture me (turn me)?			

> Surf films were produced by surfers who were influential in the subculture ... They not only articulated and represented the cutting edge of surfing and surf technology, they promoted new values and ideas ... not always adopted with enthusiasm
>
> *Ormrod, 2009, p. 38*

Analytically, multimedia research queers the boundaries and perceived linearity of research stages and processes. In my own work, my experience was documented in a variety of timespaces, in the moment of surfing, in reflections and activity immediately after, in the moments of video elicitation creating new recordings of my interpretations, memories and revisited experience, and in the moments of stitching the field texts in such a way to narrate several research texts. I would argue that various forms of analysis were more and less present in most spacetimes. Perhaps the only time that analysis of some form was not active was in the unconscious acts during the surfing session, acts that were often recorded and reflected upon during audio-visual elicitation. Returning to the work of Miles (2017) acting as filmmaker-researcher (writer, director, paddler and co-producer) "the creative, reductive formulas of film-making (following story arches with conflict and resolution, introducing people and places, highlighting epiphanies and hardships, etc.) has meant [their] introspection and reduction of raw footage is by nature an intensely analytical process" (p. 121).

Why pay attention to the senses and multimedia?

The most significant features of 'multi-' sensory, modal and media research, for me, are the spatio-temporal affordances, participatory inclusion, research being more explicitly publicly active and the extension of sensory literacies. These are captured in an investigation of cycling (Spinney, 2011), which highlighted three ways in which mobile video ethnography could contribute to research. The first was using video to feel like one was there in the experience when unable to be there – as Georgia and Beau illustrate above by producing film; secondly as a way of capturing "fleeting moments of mobile experience" that myself and Beau have illustrated above, and the third as a way to "extend sensory vocabularies" (p. 161) – something I continue to experiment with (lisahunter, 2018). Capturing, co-constructing and/or eliciting affect or emotion, to understand human experiences of the outdoors, and/or the mechanisms of learning, are common intentions of outdoor studies. Multimedia has the potential to provide a rich, multidimensional, multimodal and multisensational assemblage to communicate experience associated with outdoor research.

The availability and ease of use of audio-visual technologies facilitates participatory projects, where research participants act as co-researchers, enabled not only to create their own audio-visual images, but to create a final research text in the form of a documentary for example (see also Chapter 15). Given autoethnographic accounts, and even delving creatively in to more-than-human imaginaries, there is multimedia potential to facilitate a sense of 'multi' where unimodal forms of field, interim and research texts (written text) can be replaced with more potentially inclusive modes or choices in how an experience is documented and communicated/shared. McNutt (2013) has used participant researcher narratives to aid participants in (re)storying lives and Everett (2017) has used them as powerful pedagogical tools. The relationship between narrative and the audio-visual in the evocative, therapeutic or remembering process is not yet fully understood, but it is one that has and can be explored through Art Therapy, visual narrative inquiry and, I would also suggest in increasing number, outdoor studies.

The first move in public activism is dissemination. In the reach of audio-visual, through for example YouTube and Vimeo, we might notice a relatively large research blind-spot. A recent documentary *Out in the Lineup* (Thomas, 2014) has perhaps done more to get the message of homophobia in surfing experiences into the public eye than much of the sport-related scholarship (see lisahunter, 2018).

Multi-perspective multimedia (whether one subject's experience from different temporal, relational and spatial views or multiple subjects viewing the same 'event') reflects one's position within the research by revealing and examining one's onto-epistemologies and positioning into the context as well as how one is positioned in such contexts. In a simple example, cameras may pick up information the subject may otherwise be unaware of (from my fingers in the water to sharks swirling below the surfer's feet or bumping one's craft, as experienced by Miles). While any communication of experience is already filtered by the subject positions, perceptions and perspectives, multi-perspective multimedia can make these 'unaware events' available for further noticing and analysis.

Audio-visual narrative research reveals complex embodied and emplaced social phenomena within this field. However, there are still many questions about how we might begin to take more seriously the lived body, the phenomenological and subjective experiences of those people whose practices constitute this field.

Warnings, considerations, limitations

Filmmaking in academic research is a complex set of practices, a "creative process occurring within a complex web of social, technological, personal and economic relations" (Berkeley, 2018, p. 30) that has the potential to create new knowledge, but like all methodologies there are also limitations and challenges. Here I will highlight what is perhaps the most significant contemporary limitation and challenge of doing multimedia research in the outdoors, that of time (see Miles, 2017, for more comprehensive discussion on documentary and Sparkes, 2017b, for a discussion of the sensory). The very real imperative of 'time' to do multimedia research is captured in FitzSimons' work (2015); most neoliberal universities have not yet caught up to 'counting' creative works such as documentary. Time to work through any research process is significant for us all let alone the additional challenges of learning filmmaking: creating visual and sound files and then editing them to create a significant research text that meets the very public (artistic) performance expectations as well as the academy's research expectations. Many resort to creating both a film text and a more easily 'legitimated' written academic text, as Miles notes:

> My relationship to the technical process and skillset of filmmaking, for example, is pivotal to what I can actually tell, given the editing expertise required. Such learning is a typical component of producing and directing a film, yet most often invisible to the audience
>
> *2017, p.121*

But these tensions can offer an upside of research in the richness of sensory engagement, in the new insights the process affords, and sometimes even in the sensory pleasures of re-experiencing 'being there', the sense I have had in my developing practices with surfing.

Other points to be mindful of have perhaps changed in nature over time and with new technologies, but not in essence. Here I am referring to the standard research ethics and logistics questions: the logistics of equipment (its cost, and invasion into the scene); participant anonymity/de-identification; ethics of participant inclusion in the research process; representation; the framing of research questions, and, consideration of what is captured or left out of the image/sounds. These onto-epistemological and methodological topics have had much scholarly attention and will be an important part of your own exploration should you decide to employ sensory and/or multimedia methodologies.

Conclusion

With the explosion of digital platforms and digital technology, new forms of multimedia genres are emerging to engage with, for and as research, particularly in collaborative, activist and ecological work (see for example Aston, Gaudenzi & Rose, 2017; McClernon, 2016). Giving attention to the multisensorial and translating that to multimodal possibilities for/as research heralds exciting times for researchers attempting to move beyond ocularcentric research and written-text-based representations (the irony of writing about it here is not lost to me!). Whatever sensory texts or 'representations' are created at any of the research moments, they are each assemblages, co-produced between participant, technology, spacetime and the researcher. Clearly multisensorial and multimedia methodologies have an important role to play in outdoor studies, whether for the experiential learning and experiential education that outdoor studies purport to facilitate (see Quay & Seaman, 2016), the continuation of adventure education (Sibthorp & Richmond, 2016), or even human (re)engagement with ecologically sustainable practices through ecotourism or Forest Schools. Multimedia is an important medium for public environmental awareness and action – with quick communication on multiple levels: there are a large volume of messages in three minutes of audio-visual. This works well for those considering message systems for pedagogy and environmental action – both employed through outdoor education and outdoor studies (e.g. Sandell & Öhman, 2010). The sophistication and accessibility of multisensory, multimedia, and multimodal technologies make possible multiliterate research projects, puzzles and audiences, and are providing us with new ontologies, new epistemologies, new methodologies and new ways of being in the world.

Acknowledgements

Thanks to the generous reviewing of elke emerald, Barbara Humberstone and Heather Prince.

Notes

1 www.youtube.com/watch?v=yblB87dpJGc
2 Victorian Aboriginal Child Care Agency, "Georgia's Story: SeaCountry". www.vacca.org
3 www.beaumiles.com/films/bassbykayak/
4 www.beaumiles.com/2016/04/home-truths-bass-strait-in-a-salad-bowl

References

Ahmed, J. (2010). Documentary research method: New dimensions. *Indus Journal of Management & Social Sciences, 4*(1), 1–14.

Allen-Collinson, J. & Hockey, J. (2010). Feeling the way: Notes toward a haptic phenomenology of distance running and scuba diving. *International Review for the Sociology of Sport, 46*(3), 330–345.

Angel, S. (Director), Liang, Y. (Producer), Holliday, J. (Cinematographer), lisahunter (Sound) & lisahunter (Editor). (2017). *Reaching Higher Ground* (Documentary). Florida: Actuality Media.

Aston, J., Gaudenzi, S. & Rose, M. (Eds.). (2017). *I-Docs: The evolving practices of interactive documentary*. New York, NY: Columbia University Press.

Berkeley, L. (2018). Lights, camera, research: The specificity of research in screen production. In C. Batty & K. Susan (Eds.), *Screen production research* (pp. 29–46). London: Palgrave Macmillan.

Clandinin, D.J. & Connelly, F.M. (2000). *Narrative inquiry: Experience and story in qualitative research*. San Francisco, CA: Jossey-Bass Inc.

Coates, E., Hockley, A., Humberstone, B. & Stan, I. (2016). Shifting perspectives on research in the outdoors. In B. Humberstone, H. Prince & K.A. Henderson (Eds.) *International handbook of outdoor studies* (pp. 69–77). Abingdon: Routledge.

Everett, M.C. (2017). Fostering first-year students' engagement and well-being through visual narratives. *Studies in Higher Education, 42*(4), 623–635. doi:10.1080/03075079.2015.1064387

FitzSimons, T. (2015). "I've got to STOP writing this (adjective of choice) article and get onto my filmmaking": Documentary filmmaking as university research – some history and case studies. *Studies in Australasian Cinema, 9*(2), 122–139.

Humberstone, B. (2011). Engagements with nature: Ageing and windsurfing. In B. Watson & J. Harpin (Eds.), *Identities, cultures and voices in leisure and sport* (pp. 159–169). Eastbourne: Leisure Studies Association.

lisahunter. (2013). What did I do-see-learn at the beach? Surfing festival as a cultural pedagogical sight/site. In L. Azzarito & D. Kirk (Eds.), *Physical culture, pedagogies and visual methods* (pp. 144–161). New York, NY: Routledge.

lisahunter. (2018). Sensory autoethnography: surfing approaches for understanding and communicating "seaspacetimes". In M. Brown & K. Peters (Eds.), *Living with the sea* (pp. 100–113). Abingdon: Routledge.

lisahunter & emerald, e. (2016). Sensory narratives: Capturing embodiment in narratives of movement, sport, leisure and health. *Sport, Education and Society, 21*(1), 28–46.

lisahunter & emerald, e., (2017). Sensual, sensory and sensational narratives. In R. Dwyer, I. Davis & e. emerald (Eds.) *Narrative research in practice: Stories from the field* (pp. 141–157). Singapore: Springer.

lisahunter, Wubbels, T., Clandinin, D.J. & Hamilton, M.L. (2014). Moving beyond text: Editorial for special issue. *Teaching and Teacher Education, 37*(1), 162–164.

Mason, J. & Davies, K. (2009). Coming to our senses? A critical approach to sensory methodology. *Qualitative Research, 9*(5), 587–603.

McClernon, T. (2016). Animal outlaws: Capitalism, containment, and documentary activism. *Jump Cut: A Review of Contemporary Media, 57*(Fall), 1–9.

McNutt, J. (2013). Art therapy as a form of visual narrative in oncology care. In C. Malchiodi (Ed.) *Art therapy in health care* (pp. 127–135). New York, NY: Guilford Publications.

Miles, B. (2017). *The secret life of the sea kayaker – An autoethnographic inquiry into sea kayak expeditioning* (Unpublished doctoral dissertation). Monash University, Melbourne.

Ormrod, J. (2009). Representing "authentic" surfer identities in "pure" surf films. In B. Wheaton & J. Ormrod (Eds.) *On the edge: Leisure, consumption and the representation of adventure sports* (pp. 17–42). Eastbourne: Leisure Studies Association.

Pink, S. (2006). *The future of visual anthropology: Engaging the senses.* Abingdon: Routledge.

Pink, S. (2013). Engaging the senses in ethnographic practice. *The Senses and Society, 8*(3), 261–267.

Pink, S. (2015). *Doing sensory ethnography* (2nd ed.). London: SAGE.

Quay, J. & Seaman, J. (2016). Outdoor studies and a sound philosophy of experience. In B. Humberstone, H. Prince & K. Henderson (Eds.) *Routledge international handbook of outdoor studies* (pp. 82–92). New York, NY: Routledge.

Sandell, K. & Öhman, J. (2010). Educational potentials of encounters with nature: Reflections from a Swedish outdoor perspective. *Environmental Education Research, 16*(1), 113–32.

Sibthorp, J., & Richmond, D. (2016). Adventure education: Crucible, catalyst and inexact. In B. Humberstone, H. Prince & K. Henderson (Eds.) *Routledge international handbook of outdoor studies* (pp. 291–302). New York, NY: Routledge.

Sparkes, A. (Ed.) (2017a). *Seeking the senses in physical culture: Sensuous scholarship in action.* Abingdon: Routledge.

Sparkes, A. (2017b). Researching the senses in physical culture and producing sensuous scholarship: Methodological challenges and possibilities. In A. Sparkes (Ed.) *Seeking the senses in physical culture: Sensuous scholarship in action* (pp. 174–197). Abingdon: Routledge.

Spinney, J. (2011). A chance to catch a breath: Using mobile video ethnography in cycling research. *Mobilities, 6*(2), 161–182.

Stockall, N. (2001). Video elicitation of the semiotic self. *International Journal of Applied Semiotics, 2*(1–2), 29–37.

Sweet, E. & Escalante, S. (2015). Bringing bodies into planning: Visceral methods, fear and gender violence. *Urban Studies, 52*(10), 1826–1845.

Thomson, I. (Director) & Castets, T. (Producer). (2014). *Out in the line-up* (Documentary). Yellow Dot Productions.

Wood, M. & Brown, S. (2010). Lines of flight: Everyday resistance along England's backbone. *Organization, 18*(4), 517–539.

21

REPRESENTING EXPERIENCE

Creative methods and emergent analysis

Marcus Morse and Philippa Morse

> *Standing on the banks of the Shoalhaven River, alongside year 9 students, I was painting for the first time in quite a few years; "paint what you can see, not what you think you can see". These guiding words from the art educator provided a moment of pause – to look anew at the riverscape and see the vibrant colours and textures of the riparian vegetation and feel the relentless movement of the river.*
>
> *Author*

Introduction

In considering outdoor experiences it can be tempting to focus on what people are thinking about or how they might describe experiences – yet cognition and language can only ever describe a part of how we experience our surrounds. We come to know about the world, at least in part, through aesthetic and embodied engagement. Merleau-Ponty (1962, 1968), for example, highlights the way in which our bodily engagement orientates us within the world:

> Merleau-Ponty's approach is defined by his conviction that nature has its own meaningful configuration to which we are oriented at a level more originary than thought, at the level of our bodily engagement with the perceived
>
> *Toadvine, 2009, p. 131*

What types of experience, then, might be conducive to embodied engagement? And what might be the educational value of such experiences? In an effort to open up broader notions of what an experience might be, we suggest exploring creative methods that de-privilege language and cognition and attempt to re-present aesthetic and embodied components of experience. Such approaches can include artistic, imaginative and creative representations of lived experience.

In highlighting possibilities for such re-presentations, we also consider an interrogative form of emergent analysis. We advocate for a careful reflective process that provides opportunities for what Merleau-Ponty (1968) suggests might allow the world to say, "what in its silence *it means to say*" (italics in original, p. 39). We undertake an analysis that privileges implicit relationships between and within representations of experience. To illustrate such an approach, we offer an example of a research study conducted during a five-day canoe journey on the Shoalhaven River, Australia, that involves 18 Year 9 (aged 14–15) school students representing their experiences through creative mediums.

A phenomenological starting point

This study follows from previous research by one of the authors that describes meaningful experiences of adult participants on a river journey (Morse, 2014, 2015a, 2015b). This study, described herein, sets out to explore experiences of younger students in an outdoor context. Although we initially adopted a phenomenological approach, we did so not by adhering strictly to phenomenological methods; rather, we allowed a phenomenological understanding to guide our explorations. In beginning this project, we considered that traditionally approached interviews might not be conducive to providing the fullest opportunities for younger participants to 'return to' or describe lived experiences. We therefore decided on an outdoor journey context that included artistic possibilities for representing experience – with the primary purpose of using creative artefacts as a way of getting into descriptive interviews with students. However, although this did occur. the findings revealed further surprising possibilities for creative methods performing as critical points of departure in themselves.

Art and representing experience

Phenomenology provides a vantage point to view the world and an alternative perspective upon the nature of knowledge (van Manen, 1997). It is not a search for our reflective cognition of the world, but a search for our original perception of the world. As Hay (2003) suggests, "the point of phenomenology, after all, is to suspend theory so that Being can be met, unmediated" (p. 247). In other words, phenomenology seeks to describe the immediacy and essence of original pre-reflective lived experience. Phenomenology offers methodological elements for attempting such a task, including the 'reduction'. In phenomenology, reduction does not mean to make smaller; rather it is derived from *re-ducere*, meaning to 'lead back to' (see van Manen, 2014, pp. 215–235). Phenomenological reduction, then, means attempting to return to, or connect with, lived meaningful experience.

How, for example, might we attempt a return to a lived moment without cognitively reflecting upon it? How might we gain access to such moments when we primarily use words to describe what, for many people, are complex, mysterious and/or ineffable experiences that are recalled post-reflection? Indeed, participant descriptions of experiences must not only be incomplete, but are mediated by personal and social/cultural preconceptions, despite deliberate phenomenological bracketing (or epoche; a process of suspending judgement and preconceptions (Moustakas 1994). We are, therefore, required to reflect not only on what is said, but also that which is not said, but may be made discernible through careful interrogative reflection. As Merleau-Ponty (1968) suggests of this reflective process:

> It must question the world, it must enter into the forest of references that our interrogation arouses in it, it must make it say, finally, what in its silence *it means to say* ... We know neither what exactly is this order and this concordance of the world to which we must entrust ourselves, nor therefore what the enterprise will result in, nor even if it is really possible
>
> *p. 39*

Although the potential impossibility of this task on the surface may appear a hindrance, it also provides the opportunity to search at a deeper than surface level, to not be satisfied wholly with the words or categories used, and to interrogate individual experiences in light of creative responses and collective recollections. In other words, to allow participant descriptions and representations to say more than might originally appear to be said via a process of careful interrogative reflection.

Liberman (2000, p. 39) reflects on Merleau-Ponty's position above: "Merleau-Ponty's doubt here is humbling, but our task here is such that we can put that humility to good use". We argue that such interrogative reflections can be strengthened by creative representations of lived experience. As Noë (2000) argues: [art] "can teach us about perceptual consciousness by furnishing us with the opportunity to have a special kind of reflective experience. In this way, art can be a tool for phenomenological investigation" (p. 124). In representing experience through creative responses, we highlight the possibilities in the act of 're'-presenting. While it is not possible to fully return to the lived experience – art can offer possibilities for the intensifications of experience:

> The everyday notion of representation could mean 'to depict,' or 'to present again' (re-present), but Jean-Luc Nancy asserts that the "re- of the word representation is not repetitive but intensive ... mental or intellectual re presentation is not foremost a copy of the thing," but an intensified presentation. It is "a presence that is presented" (Nancy 2007, p. 36). The re- in represent is, in other words, an amplification; to represent is to present more of what is.
>
> *Stern, 2018, pp. 18–19*

The study

Rising in the southern tablelands of New South Wales, Australia, the Shoalhaven River has been at the heart of life in the region for thousands of years, flowing as it does through lands for which the Wodi Wodi, Wandi Wandian and Yuin nations are traditional custodians. The choice of river context was based on insights previously gained (Morse, 2014, 2015a) and the way in which river journeys provide unique and rich contexts for exploring participant perceptions of surrounding environments (Djohari, Brown & Stolk, 2018; Stewart, 2004; Wattchow, 2007). The five-day journey, involving 18 students from a Sydney-based Rudolf Steiner School, Glenaeon, was a canoeing experience, with an arts focus, that included time spent at the Bundanon Art Centre on the Shoalhaven River. Bundanon is the previous home of iconic Australian artist and painter Arthur Boyd (1920–1999). Boyd was a leading Australian painter of the late 20th century who, along with his wife Yvonne, donated the Bundanon properties to the Australian Government in 1993. Boyd's paintings of the Shoalhaven river display an incredibly inspiring and long-held connection with place (see McGrath and Boyd, 1982).

The first day of the canoe journey saw the students familiarise themselves with the place and craft by spending time at the put-in. Unusually for a river journey, the next day was spent paddling *upstream* into a wilder section of the river. Upon arrival at the higher camp students participated in a two-hour solo session that included possibilities for creating sketches, poems and/or paintings focused on the place. The following morning students travelled downstream, past the original put-in and then, after travelling through a variety of changing landscapes, arrived at Bundanon on the fourth day. The remainder of the fourth and fifth days were spent at Bundanon, engaged in creating art-based responses to the riverscape and journey experienced (whilst still being based next to the river). In particular, students were involved in producing sketches, printmaking (often based on creative work produced from earlier sketches) and painting. The Glenaeon students had been engaged in a school-based art programme focused on Australian landscape artists (including Boyd) prior to coming on the river.

Creative artefacts as method

During this research study we collected data via a number of methods including researcher observations, creative artefacts, journals and interviews. Although the creative artefacts were originally designed as a way into conversations with students, they proved to be a crucial element of the research design in themselves; revealing possibilities for meaningful experience that might otherwise have gone unnoticed. We do not suggest that the methods described provide direct access to pre-reflective lived experience; rather, that by including carefully considered creative methods, aesthetic and embodied ways of knowing might be brought to the fore. For all forms of data, including creative artefacts we used emergent thematic analysis (Patton, 2014).

Whilst on the journey students were encouraged to represent their experience of the river journey via three key creative forms. First, students sketched, wrote poems and painted at a variety of times within a sketchbook (planned and unplanned) as they travelled down the river. Second, students were engaged in creating *collagraphs,* a form of printmaking in which materials are glued to a substrate (such as cardboard), inked, and then run through a printing press to produce an image on a secondary sheet of paper onsite at Bundanon. The collagraphs produced often involved work derived from previous sketches. And third, students were engaged in painting, and in particular *en plein air* painting (an outdoor form of painting looking and sensing the subject directly).

In providing opportunities for themes to emerge from the creative artefacts, we took care as far as we could to bracket our own assumptions with regard to student representations and meaning. We undertook emergent thematic analysis across a range of data including the creative artefacts to describe the phenomenon under inspection. In many cases, however, it was the artefacts themselves from which the key-themes emerged, with the place itself appearing to be perceived, in some ways, on its own terms. Below we describe three of these emergent themes.

1) Texture and complexity

One of the most striking themes available from the art pieces was a focus on texture and complexity (see Figure 21.1). With glue and string in hand, the students created relief prints, that once dry were pushed through the large ink press. These prints revealed a particular focus on the textural elements of the surrounding environment that was consistently revealed. For example, prints commonly involved an attention paid to combinations of textured elements within the surrounds such as smooth river rocks, wind-formed ripples on the surface of the water, currents in the water, shiny grasses and rough overhanging branches.

FIGURE 21.1 Collagraph – river textures

We further explored these textured elements within interviews and subsequent written reflections, revealing a focus of attention on the perceived beauty of the place. Yet the way in which students described this was often via texture and complexity (commonly perceived as hidden); the interweaving of the elements suggested something more than could be described through order, categorisation or language. It appeared to be the texture and complexity which attracted and held attention, with many of the collagraphs based on the sketches or recordings made during the solo or quieter times.

> With the collagraph you are able to add kind of texture and meaning to it. Like different meshes and cardboard and tissue paper. I felt like that described the experience, like when we camped we were on dirt we were on sand we were on rocks.
>
> *Alison*

> I loved seeing how each time you put your paddle in the water, to do a stroke, it created a miniature whirlpool, that rippled on even after your paddle had left the water.
>
> *Larni*

Such a focus of attention on textured material objects (like smooth river stones) perceptually acknowledges the stones themselves as playing a role in attracting attention. As Rautio (2013) asserts, "tiny black inanimate pebbles can invite us into interaction by virtue of existing, guide the nature of this interaction by virtue of their physical form, support our activity through lending themselves to be investigated and engaged with" (p. 402). In this way, attending to material surrounds might provide opportunities for acknowledging a quality of relationship with(in) the place.

> Texture is patterned, full of contrast and movement, gradients and transitions. It is complex and differentiated … the organic and the inorganic, color, sound, smell, and rhythm, perception and emotion, intensely inter-weave into the 'aroundness' of a textured world, alive with difference.
>
> *Manning, 2016, p. 4*

The nuance of sensuous texture was also revealed through the medium of painting – with students observed applying thickened paint with knives or sticks. We were intrigued by this vigorous and abundant layering of vibrant colour in this way. When asked about this texturing many students described attempting to (re)present the *sensed* feeling of being on the river; for example, a feeling of the wind against one's face, the slap of the water on the canoe, the sensed resistance to paddle strokes and/or the sounds of waves. In this way the texture was a theme which was commonly linked to sensuous experience, with some students describing being transported back to feelings and/or emotions as they painted (see Figure 21.2).

FIGURE 21.2 Painting – a textured riverscape

2) Time and attentiveness

The art works highlighted a quality of attention that was often related to time. For many students the focus of their work was derived from the solo-time or perceived quieter moments on the river. In such moments, students' attention was seemingly easily garnered and held, with the experience being able to run its course, at its own pace – adding a sense of fullness or wholeness to the experience (see Dewey, 1934). By reflecting on this theme emerging from the creative pieces and subsequent stories that students told about their work it became clear that a good deal of meaning was derived from such moments (see Figure 21.3). Djohari et al. (2018) describe the way in which meaning might flow for some participants in moments of silence with others on a riverbank; "their 'mattering' comes from silent companionship of co-weaving the bankside world into being together" (p. 365). Whilst students did not need to be alone as such, the time/ing often involved a lack of social and/or temporal intrusion.

> I think it was good to be separate from everything else to be able to focus. I think just being alone in such an amazing place. You really get to take it in. If there is heaps of noise and distraction around you, then you don't get to focus.
>
> *Amy*

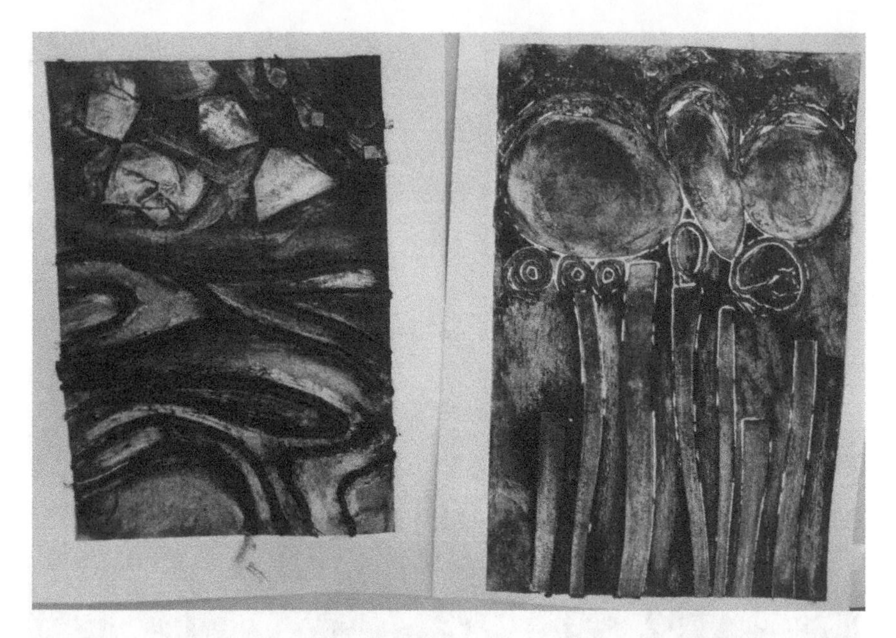

FIGURE 21.3 Collagraphs – banks of the river

There was a sense within the pieces, and subsequently expressed by students, that time and the lack of distractions provided opportunities to attend in a way that enabled, for example, to draw what felt right to draw. Such interactions involved a sense of comfort in being able to attend and possible freedom from judgement in terms of what they paid attention to and/or how their focus of attention was recorded. There was, then, an expression of freedom through art to attend to the place in a way they felt called to; to a seemingly reciprocal relationship (see Figure 21.4).

It is worth noting that the solo-time on the river had some embodied peculiarities that appeared to impact students profoundly. When students chose a spot for their solo-time, all but one of the students sat on/near the river bank. This position of comfort, with one's back against the forest and feet slightly lower than the back, looking across the river and relatively free from social distractions, was perceived by students to be a welcomed chance to attend to the place. In this way many creative responses highlighted a careful, and valued, attention to surrounding environs. Sensorial peculiarities of travelling within a river environment, as well, include the way in which everything attended to (visually, auditorially and physically) resides within that valley. And to travel by canoe can add a sense of comfort in having someone else in the craft – yet with the lack of face-to-face interaction, also a sense of solitude.

3) *Image-ing one-self with(in)*

It appeared for many students that their experience of the river and artwork were in some ways interwoven or melded together – that by placing themselves *within* their

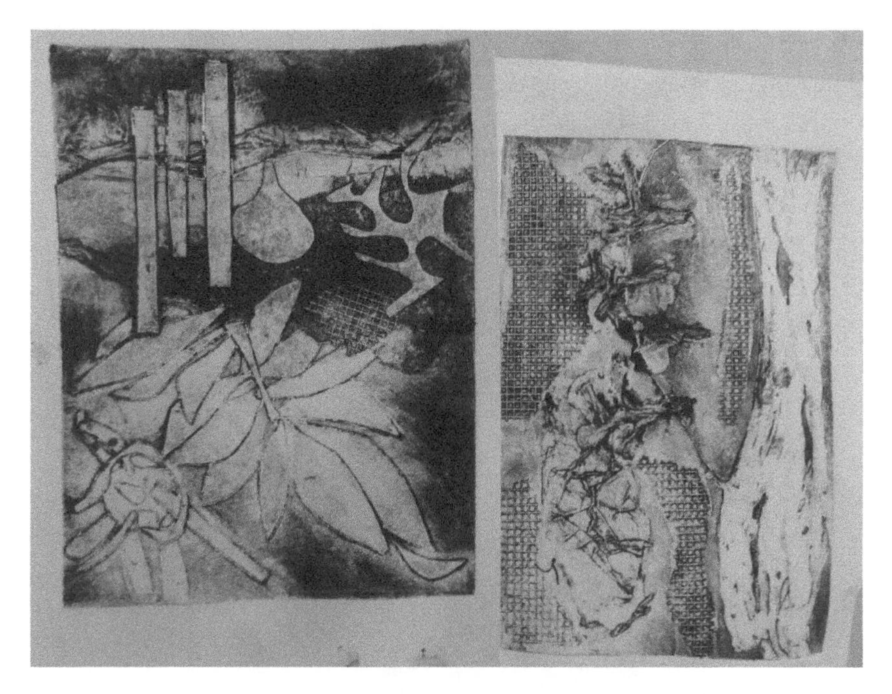

FIGURE 21.4 Collagraphs – attending to detail

work they were also within their experience. In this way there was a sense that the student's experience of the river was the art, and vice versa; art was, then, an expression of lived experience. In discussing experiences many students acknowledged it was difficult to describe what the experience had meant for them – yet their art provided an opportunity for expressing or representing experience without perceived judgement from others. Art, experience and dialogue, then, melded together to form an intimate interaction (see Figure 21.5).

While students described the physical features of the journey as providing inspiration for their work, such as the texture and complexity of the rock pools and the droplets of water falling from the trees – more surprising were the numerous comments around the way in which they felt their artwork conveyed feelings of *movement* down the river. Students described painting or expressing the river by how it felt for them to be on the river – attempting to draw the movement and power of the water, the way the currents moved the canoe and/or how the wind felt moving across one's face. There was also a sense of physically difficult work and emotional response; of being immersed in the river and the surrounding environment.

> *Although we paint, kind of what we see, like we paint a tree, when you are canoeing in where you are, what you're painting, you're in your painting … when you are in the river that you're going to paint, you get a better understanding of what it really is. When you are painting the river, you can really feel what it's like to be in the river.*
>
> *Robert*

FIGURE 21.5 Painting – an expression of experience

> *As you are going down the river you see so much more of the landscape than you usually would, it's just a different feeling … it had a meaning to it, to paint. Rather than just a river, it was a river we had been down, and paddled, and experienced.*
>
> *Joel*

> *We got to interact with the water, you get to understand how it works. When you are out on the water you're battling, then you know the hard stuff so then you go and take that with you when doing your art. It was nice to able to work with it on paper.*
>
> *Amy*

This melding of the river and art experience appeared to work in both directions, so that the knowledge that they might express their experience via art seemingly enhanced their experience of the place. Many students reflected that felt like they had to paint the river differently after being on the river; that the river experience changed their representations and how they went about them. Most notable perhaps were the moments when students sketched, wrote or recorded in some way their impressions of the place. This revealed the way in which creative responses reflected aesthetic elements born from lived experience. Dewey (1934), in considering aesthetics, reminds us that the aesthetic of art begins with lived experience: "in order to understand the aesthetic in its ultimate and approved forms, one must begin with it in the raw; in the events and scenes that hold the attentive eye and ear of man, arousing his interest and affording him enjoyment as he looks and listens" (pp. 4–5). Art, then, might be considered to be preceded by lived experience and the imagination of the artist in its process of creation. Through this research project it became clear to us that elements of the raw eventing of the environs was interwoven within participant sketches, prints and paintings.

Potential value of such methods

What might be the value of attending to aesthetic components of experience via creative methods? What might be the value of describing, highlighting and/or acknowledging perceptual or embodied components of experience? Quay (2013) asserts the crucial importance of pre-reflective experience in providing opportunities for building *ways of being* in the world, by highlighting that any reflective experience must always have foundational components of aesthetic experience. In other words, when we reflectively position ourselves in relation to others (including human, more-than-human and material) we are, at least to some degree, reliant on aesthetic components of experience. Considered in this way, both facilitating opportunities for aesthetic experience and representing ways of coming to know the world are critical.

Quay (2013) argues that more emphasis should be placed on the aesthetic lived experience of students because to do so is to provide opportunities for students to gain a sense of being within the world. This is not to suggest that embodied preconceptual components of experience are separate from cognitive thoughts or intellectual processes – they are inextricably intertwined and necessarily linked (O'Laughlin, 2006) – yet, at times, it is easy to take for granted the former and prioritise the latter. A focus on pre-reflective components of experience, then, is important for expanding our sense of not only what experience might be, but also for considering the qualities of our relationships with(in) the world. Kohak (1992) asserts the importance of pre-reflectively perceiving worth within the world, rather than simply post-reflectively conceiving of worth – because ultimately the latter is not possible without the former being implicated: "I am persuaded that the ability to formulate an adequate and efficacious *conception* of value is contingent on a prior, prereflective *perception* of value" (p. 173). Attempting to return to perceptual lived experience is never entirely possible, yet through creative methods additional insights may be revealed. By focusing on participant experiences in the Shoalhaven River landscape we have described in this chapter how affectual places are continuously (re)made present through people's active engagement with(in) the world and have highlighted possibilities for creative methods, for both research and practice.

References

Dewey, J. (1934). *Art as experience*. London: George Allen and Unwin.

Djohari, N., Brown, A. & Stolk, P. (2018) The comfort of the river: Understanding the affective geographies of angling waterscapes in young people's coping practices. *Children's Geographies*, 16(4), 356–367.

Hay, P.R. (2003). Writing place: Unpacking an exhibition catalogue essay. In J. Cameron (Ed.) *Changing places: Reimagining Australia* (pp. 272–285). Double Bay, NSW: Longueville Books.

Liberman, K. (2000). An inquiry into the intercorporeal relations between humans and the earth. In S. Cataldi and W. Hamrick (Eds.) *Merleau-Ponty and environmental philosophy: Dwelling on the landscapes of thought* (pp. 37–50). Albany, NY: State University of New York Press.

Kohak, E.V. (1992). Perceiving the good. In M. Oelschlaeger (Ed.) *The wilderness condition* (pp. 173–187). San Francisco, CA: Sierra Club Books.

McGrath, S. & Boyd, A. (1982). *The artist and the river: Arthur Boyd and the Shoalhaven*. Sydney, NSW: Bay Books.

Manning, E. (2016). *The minor gesture*. Durham, NC: Duke University Press.

Merleau-Ponty, M. (1962). *Phenomenology of perception*. London: Routledge & Kegan Paul.

Merleau-Ponty, M. (1968). *The visible and the invisible: Followed by working notes*. Evanston, IL: Northwestern University Press.

Morse, M. (2014). A quality of interrelating: Describing a form of meaningful experience on a wilderness river journey. *Journal of Adventure Education and Outdoor Learning, 14*(1), 42–55.

Morse, M. (2015a). Being alive to the present: Perceiving meaning on a wilderness river journey. *Journal of Adventure Education and Outdoor Learning, 15*(2), 168–180.

Morse, M. (2015b). Paying attention to experience within nature. In M. Robertson, R. Lawrence and G. Heath (Eds.) *Experiencing the outdoors: Enhancing strategies for wellbeing* (pp. 113–122). Rotterdam: Sense Publishers.

Moustakas, C.E. (1994). *Phenomenological research methods*. Thousand Oaks, CA: SAGE Publications.

Nancy, J. (2007). *The ground of the image*. New York, NY: Fordham University Press.

Noë, A. (2000). Experience and experiment in art. *Journal of Consciousness Studies*, 7(8–9), 123–136.

O'Laughlin, M. (2006). *Embodiment and education: Exploring creatural existence*. Dordrecht: Springer.

Patton, M.Q. (2014). *Qualitative research and evaluation methods* (4th ed.). Thousand Oaks, CA: SAGE.

Quay, J. (2013). More than relations between self, others and nature: Outdoor education and aesthetic experience. *Journal of Adventure Education and Outdoor Learning, 13*(2), 142–157.

Rautio, P. (2013). Children who carry stones in their pockets: On autotelic material practices in everyday life. *Children Geographies, 11*(4), 394–408. doi:10.1080/14733285.2013.812278

Stewart, A. (2004). Decolonising encounters with the Murray River: Building place responsive outdoor education. *Australian Journal of Outdoor Education, 8*(2), 46–55.

Stern, N. (2018) *Ecological aesthetic: Artful tactics for humans, nature and politics*. Hanover, NH: Dartmouth College Press.

Toadvine, T. (2009). *Merleau-Ponty's philosophy of nature*. Evanston, IL: North Western University Press

van Manen, M. (1997). *Researching lived experience: Human science for an action sensitive pedagogy* (2nd ed.). London: Althouse Press.
van Manen, M. (2014). *Phenomenology of practice: Meaning-giving methods in phenomenological research and writing*. New York, NY: Left Coast Press.
Wattchow, B. (2007). Playing with an unstoppable force: Paddling, river-places and outdoor education. *Australian Journal of Outdoor Education, 11*(1), 10–21.

PART IV

Quantitative and mixed methods

22

DERIVING METRICS AND MEASURES IN OUTDOOR RESEARCH

Roger Scrutton

Introduction

There is a widespread view that spending time outdoors for the purposes of recreation or learning has a positive effect on people's personal and social development, health and wellbeing, academic achievement and related areas (e.g. Malone & Waite, 2016). Nevertheless, rightly or wrongly, stakeholders in outdoor activities – educators, public sector and commercial providers, grant awarding bodies, government departments – like to have evidence of the effectiveness of these activities in numbers as well as from narrative and anecdote. Metrics and measures of effectiveness are therefore important. Outdoor research to derive metrics and measures is aimed at finding out what learning experiences work and by how much, and has predominantly come from the education sector through the measurement of the effectiveness of an intervention, such as a residential week or an expedition. Some major reports from national agencies provide quantitative evidence, however, such as the Getting Active Outdoors review of recreational use of the outdoors in England (Sport England, 2015) and, following on from that, the Reconomics exercises (Sport and Recreation Alliance, 2017) to evaluate the cash value of outdoor recreation to public and commercial purses.

The terms 'metric' and 'measure' are often used interchangeably across a range of disciplines. 'Metric' as a noun (as opposed to an adjective) has become more widely used in recent years to refer to a measure or group of measures of performance against a target, with or without an intervention or experiment. In the Getting Active Outdoors review, recreational use was characterised by several variables – age, wealth, ethnicity, relationship to the outdoors, etc. Studies of a cross-sectional nature, such as this, might be said to generate metrics. A more extensive statistical analysis of metric data has been published by Hinkley, Brown, Carson and Teychenne (2018) to show that screen time may be adversely, and outdoor play favourably, associated with preschool children's social skills. However, in outdoor research using quantitative methods, we are more familiar with measures obtained against a scale associated with a questionnaire to measure the impact of an intervention. For the purposes of this chapter I will not pursue a rigorous distinction between metrics and measures; I will be focusing on experimental research that uses an outdoor learning intervention in anticipation of detecting a measurable benefit for the participants. There will not be a detailed description of quantitative methods or statistical analysis of data, for which there are a number of excellent texts, e.g. Cohen, Manion and Morrison (2018), Field (2013), rather an outline of methodologies and their pros and cons.

Theoretical framework for outdoor research

The outdoor, or natural, environment is seen to offer learning experiences for young people and others that are deeply experiential, more memorable and, therefore, more effective than learning in other environments. Thus, the rationale for conducting quantitative studies in outdoor research is to discover what outdoor learning experiences work best and by how much and through this justify and optimise programmes of learning across the sector for the benefit of both participants and providers.

This is the philosophical position from which the specific research methodologies discussed in this chapter derive. The methodologies are designed to be objective, and to challenge and test the concept that outdoor learning is particularly effective. However, in doing this we discover that conducting quantitative research in the natural environment presents practical difficulties, which might be hampering us in deriving a true measure of effectiveness.

Reviews and critiques of measures in outdoor research

Primary studies to derive metrics and measures of the impact of outdoor education markedly increased in number about 50 years ago, and many of these focused on the impact of interventions provided by the Outward Bound Trust (Hahn, 1961). Experimental studies of sufficiently good quality to be included in more recent meta-analyses date from about this time. However, the measurement of impact by quantitative methods has not always had a 'good press'. The outdoor research community recognises that it is difficult to acquire high-quality metrics and measures that might influence stakeholders. In their 1997 meta-analysis of primary studies, Hattie, Marsh, Neill and Richards (1997) said, "Where there was some attempt at evaluation beyond the anecdotal, the analyses were rarely more than correlational" (p.45). Nevertheless, it was possible for them to calculate a mean Cohen's effect size[1] of benefit for a range of personal skills of about +0.3 to +0.4 ('small' to 'medium'). At more or less the same time, a review by Barrett and Greenaway (1995) heavily criticised the quality of evaluations of the effectiveness of outdoor adventure to develop personal and social skills as "isolated", "inconclusive", "over-ambitious", "uncritical", "not of a high standard" and "difficult to locate". Cason and Gillis (1994), in a meta-analysis of measurements of the effectiveness of adventure programmes, specifically commented on a lack of equivalent control groups and lack of randomised participant assignment. In 2008, the quantitative approach to measuring impact from outdoor programmes for personal effectiveness was again reviewed (Neill, 2008). Despite finding that as many as 35% of primary studies yielded zero or negative effect sizes, and bemoaning the fact that quantitative research using an experimental approach was "limited by a lack of appropriate dependent measures, low statistical power, over-reliance on inferential statistics, a lack of control and comparison groups, a lack of longitudinal data, and a lack of investigation of independent variables" (p.xxiv), the review confirmed that a small to medium positive effect size was typical.

Figure 22.1 shows the distribution of mean effect sizes from meta-analyses of outdoor research as reviewed by Neill (2008) compared to mean effect sizes in cognate fields: their range from about +0.1 to about +0.5 coincides with the peaks in distribution in related research areas. Note, however, how few meta-analyses there are in outdoor research, reflecting a relatively new and small research discipline.

This distribution appears reassuring, but it also demonstrates that as it is evaluated quantitatively outdoor education is only as effective as other educational interventions. Laidlaw (2000) conducted meta-analytic research to understand the importance of study design on effect size from outdoor programmes, concluding that some 17% of the variance in the data can be accounted for by study design factors and the quality of the sample in particular. His results are summarised in

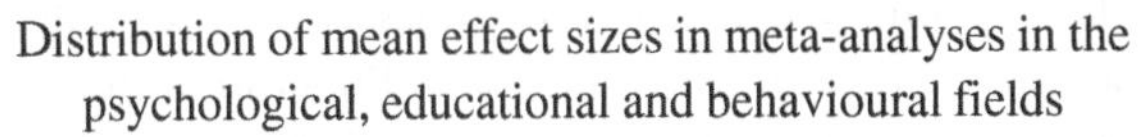

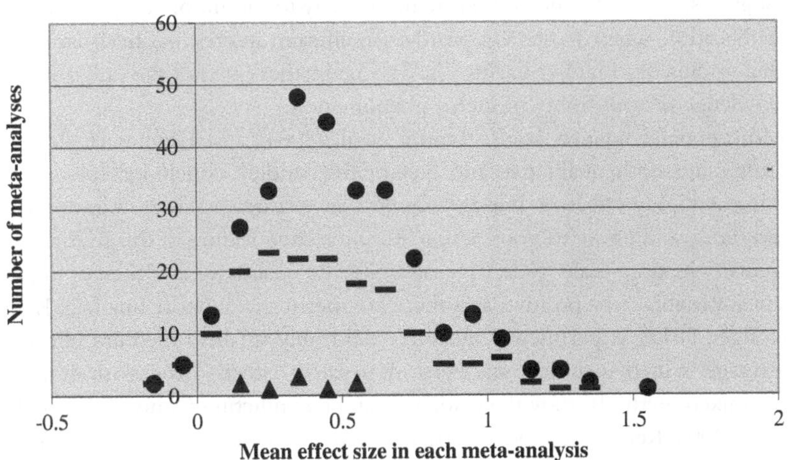

FIGURE 22.1 The distribution of Cohen mean effect sizes in meta-analyses in outdoor adventure education and cognate fields. Triangles – outdoor adventure education (Neill, 2008); dots – psychological, educational, and behavioural interventions (Lipsey & Wilson, 1993); bars – educational only

Source: Reproduced with permission from Scrutton and Beames, 2015

TABLE 22.1 Mean effect sizes for different study designs in a meta-analysis (adapted from Laidlaw, 2000)

Study design	*Mean effect size*	*Standard deviation*
Pre-test/post-test, no control	0.55	0.24
Quasi-experimental with a control group	0.45	0.20
Multiple-time series	0.30	0.16
Experimental (attempting to conduct the 'true experiment' – next section)	0.17	0.13

Table 22.1, in which the mean effect size for different methodological approaches ranges from +0.17 (most sound methodologies) to +0.55 (least sound methodologies). Other authors have also suggested that effect sizes tend to increase as study design deviates further from a truly experimental one. Nevertheless, Laidlaw found a mean effect for all the primary studies he used of nearly +0.5, comparable with other meta-analyses, the full range of effect sizes being -0.38 to +1.38. In longitudinal studies there is a similar spread of effect sizes for the retention of benefit (Hattie et al., 1997), with a tendency towards lower values over time (Fiennes, Oliver, Dickson, Escobar, Romans & Oliver, 2015). Here the attrition of sample size is an additional challenge for researchers.

Another area of concern in relation to the distribution of effect sizes in outdoor research is the phenomenon of publication bias. Bowen & Neill (2013), in a meta-analysis of adventure therapy outcomes, said "Publication bias may occur because investigators, reviewers, and editors are more likely to submit or accept manuscripts for publication when results are positive, significant, interesting, from large well-funded studies, or of higher quality" (p. 31). Again, these are not the only authors to acknowledge the possibility of such a phenomenon.

More general reviews by Rickinson et al. (2004) and Fiennes et al. (2015), encompassing both qualitative and quantitative studies, concluded that outdoor learning is highly effective, but also found that research is confusing, unfocused, spread thinly and prone to poor design. An interesting feature of this research field, particularly in large-scale studies, is that whilst the qualitative evidence of impact is almost invariably very positive, and there are metrics to support this (e.g. Kendall & Rodger, 2015), experimental measures yield only small to medium effect sizes on average, which include some zero or negative effects. Even within an individual mixed methods study the evidence can be conflicting (Amos & Reiss, 2012; Christie, 2004; Kendall & Rodger, 2015): whereas some outcomes assessed qualitatively through teachers and parents, such as pupils' improvement in attitudes towards school work, are substantially positive, the same outcome measured quantitatively is small or non-existent. This tension between qualitative and quantitative results highlights the need for researchers to be very careful with their methodologies, particularly in large-scale studies, where a sample of thousands for the quantitative work, while in theory providing high statistical power for the experiment, might be introducing a high level of scatter in data, i.e. high standard deviations, and an associated reduction in the measured effect size.

It is important to note that these critiques of research quality are based chiefly on studies of adventure education interventions to benefit personal and social development, where the dependent variable is often a latent variable[2] and extraneous variables are quite difficult to control. Interventions aimed at benefiting cognitive development and academic achievement, which generate directly measurable outcomes with better control of extraneous variables, often yield larger effect sizes, e.g. Nundy (1999), in which an effect of +1.6 was found for cognitive gain in a study of the impact of residential geography fieldwork on upper primary school pupils. More generally, interventions in which the outcome measure is closely aligned with the programme intent tend to yield larger effect sizes (Laidlaw, 2000).

Perhaps in response to these critiques of quantitative outdoor studies, it is now common practice for researchers to recognise the methodological limitations of their study. In the process of collecting metrics, Williams (2012) acknowledged constraints on his sampling strategy that might have compromised his ability to generalise results to a larger population. Sheard & Golby (2006) confess to using too small a sample of about 27 in their experimental and control groups, thus threatening statistical power in their study of the impact of an outdoor adventure education curriculum on aspects of positive psychological development. Gustafsson, Szczepanski, Nelson and Gustafsson (2012), while using adequate sample sizes in the experimental and control groups for their study of the effects of outdoor education interventions on the mental health of schoolchildren, bemoaned the fact that they could not randomly allocate participants to these groups. Are such challenges to quantitative research in outdoor studies inhibiting our ability to demonstrate the full value of outdoor education?

Aspiring to the 'true experiment', and alternatives

To measure the impact of an intervention we might aspire to the 'true experiment' (Cohen et al., 2018). This is in effect a randomised control trial (RCT), advocated for outdoor studies by Fiennes et al. (2015) and for education more generally by Goldacre (2013), in which participants are randomly allocated to an experimental or control group with similar demographic properties and representative of the population to which the results of the experiment will be generalised. They should also be large enough to provide sufficient statistical power. Ideally, each group should be ignorant of the fact that there is another group participating in the experiment. Using a questionnaire that is appropriate to the measurement of the dependent variable, tests are applied to both groups shortly before and shortly after the intervention (and later, in longitudinal studies) so that the test scores for the two groups can be compared and the impact of the intervention calculated. Studies in outdoor education experience difficulties in meeting these criteria in several ways. Cohen et al. (2018) recognise this when they point out that what happens in the artificial world of the laboratory (where RCTs are commonplace) is unlike what happens in the 'real world'. They describe numerous variables that are difficult to control in natural settings, from the difficulty of replicating interventions to participant-participant interactions and the complex outcomes that arise, in our case from the interaction of the multiple dimensions of an intervention (McKenzie, 2003). Williams (2012) suggested an analogy with complex systems in the natural and social sciences.

So, to be realistic, whilst the outdoor researcher might aspire to a true RCT, it is very unlikely they will be able to conduct one using an intervention in which the participants within the experimental and control groups interact with each other and with their respective environments as a matter of course (indeed, we would encourage them to interact to promote personal and social development, group learning etc.). Encouragingly, many researchers have found it possible to conduct an experiment that is very close to a RCT, one in which participants are assigned to experimental and control groups, both groups are tested with the same questionnaire at an appropriate time with respect to the intervention undertaken by the experimental group and best efforts are made to identify and control extraneous variables (Ewert & Sibthorp, 2009). What is missing from the true experiment might be the random assignment of participants to groups and the isolation and controlled environments in which the two groups and their members function. Additionally, in outdoor studies the dependent variable is often a latent variable. Researchers in outdoor studies frequently talk of having conducted such a quasi-experiment.

The study by Nundy (1999) referred to earlier was a quasi-experiment rather than a true experiment. Participants were not allocated randomly to experimental and control groups, and almost certainly he could not perfectly control some confounding variables. Quasi-experimental is the most common form of design found in the outdoor research literature – the so-called 'pre-test post-test non-equivalent groups design'. Other examples of the quasi-experimental approach are by Sheard and Golby (2006), and Ewert and Yoshino (2011). It is predominantly this design that the criticisms reviewed above are directed towards. However, prior knowledge and understanding of potential limitations should allow the researcher to address them at the project planning stage and reduce their impact on the validity of the experiment.

In addition to the quasi-experiment, what are possible alternative study designs? A second quasi-experimental design is the 'one-group pre-test post-test design'. As the name implies, there is no control group in this experiment, thus threatening the control of extraneous variables that might confound the measurement of impact from the intervention. Under certain circumstances it might be possible to control some extraneous variables, such as an intervention involving sole use of a residential centre and pre- and post-tests conducted on the day of arrival and the day of departure – the experiment is essentially a 'closed system'.

Another design is the one-group time series design. This was employed by Hayhurst, Hunter, Kafka and Boyes (2015) to measure the impact of a ten-day developmental sail training voyage on the resilience of a sample of 72 high school pupils, who were tested one month prior to the voyage, on the first day of the voyage, the last day, and five months after the voyage in order to establish the stability of resilience levels before and after the intervention. In this case, a control group of 74 high school pupils was employed who were tested only at one month prior to the voyage to ensure that the baseline resilience of members of the experimental group was not unusual. The experiment was successful and a stable, statistically significant improvement in resilience was observed in the experimental group. A limitation of this study was that not all participants completed all tests, suggesting that some of the statistical results were not based on repeated measures on the same cohort of participants.

Another approach, which might qualify as truly experimental as opposed to quasi-experimental because it incorporates the essential component of random allocation of participants to experimental and control groups, is the matched-pairs design. In this design, participants are allocated to groups following the pre-test by creating matched pairs of participants on the basis of their scores in the pre-test and other measures and then allocating them randomly to the two groups.

Looking forward: The importance of project planning

I have equated metrics with measurements taken during cross-sectional studies or on progress towards a target, but then focused the bulk of the discussion in this chapter on experimental approaches to measuring the impact of outdoor interventions. For both metrics and measures, good project planning is essential for a successful outcome to the research. Despite the discouraging comments in the literature concerning quantitative methods, researchers should not be deterred from collecting metrics and measures as indicators of what works and by how much. This final section provides some encouragement for that.

We have seen that two aspects with which care must be taken when collecting metrics are the size and composition of the sample and the control of confounding variables. The sample is of particular concern in longitudinal studies where attrition and conditioning are likely with the possible loss of being representative of the larger population. A forward look whilst planning will alert researchers to potential weaknesses such as these. We have also noted that cross-sectional studies offer the opportunity for some quite sophisticated statistical analysis, but inference of cause and effect from this is rarely possible unless there is planning for sound longitudinal evidence.

With regard to experiments, 'true experiments' (RCTs) are very difficult to conduct 'in the real world'. Nevertheless, several experimental designs are available to obtain good-quality measurements – as long as critical concerns about technical aspects of outdoor studies are addressed at the planning stage. The planning stage is the most important stage of a research project. It should be thoroughly worked through before any data are collected, because once data are collected it is difficult to correct omissions, and it is the data set that will determine what statistical tests can subsequently be carried out to address project objectives.

An aide-memoire for project planning is offered here. It places the challenges to experimental outdoor research identified by Neill (2008) against steps in the implementation of an experiment as suggested by Cohen et al. (2018, p. 411) (Table 22.2). The table highlights the fact that there are three broad areas requiring

TABLE 22.2 Challenges to experimental validity (Neill, 2008) set against steps in the implementation of a research project (Cohen et al, 2018)

Steps in the implementation of an experiment (Cohen et al., 2018)	*Challenges to experimental validity in outdoor studies (comments from Neill (2008) in italics)*
Identify the purpose of the experiment	
Select or identify the relevant variables	"*a lack of appropriate dependent measures*". (1) Choose appropriate dependent variables that meet the purposes of the project (including measurable proxies to quantify latent variables); (2) Use questionnaires that are proven to have good validity and reliability and are appropriate to the outdoors, demographics of the sample and the level of intervention. "*a lack of investigation of independent variables*". Apart from those manipulated by the researcher (fundamentally, participation in the intervention or not), be aware of the influence of other independent variables not manipulated for experimental purposes (e.g. gender, age, previous outdoor experience).
Specify the level of the intervention (low, medium, high)	
Isolate and control the experimental conditions and environment	"*a lack of control and comparison groups*". Recruit matched control and experimental groups to exert a level of control on potential threats to research validity.
Select the appropriate experimental design	"*a lack of longitudinal data*". Use a delayed test to assess transfer of learning to everyday life or the classroom – an important objective of outdoor studies.
Administer the pre-test Sample the relevant population and assign the participants to the groups	In outdoor studies, normal practice is to sample the population to create matched experimental and control groups before testing starts. Alternatively, use the matched-pairs experimental design mentioned above. "*low statistical power*". Cohen (1988) published tables of desirable sample size to give adequate statistical power for the adopted statistical significance criterion (e.g. 0.05) and target effect size (e.g. small/medium/large).
Conduct the intervention	
Conduct the post-test	"*a lack of longitudinal data*".
Analyse the results	"*over-reliance on inferential statistics*". Examples are reliance on a simple correlation between two data sets to infer cause and effect, and scaling up the results of an experiment to a population that the sample does not really represent.

particular care: identification, use and control of variables, selection of a reliable and valid questionnaire and recruitment of an appropriate sample (see Scrutton & Beames, 2015, for further discussion).

Results from meta-analyses have featured quite prominently in this chapter. In looking forward, a quote from Suri (2018) raises again the fact that stakeholders in outdoor activities like to have evidence of the effectiveness of these activities for their own purposes in numbers as well as from narrative and anecdote. Suri says, "Making sense of such complex domains of literature to inform policy, practice or further research can be challenging .... research syntheses are increasingly gaining prominence as valid methods for knowledge generation in their own right". Ongoing research synthesis (meta-analyses) of outdoor studies is confirming that, quantitatively speaking, effect sizes from outdoor interventions appear to be rather 'normal' and around +0.35 (Figure 22.1). However. the success of the meta-analyses that stakeholders are increasingly turning to depends on there being a large number of good-quality primary studies of metrics and measures that address the issues discussed here of variables, questionnaire and sample. It is important that we maximise the quality of primary studies.

Notes

1 Effect size. "An objective and (usually) standardized measure of the magnitude of an observed effect. Measures include Cohen's *d* ..." (Field, 2013, p. 874). Expressed mathematically, Cohen's *d* is the difference between two means (such as (post-test mean – pre-test mean)) divided by the pooled standard deviation for the two data sets.

2 Latent variable. "A variable that cannot be directly measured but is assumed to be related to several variables that can be measured" (Field, 2013, p. 878). The measured variables are used to derive a score for the latent variable. Examples in outdoor education might be social efficacy or leadership ability.

References

Amos, R. & Reiss M. (2012). The benefits of residential fieldwork for school science: Insights from a five-year initiative for inner-city students in the UK. *International Journal of Science Education, 34*(4), 485–511.

Barrett, J. & Greenaway, R. (1995). *Why adventure? The role and value of outdoor adventure in young people's personal and social development.* Coventry: Foundation for Outdoor Adventure.

Bowen, D. & Neill, J. (2013). A meta-analysis of adventure therapy outcomes and moderators. *The Open Psychology Journal, 6*, 28–53

Cason, D. & Gillis, H. (1994). A meta-analysis of outdoor adventure programming with adolescents. *Journal of Experiential Education, 17*(1), 40–47.

Christie, E.M. (2004). *"Raising achievement" in secondary schools? A study of outdoor experiential learning* (Unpublished doctoral dissertation). University of Edinburgh, UK.

Cohen, J. (1988). *Statistical power analysis for the behavioural sciences* (2nd ed.). Hillsdale, NJ: Lawrence Erlbaum Associates.

Cohen, L., Manion, L. & Morrison, K. (2018). *Research methods in education* (8th ed.). Hoboken, CT: Taylor & Francis.

Ewert, A. & Sibthorp, J. (2009). Creating outcomes through experiential education: The challenge of confounding variables. *Journal of Experiential Education, 31*(3), 376–389.

Ewert, A. & Yoshino, A. (2011). The influence of short-term adventure-based experiences on levels of resilience. *Journal of Adventure Education and Outdoor Learning, 11*(1), 35–50.

Field, A. (2013). *Discovering statistics using IBM SPSS statistics* (4th ed.). London: SAGE.

Fiennes, C., Oliver, E., Dickson, K., Escobar, D., Romans, A. & Oliver S. (2015). *The existing evidence base about the effectiveness of outdoor learning.* London: Institute of Education, University College London.

Goldacre, B. (2013). *Building evidence into education.* Retrieved from: www.gov.uk/government/news/building-evidence-into-education

Gustafsson, P., Szczepanski, A., Nelson, N. & Gustafsson, P. (2012). Effects of an outdoor education intervention on the mental health of schoolchildren. *Journal of Adventure Education and Outdoor Learning, 12*(1), 63–80.

Hahn, K (1961). Origins of the Outward Bound Trust. In D. James (Ed.) *Outward Bound* (pp. 1–17). London: Routledge & Kegan Paul.

Hattie, J., Marsh, H.W., Neill, J.T. & Richards, G.E. (1997). Adventure education and Outward Bound: Out-of-class experiences that have a lasting effect. *Review of Educational Research, 67*, 43–87.

Hayhurst, J., Hunter, J.A., Kafka, S. & Boyes, M. (2015). Enhancing resilience in youth through a 10-day developmental voyage. *Journal of Adventure Education and Outdoor Learning, 15*(1), 40–52.

Hinkley, T., Brown H., Carson V. & Teychenne, M. (2018). Cross sectional associations of screen time and outdoor play with social skills in preschool children. Retrieved from: http://journals.plos.org/plosone/article/metrics?id=10.1371/journal.pone.0193700

Kendall, S. & Rodger, J. (2015). *Evaluation of Learning Away: Final report.* London: Paul Hamlyn Foundation.

Laidlaw, J.S. (2000). *A meta-analysis of Outdoor Education programs* (Unpublished doctoral dissertation). University of Northern Colorado, US.

Lipsey, M.W. & Wilson, D.B. (1993). The efficacy of psychological, educational and behavioral treatment. *American Psychologist, 49*(12), 1181–1209.

Malone, K. & Waite, S. (2016). *Student outcomes and natural schooling.* Plymouth: Plymouth University. Retrieved from: www.plymouth.ac.uk/research/oelres-net.

McKenzie, M. (2003). Beyond "The Outward Bound Process"': Rethinking student learning. *Journal of Experiential Education, 26*(1), 8–23.

Neill, J. (2008). *Enhancing life effectiveness: The impacts of outdoor education programs* (Unpublished doctoral dissertation). University of Western Sydney, Australia.

Nundy, S. (1999). The fieldwork effect: The role and impact of fieldwork in upper primary school. *International Journal of Geographical and Environmental Education, 8*, 190–198.

Rickinson, M., Dillon, J., Teamey, K., Morris, M., Choi, M., Sanders, D. & Benefield, P. (2004). *A review of research on outdoor learning.* Shrewsbury: National Foundation for Educational Research and King's College London.

Scrutton, R. & Beames, S. (2015). Measuring the unmeasurable: Upholding rigour in quantitative studies of personal and social development in outdoor adventure education. *Journal of Experiential Education, 38*(1), 1–18.

Sheard, M & Golby, J. (2006). The efficacy of an outdoor adventure education curriculum on selected aspects of positive psychological development. *Journal of Experiential Education, 26*(2), 187–209

Sport and Recreation Alliance (2017). *Reconomics Plus.* Retrieved from: www.sportandrecreation.org.uk/pages/reconomics-plus

Sport England (2015). *Getting Active Outdoors*. Retrieved from: www.sportengland.org/media/3366/outdoors-participation-report-v2-lr.pdf

Suri, H. (2018). Meta-analysis, systematic reviews and research syntheses. In L. Cohen, L. Manion & K. Morrison (pp. 427–439). London: Routledge.

Williams, R. (2012). *The impact of residential adventure education on primary school pupils* (EdD dissertation). Retrieved from: http://hdl.handle.net/10036/3518.

23

SCIENTIFIC INVESTIGATIONS IN OUTDOOR ENVIRONMENTS

Lois Mansfield

The outdoors provides a plethora of opportunities to devise research projects exploring many aspects of the physical environment. Indeed, this area of geography is typical of most national secondary education systems, and embedded in many tertiary programmes in outdoor-related themes.[1] In some university departments specialising in outdoor provision few students opt to do physical environment-based, positivist research projects. This trend suggests several possible reasons: the thought of mathematical or statistical data manipulation disengages many outdoor undergraduates; teaching and research staff are mainly constructivist advocates; or that outdoor curricula relegate the importance of the natural environment in the creation of outdoor experiences to early stages of programmes and research projects follow later. The purpose of this chapter, is, therefore, to demystify the character of positivist research through the lens of physical geography and to provide some pragmatic design principles needed to plan, run, analyse and interpret such research projects applicable in schools or for undergraduates.

Process overview

Completing research relating to physical environments requires researchers to follow a positivist approach. Positivism was first proposed by Auguste Comte (1798–1857) to draw a distinction between science and both metaphysics and religion, through a commitment to empiricism; that is, making inference through real world observation and/or derivation of theories capable for verification (Cresswell, 2013). For physical geography, these ideas were adopted in the Quantitative Revolution (1950s and 1960s), when geographers sought to explore and explain spatial and temporal patterns through the application of numerical, analytical and inferential techniques. Recent philosophical developments revolve

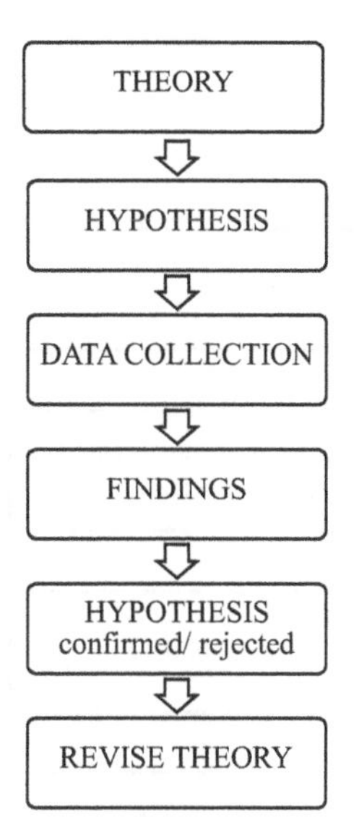

FIGURE 23.1 Research design: Critical rationalism or hypothetico-deductive

around critical rationalism, which develops theories, hypotheses and experimentation to falsify our suppositions through a hypothetico-deductive approach (Popper, 1975; Richards, 2009). In other words, geographers like to interrogate theories by testing them; in this way our knowledge expands and improves on a subject. We do this by developing a *hypothesis* (or hypotheses) and then designing a research strategy to test it (Figure 23.1). This research process is part of wider scientific method, which positivists adopt to create an accepted structure to complete their investigations. Consequently, a well-considered scientific enquiry couched in positivism, in which results will be acceptable to the scientific community, requires us to not only consider process, but also method with several feedback loops in terms of planning properly (Figure 23.2).

This chapter will explore how we can accomplish the planning process in relation to the outdoor environment by drawing on a range of case studies of research projects completed by students and staff.

Choosing a topic

Probably one of the most difficult decisions researchers have to address is selection of a topic of study (see Chapter 3). The physical world is extremely complex, with many variables operating simultaneously in varying degrees. A minute adjustment in one variable can cause effects across systems leading to very different outcomes. Our first task therefore, is to read widely to understand the current knowledge within the relevant discipline area. We may discover contradictions in materials; this is important, as it allows us to develop our critical faculties to judge the veracity of

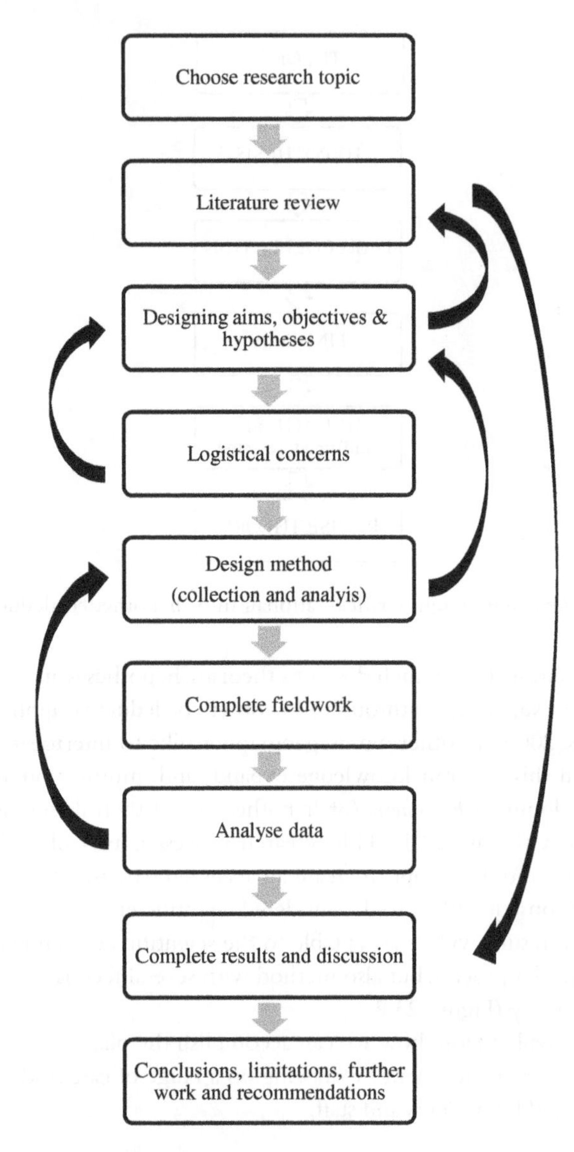

FIGURE 23.2 The positivist research process
(Note the loops of information flow, which can influence decision-making and analysis)

the research. Furthermore, reading provides us with an opportunity to triangulate the range of methods available to collect our data, allowing us to weigh up, and acknowledge, the strengths and weaknesses of each (Clifford, Cope, Gillespie & French, 2016).

For those interested in the outdoors this can lead them down a very long detailed path into areas of which they have little knowledge and may only have a limited initial understanding. The other challenge is that areas in which outdoor researchers become interested often cross a number of sub-disciplines, where the interface between people and the natural world is crucial to understanding what is actually going on. A good example of this is footpath erosion, which I have explored with many students. It is an extremely common problem in areas of high tourism and recreation.[2] Our knowledge of variables responsible is comprehensive as the subject has been studied by many people from the 1980s to present. Walker behaviour, psychology of route section, land use, vegetation studies, how soils react to compaction and the impacts of the wider environment such as weather patterns, climate change, angle of slope, geological variation etc., all play a role (e.g. Bayfield & Aitken, 1992; Coleman, 1981;Morrocco & Ballantyne, 2007).

Reading widely also provides the researcher with an opportunity to organise their thoughts, plan the structure of their literature review and develop a research design that focuses down to the area in which they are specifically interested. A typical problem with many students is that they ask too broad a research question which leads to a piece of work that is multi-limbed, overly complex, poorly structured and often fails to get to the central point of what is going on. Parsons and Knight (1995, p. 113, citing Kennedy, 1992) beautifully refer to this as the "splodge" research strategy rather than the elegant "wine glass" and thus, I find using spider diagrams, and analogies such as putting your idea through a nest of finer and fine meshed sieves, helps to focus the research.

Logistics

The next consideration is to ensure the research location can be accessed and the method employed is permissible. This is our first feedback loop, for the choice of method for data collection is interrelated to what we can actually do on site. Typical points needing clear thought are shown in Table 23.1. Some countries have very strict rules regarding research permits (e.g. Iceland) or the importation of animal parts can be prohibited under legislation such as CITES (Convention on International Trade in Endangered Species of Wild Fauna and Flora).

Operational issues needing close consideration include health and safety, and the management of risk (Higgitt & Bullard, 1999). Tidal patterns, use of boats and those projects involving rock climbing can be particularly problematic, but not impossible if properly managed. Ethical issues are less influential for physical environment research; but we still need to consider access to others' land: considering the impact we might have from physical disturbance or keeping locations undisclosed if dealing with protected species. All projects should thus be risk assessed and approved before actual commencement.

Formulation of aims, objectives and hypotheses

Once we have worked out what we are interested in, our focus turns to developing an, or some, aims. An aim is what we are going to investigate, our ultimate goal.

TABLE 23.1 Logistical considerations for a positivist approach to research

Issue	*Points to consider*
Access	Land ownership Protected status limiting activities or revealing locations The need for a research permit
Timing	Weather conditions Seasonal fluctuations
Cost	Repetitive visits Distance from base Equipment expense
Repetition	Is this a monitoring project requiring multiple visits? Can you get back there?
Equipment	Durability Left to 'run'– consider vandalism or theft Personal protective equipment
Hazard/Risk	Habitats, animals, disease, environment, other people, chemical usage, exposure to UV, working in extreme/hostile/remote environments, dysfunctional body
Sampling	Permanent removal of samples from site Temporary disturbance Legal requirements/permits You need more than one person to run the experiment

From this, we will derive a set of stepping stones of tasks to accomplish (our objectives). I often use the analogy of eating a jam sandwich with students. The aim is to eat the sandwich; the objectives are: 1) select ingredients, 2) buy ingredients, 3) make sandwich, 4) eat it. Note how the objectives are all verbs.

Designing and wording hypotheses is a tricky topic for many researchers. A hypothesis is a prediction we make that can be testable in the real world (laboratory or outdoors). Using Popper's (1975) ideas of critical rationalism and hypothesis testing, the best line of enquiry is to employ *falsification*. Here, we attempt to 'deny the consequent', in other words, we try to disprove our idea. Two deductive arguments can be employed as shown in Table 23.2. *Affirming the consequent* leads us to what is termed as a fallacy; *denying the consequent* is a more powerful scientific method as it is more likely to eliminate the probability that it was by chance.

Further useful tips for setting up hypotheses is to try and have two concepts in the statement which are clear, identifiable and measureable along with some form of relationship between them, examples being:

- Width of a footpath is related to erosion depth;
- Gryke depth on limestone pavements gets deeper as width gets wider;
- Tree girth decreases with altitude;

TABLE 23.2 Setting up hypotheses

	Affirming the consequent	*Denying the consequent*
Hypothesis	Rain causes rocks downslope to be wet	Rain causes rocks downslope to be wet
Consequence	If it rains, rocks get wet	If rain does not fall, rocks will NOT be wet
Test	Let it rain	It does not rain
Result	Rocks are wet	Rocks are wet
Conclusion	The hypothesis is correct FALLACY	The hypothesis is NOT correct VALID

- Boulders at the top of a scree slope are smaller than those at the base;
- Species diversity is greatest on the least accessible ledges on a crag.

Note how some of these hypotheses are directional (some form of vector is employed – larger, bigger, deeper) and others are not (a relationship to be determined). Which we use is very much dependent on our earlier investigations as part of the literature review. Some researchers subscribe to the view that one should know what the outcome will be and thus employ directional hypotheses, whereas others do not and leave it open to see what the results deliver.

Once we have created a testable hypothesis we then create a pair of two contrasting statements; the null (H_0) hypothesis and the alternative or research hypothesis (H_1). Setting these up correctly at this point is dependent on the type of statistical analysis we will want to do in relation to the research, which, in turn, is governed by the inferential statistic we use AND how we measure our data. For example:

H_0 – width of a footpath is not related to erosion depth
H_1 – width of a footpath is related to erosion depth
And:
H_0 – species diversity is not the greatest on the least accessible ledges on a crag
H_1 – species diversity is greatest on the least accessible ledges on a crag

Undergraduates find this particular part of the scientific method difficult, as they have to forecast what sort of data they will collect and in what format they are needed, in order to run a specific statistical process properly. Many simply collect data and then say "what can I do with this?" with the onus on supervisory staff to guide them through the process.

Using Popper's ideas, new knowledge can be fed back into our research and then the experiment run again, until eventually we may find a hypothesis which is not falsified, suggesting this is actually what is going on. This knowledge can then be used to formulate 'laws' or 'theories' about how the real world works.

Scientific method

Now we have some idea as to what we want to find out, this allows us to focus on the scientific method itself. Elements of this include: selecting an experimental design, setting a sampling regime, data collection, and completing our data analysis.

Experimental design

Experimental design in positivist science follows five options:

1. Classic – comparing a control site to experimental one, the latter of which has some form of intervention to differentiate it from the former. The idea is that it should be easy to spot causal changes. The downsides are that it requires several runs to ensure results do not occur by chance.
2. Cross-sectional/ correlational – collection of a variety of variables (measurements) from at least two locations at one point in time. We compare the extent to which the locations differ variable by variable. The key factor here is to eliminate any matched (similar) variables leaving only variables that have the greatest difference and are therefore telling us something. Nevertheless, negative results are just as good as positive ones. Other challenges relate to the measurement of the right variables as we cannot return to collect more data another time.
3. Panel – we only use the experimental group over a set time. We inject an intervention and watch what happens. The downsides are that it can be hard to ascertain the cause of a change and it is expensive as they have to be long-term experiments.
4. Longitudinal – where we follow certain variables/ locations for very long periods of time in terms of their evolution and reaction to events (natural or human). Whilst expensive in time and labour, these types of study are common in the ecological sciences in relation to animal studies and for physical environments to see evolutionary changes in geomorphological systems.
5. 'One shot' (case study) – collect a range of data for one location at one point in time. This can lead to a descriptive outcome, rather than explaining what is going on; we have eliminated the possibility of identifying causality. This can make the researcher think they have found the answer, when they have not, but the design is very cheap in terms of time and labour.

As many undergraduate or school projects tend to fall into the latter category due to time and access constraints, it is important to ensure multiple sets of readings of the phenomenon are taken (sampling regime). Recording three or four sets of readings is not enough in natural or medical sciences, experiments need to be run many times to eliminate the one-off chance results as well as fulfil the criteria for running certain statistical operations. This is very different from accepted norms in qualitative methodology and can be new to many students.

Devising sampling regimes

The first step is to ensure we have identified exactly what it is we need to measure. Measurement is an entire topic in its own right, which needs careful consideration to include:

1. Type? – what do we want to measure, this will have derived from our reading, research objectives and hypotheses construction.
2. How? – which techniques will allow us to collect the data we need?
3. When? – are there seasonal or other constraints affecting data quality or even availability?
4. Level? – will the data we collect be: Nominal, Ordinal, Interval or Ratio. This is crucial, as it will control how we analyse our data; so we need to think about how we will summarise and describe as well as which statistical tests we wish to use.

From this, we may need to consider if we want to control certain variables and watch what happens to others in response. The former is known as the independent variable and the latter as the dependent; knowing which one is which is important for later data manipulation. A good example here is changing the number of walkers on a path (independent) to see what happens to erosion levels (dependent).

Next we need to decide if we are employing a probability strategy (whereby locations are chosen at random or through stratified sampling) or a non-probability strategy (which is systematic). How we do this very much depends on the discipline we are working in and the approach to sampling. Again, reading helps determine the preferred choice of a particular research field.

Finally we need to decide how many. The sample we take should be both representative of the entire population of X, with as small a sampling error as possible: A large sample reduces the over representation of a particular situation. What this means in reality is that as sample increases in size so the sampling errors get smaller; however, the relationship is not linear but exponential. Thus once we have about 150 samples, the precision of our sampling does not get any better if we double or even triple the sample size (Silk, 1979). This sounds quite daunting for younger researchers, so these numbers can be adjusted downwards appropriately to perhaps 50, for unpublished works.

Data collection

Data collection needs good planning not only to ensure we capture all we need, but that what we collect is intelligible and useful. Pre-prepared data collection tables are a must to keep data tidy, with clearly labelled locations (this is crucial for longitudinal studies where we need to keep going back to the same spot), dates, times and weather conditions.

Precision and accuracy have specific meaning within scientific method. Precision measures the quality of our data and thus needs repetition to ensure measurements are close to each other. Accuracy, in contrast, is about ensuring our measurements are close to the correct (accepted) value, but it does not measure quality. An example of this is the analysis of water moving into and through cave systems for carbonate and bicarbonate concentrations. We employ a technique known as titration, using a glass burette to add acid into a prepared water sample. The burette is graduated into 1mm divisions and can be subject to misreading (precision). At the same time if acid is added too fast, the end point of the chemical reaction is exceeded and the experiment has to be repeated until at least two concurrent results are obtained (accuracy).

Finally, we need to check our equipment is field-fit for the conditions, e.g. avoiding battery-driven equipment in work on active glaciers as the batteries can run out within ten minutes due to the cold. Mechanical pieces are thus more reliable, easier to fix and, if necessary, we can actually build them. A piece of equipment I often build with students is the pebbleometer (giant calipers). It is designed to measure the three dimensions of sediments over 20mm in size, the data of which can be manipulated in many ways to explore environments of deposition. The simplicity of design allows easy use by primary- and secondary-age children and their construction employs multiple curricula concepts in mathematics and design and technology.

Of course, some forms of data collection lend themselves to modern technology through automated systems, whereby data can be downloaded from monitoring stations on site, or remotely via satellite or mobile phone relay. This is particularly prevalent in river catchment studies, where data need sampling every few minutes, or even after every 0.2mm of precipitation, especially during a flood event.

Data analysis

Often it is the maths (statistical) component of scientific method that initially puts researchers off, which is a pity, for, as we have seen, the actual process is very straightforward, well tested and easy to follow. Data analysis for physical geography seeks spatial patterns and temporal trends, and goes on to explain them; to demonstrate if X causes Y, or A is different from B, and in what way. We can even use statistics to make predictions, which is useful, for example, in flood management.

The analysis phase requires us to complete two or three steps depending on the nature of our research. Our first analysis task is to describe our data using descriptive statistics, tables, graphs and/or maps (Harris, 2016). These techniques are designed to summarise our data and depend on level of measurement (Table 23.3). Key characteristics to note here as we move <u>from</u> nominal through <u>to</u> ratio:

- Level of understanding by the public reduces significantly;
- Level of acceptance by the scientific community increases;
- Number of techniques available to us increases;
- Statistical and mathematical complexity increases.

TABLE 23.3 How to describe a data set

Level of measurement	*Descriptive statistics*	*Possible graphical techniques*
Ratio	Arithmetic mean	Histogram
	Standard deviation	Scattergraph
	Variance	Triangular plot
Interval	Range	Kite diagram
	Interquartile range	Choropleth
	Skewness	Frequency polygon/curve/ distribution
	Kurtosis	Cumulative frequency curve
		Box and whiskers
		Radial plots/rose diagrams
Ordinal/ ranked	Median	Bar chart
		Pie chart
Nominal/ classificatory	Mode	Pie chart
	Variation ratio	Bar chart

The researcher should calculate any relevant descriptive statistics and construct graphics with prose to outline to the reader what the data are telling us. Remember we are looking for patterns and trends to explain our observations.

The second step in data analysis allows us to test hypotheses using analytical or inferential statistics. We may be looking to prove data comes from different or similar populations, that there is some form of interrelationship/association or causality. First we need to determine whether the data follow a normal distribution (a bell shaped curve/ parametric) by looking at various properties:

- Mean, mode and median are very similar;
- The data are symmetrical about the mean
- Data are at least the interval scale
- Data form a straight line using % cumulative frequency on Arithmetic Probability paper
- Skewness = 0
- Kurtosis = 3.0

Proving or disproving normality determines which analytical statistics we can use. There is a plethora of these, which can be overwhelming. To help researchers get to grips with their options, various authors have designed decision trees to ensure the right statistical test is selected (e.g. St John & Richardson, 1996). The fundamental choices are shown in Table 23.4 and the main steps to running a test in Figure 23.3 (Harris, 2016; Siegel & Castellan, 1988; Walford, 2011).

TABLE 23.4 Choosing a statistical test

	Parametric	*Non Parametric*
Two similar or different populations	t-test (samples less than 30)	Chi-squared (nominal data)
	Z-test (samples more than 30)	D-test [Kolmogorov-Smirnoff] (ordinal data)
	ANOVA (analysis of variance)	Mann-Whitney U-test
Three or more similar or different populations	ANOVA	Chi-squared
		Kruskal-Wallis
		Friedman
Relationships	Pearson product-moment	Spearman's rank
	Regression analysis	Regression analysis
Making predictions	Z scores	Poisson
		Binomial

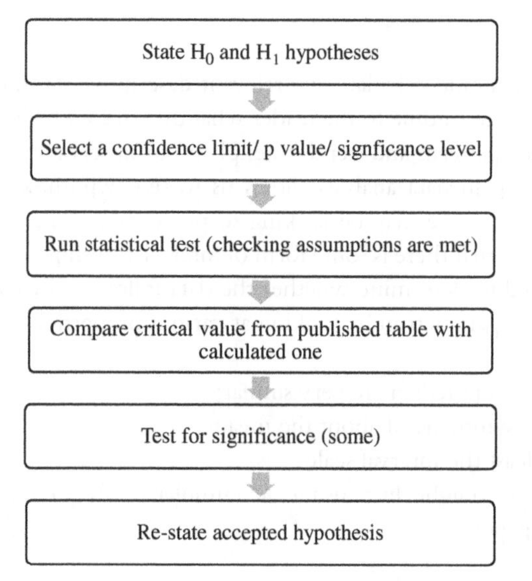

FIGURE 23.3 Conducting a statistical test

The final stage of analysis is to conduct any specialist statistical manipulations This may include tools such as Principal Components Analysis (PCA), which seeks to determine if several variables are more important than others amongst a host we may have measured (footpath erosion) or Cluster Analysis, which can suggest a classification system (soil types). Other options may be simple statistics related to very specific fields, such as sorting coefficients used as part of environment of deposition analysis in geomorphological investigations (Briggs, 1979). These often come with related specific software to use to perform the calculations. For more general statistical manipulations SPSS™, R™ and Minitab™ are most effective, and whilst many students may be more familiar with Excel, it has its drawbacks in relation to the background maths used to calculate various statistics (Dytham, 2011; Harris, 2016). Calculating inferential statistics manually can aid understanding before employing software although time might mitigate against this.

The last stage of any hypothesis testing is to ensure the results are reported. Students find how to express this difficult, but a statement such as; "gryke width is associated to gryke depth ($r_s = 0.753$ @ 95% confidence)" is acceptable. If numerous statistics tests are run, tables can be created and referred to in the results.

Interpretation

The final stage of the scientific method is to draw together the data manipulation. The positivist approach usually requires this to be written up as first, results (focusing on description of our data only) and then, discussion (seeking explanation and putting back into the broader context of the research body as known to date). A particular challenge is that students want to tell us everything in the results, but keeping to just description allows the building of clear patterns in the data as well as a cogent argument. Another challenge is helping researchers understand that negative results are just as valuable as positive ones. If the experiment did not work and there was not a difference, this allows us to eliminate ideas from our inquiry and in fact provides opportunity in the discussion for a broader debate of 'why not?' Of course, there are others who think that such a situation suggests the researcher did not properly complete their literature review and should have predicted this; but that is the real world and is what makes research fun.

Tying it all up – Conclusions

The final task for positivist research is to summarise what we have found with some succinct conclusions by going back and answering our research question, objectives and hypotheses. It is also important to reflect critically on where methodology could have been improved, and suggest further lines of inquiry outwith the scope of the current research. Increasingly I am finding undergraduates like to criticise their method in their method, which is quite bizarre as it suggests they have no idea what they are doing in the first place!

Finally, with all physical environment research we might like to consider some recommendations for management, if the subject is applied. With outdoor students, they tend to like this as it helps them understand the point of research. It may be something to get them to consider as part of the discussion and even as far back as the objectives, particularly with respect to topics related to recreational ecology/physical impacts of various outdoor adventure activities on the environment.

Concluding remarks

Whilst this chapter has explored positivism in the context of physical geography, the processes and methods are just as applicable to other areas of outdoor research. Sports physiology, functionality of outdoor equipment and questionnaire survey are all feasible using a positivist approach.

Notes

1 For example, https://australiancurriculum.edu.au/f-10-curriculum/humanities-and-social-sciences/geography/ and www.canadian-universities.net/Universities/Programs/Geography_and_GIS.html

2 www.thetimes.co.uk/article/footpath-erosion-threatens-wildlife-of-lake-district-6t60n79lmlf

References

Bayfield, N. & Aitken, R. (1992). *Managing the impacts of recreation on vegetation and soils: A review of techniques.* Banchory: Institute of Terrestrial Ecology.

Briggs, D. (1979). *Sediments.* London: Butterworths.

Clifford, N., Cope, M., Gillespie T. & French T. (2016). *Key methods in geography* (3rd ed.). London: SAGE.

Coleman, R. (1981). Footpath erosion in the English Lake District. *Applied Geography, 1*(2), 121–131.

Cresswell, T. (2013). *Geographic thought: A critical introduction.* Chichester: John Wiley & Sons.

Dytham, C. (2011). *Choosing and using statistics: A biologist's guide.* Chichester: Wiley-Blackwell.

Harris, R. (2016). *Quantitative geography – the basics.* London: SAGE.

Higgitt, D. & Bullard, J.E. (1999). Assessing fieldwork risk for undergraduate projects. *Journal of Geography in Higher Education, 23*, 441–449.

Morrocco, S. & Ballantyne, C.K. (2007). Footpath morphology and terrain sensitivity on high plateaux: The Mamore Mountains, Western Highlands of Scotland. *Earth Surface Processes and Landforms, 33*(1), 40–54.

Parsons. T. & Knight, P.G. (1995). *How to do your dissertation in geography and related disciplines.* Cheltenham: Nelson Thornes.

Popper, K. (1975). *The logic of scientific discovery.* Hutchinson: London.

Preece, R. (1994). *Starting research: An introduction to academic research and dissertation writing.* London: Pinter Publishers.

Richards, K. (2009). Geography and the physical sciences tradition. In N.J. Clifford, S.L. Holloway, S.P. Rice & G. Valentine (Eds.) *Key concepts in geography* (pp. 21–45). London: SAGE.

St John, P. & Richardson, D. (1996). *Methods of statistical analysis of fieldwork data* (2nd ed.). Sheffield: Geographical Association.

Siegel, S. & Castellan N.J. Jr. (1988). *Non parametric statistics for the behavioural sciences.* London: McGraw Hill International.

Silk, J. (1979). *Statistical concepts in geography.* London: George Allen & Unwin.

Walford, N. (2011). *Practical statistics for geographers and earth scientists.* London: Wiley Blackwell.

24

MIXED METHODS RESEARCH IN OUTDOOR STUDIES

Paradigmatic considerations

Kass Gibson and Mark Leather

Using mixed methods research techniques and approaches in a specific project is an intuitively appealing way to develop robust answers to our research questions, and has a long history. Mixed methods research is predicated on the belief that there are different and multiple valuable and legitimate techniques for developing knowledge and therefore using more than one method in research is advantageous. Nonetheless, even with specific nomenclature, "we are 30 to 40 years deep into a multiple, mixed methods discourse, and we still can't define the method or be clear on its benefits" (Denzin, 2012, p. 82). More specifically, core 'controversies' continue to be debated (Sparkes, 2015; Teddlie and Tashakkori, 2010) including definitional issues ranging from what counts as mixed methods research to specific concepts and terminology used, (in)commensurability of paradigms, methodological hierarchies, and the overall value of mixed methods research. We do not advocate any specific definition of what counts as mixed methods research by virtue of the methods used and/or when, where, and how mixing is undertaken in research. Guides for such issues can be found, for example, in the work of Creswell and Clark (2017); Creswell and Creswell (2017); Hesse-Biber (2010); Hesse-Biber and Johnson (2015) and Morgan (2014) (see also Chapter 25). Rather, we seek to show the need for informed decision making by researchers and concomitantly articulation and justification of those choices. The purpose of this chapter, then, is to support researchers to carefully and considerately reflect on the choices they make in designing and undertaking research.

These controversies are characterised by "terminological slippage" (Sparkes, 2015) regarding paradigm, methods and methodology. As Gibson (2016) demonstrates, such slippage is compounded by disciplinary hierarchies and ignorance. The fundamental problems and challenges faced in mixed methods research are not the result of appropriating research methods from the fields in which they originated or developed. We encourage using any and all approaches that develop greater understanding of a research problem (Gibson, 2012). Problems and challenges in mixed methods research are the result of failing to adequately consider how different research methodologies present different ways of knowing, and accompanying evaluation, quality and evidentiary criteria are not integrated in careful and respectful ways. Specifically, researchers in outdoor studies must avoid reproduction of celebratory and simplistic rhetoric of mixed methods research as a panacea to shortcomings in any given data collection and analysis process, and must engage carefully and considerately with research questions and data.

Research paradigms and the (im)possibility of mixed methods research

Developing a robust understanding of what is meant by paradigms is essential given the debates regarding not only paradigmatic differences (e.g. Sale, Lohfield & Brazil, 2002; Sparkes, 2012) but also their significance (Gibson, 2016) (see also Chapter 3) and is the cornerstone of methodological debates in mixed methods research. For us, paradigm refers to the fundamental assumptions – explicit or otherwise – regarding the nature of the world and the relationship between the world, ourselves, and other subjects, objects, beings, processes, events and happenings. Paradigms are defined by shared answers to questions posed about the nature of reality (ontology); the extent to which we can know that reality (epistemology); and simultaneously the most valuable and effective processes for gaining knowledge of that reality (methodology). It should go without saying, but is important to remember, that how researchers answer these questions determines what research questions they ask and what evidence they seek to answer them.

Many of our colleagues will not be convinced of the need to start empirical research projects – including mixed methods – with philosophising on knowledge per se. However, paradigm also refers to shared beliefs of research communities. Indeed, our aforementioned 'just-get-on-with-it' colleague will hold opinions regarding what are appropriate research questions and what kinds of evidence is persuasive as defined by the (social) norms of their research community. In other words, paradigms inform and guide our views on knowledge and evidence. This can be through philosophical engagement with our underlying assumptions, or through socialisation into a research community. The impact of paradigmatic assumptions on the research processes should be brought into sharp relief in mixed methods research because researchers seek to cross paradigmatic boundaries of research philosophies and/or communities. Therefore, we briefly outline key paradigmatic issues below.

Quantitative research is usually conducted assuming one regular and predictable reality governed by laws, independent of and external to individuals' experiences and perceptions. Researchers assuming reality is independent means they must seek to conduct research without influencing, or being influenced by, the objects of their study. Quantitative researchers must minimise and mitigate personal values and perceptions (i.e. bias). Such assumptions require, wherever possible, following standardised (ideally gold standard) research procedures. In the technical language of paradigms, such a perspective is usually defined as post-positivism – although there are subtleties and differences within this perspective – as defined by a realist ontology, objectivist epistemology, and experimentally driven and/or manipulative methodologies. Importantly, post-positivists claim reality and the laws that govern it are knowable, but only imperfectly, probabilistically and approximately. Therefore, their goal is robust approximation achieved through research practices that manipulate and observe events and actions in order to predict and control.

Qualitative research is often used to denote any research that rejects the goals and procedures of post-positivists to pursue contextual understandings of meaning and lived experience as deemed significant and important by people. The multiple different paradigmatic positions opposing post-positivism have been reviewed and defined elsewhere (Lincoln et al., 2011; Smith et al., 2012; Sparkes & Smith, 2014. In the context of this chapter we use *interpretivism* as a catch-all. Such interpretivist research is conducted assuming reality is dependent on meanings given to objects and events. In other words, reality is fundamentally influenced by our interpretations, which is why interpretivists often refer to *multiple realities*. Importantly, to say reality is multiple does not deny physical and social objects, events, processes exist independently of us. Rather it denies the ability of researchers to completely separate themselves from the researched because knowledge is co-constructed by researchers and participants. Therefore values inevitably mediate understandings. In paradigmatic terms, their perspectives are based on, to varying degrees, levels of emphasis, and in various combinations; relativist ontologies, subjectivist epistemologies, and hermeneutic methodologies. Combined, this means such research cannot claim to provide the truth, or even approximations of the truth, which has obvious and significant impact on not only what researchers can claim they know, but also how we should evaluate such research.

For example, the post-positivist will use particular methods and procedures in an attempt to 'control' (i.e. remove) values, biases and other variables from their research. Good research, according to post-positivists, is valid, reliable and presented in a dispassionate, objective and formulaic manner. For post-positivists the purpose of research is to know the world as it really is, corresponding to a belief that such knowledge is possible. Based on these assumptions, knowing constitutes statistically generalisable causal explanations and, in ideal scenarios, knowledge will be theory-free and timeless. However, interpretivists will respond: no method is theory-free, value-free or timeless. Similarly, judging what is good research is not theory-free, value-free or timeless (much ink has been spilled in defining good interpretivist research; see, for example: Sparkes & Smith, 2009; Smith & McGannon, 2017). Knowledge claims are contextual and appreciation of difference and of complex nuances based on discerning and sagacious distinctions of how methods are deployed practically and the implications of methodological decisions. The fundamental task of mixed methods research is establishing how, or indeed if at all, these differing assumptions can be combined within a single research programme.

Hopefully this brief review demonstrates that even though many researchers will not reflect on their paradigmatic assumptions, paradigms – as research philosophy and research practice – shape the research process from beginning to end through types of questions we ask as well as the way we seek to ask them. Many researchers – including those using mixed methods – have conducted research without explicit consideration of research philosophy. However, this does not prove that researchers can go successfully about their work without considering paradigms, "it merely means that they are operating with unexamined assumptions" (Mertens, 2010, p. 9). In other words, assumptions are often things we do not know we have. Therefore, for research 'purists' paradigmatic assumptions are of utmost importance in research. Resultantly purists argue, primarily, ontological and epistemological positions in (mixed) methodology must be identified, considered and explained. Further, mixing (certain) methods is possible, but mixing methodologies is highly problematic to the extent some methodological combinations are incompatible.

So-called mixed methods 'pragmatists' respond that philosophical reflection is unnecessary. For pragmatists the logistical demands and practical utility of methods matters far more than any epistemological or ontological differences, and dictate not only which research methods are used but also whether mixing methods is required. Projects begin by careful delineation of the research problem and purpose of their study then stepwise selection of suitable methods; assuming combining certain methods will best enable researchers to achieve their ends. Pragmatists' belief in excellence in procedure as the hallmarks of research quality and utility is the dominant, but not only (nor for us the best), approach to mixed methods research. For us, not only is it questionable whether pragmatists meet either the potential, or even the proclaimed, qualities of mixed methods research, but has also reinforced unhelpful, ill-informed and disempowering epistemological hierarchies.

Reasons for mixed methods

The aforementioned dominance of the dubiously named pragmatism in mixed methods research has resulted in a number of "purpose typologies" for mixed methods research. Greene, Caracelli, and Graham (1989) developed a popular and enduring set of five – problem focused – purposes for mixed methods research.

First, *methods triangulation*, taken analogously from surveying, refers to the use of different methods to check the veracity of results, control for biases and/or blind spots, and compensate for unexplained/irrelevant variance and error of individual methods to better understand a single concept. Convergence of findings is viewed as enhancing credibility of conceptualisations. The aggregation of analyses via methods triangulation is intuitively appealing, yet poses numerous difficult and overlooked questions regarding how, when, and if triangulation enables such understanding (see Mertens & Hesse-Biber, 2012). For example, interpreting agreement or disagreement between different kinds of data is not straightforward, unambiguous, or unproblematic in practical terms let alone when seeking to maintain epistemological and ontological coherency.

Complementarity mixed methods designs use different methods to develop broader understanding of a phenomenon as a whole, rather than a single concept. If surveyors use triangulation to establish the height of a mountain, then a *complementarity* method uses a range of approaches to understand aspects beyond the height of the mountain including, for example, geology, ecology, and its social and cultural value. Complementarity emphasises aggregation of the strengths of individual methods vis-à-vis different aspects of the research problem rather than compensating for weaknesses and blindspots of individual methods, and mutual reinforcement on points of agreement. Nonetheless, epistemological and ontological coherence is still questionable. Such questions are usually side-stepped in complementarity designs by conducting and reporting research in parallel rather than integrated, which is reminiscent of differences in collaboration and integration between multi- and interdisciplinary research.

Developmental mixed methods designs use different methods sequentially. Some will quibble whether using data generated by one method to inform subsequent phases of research using different methods is actually mixed methods research. We are not interested in policing what counts of mixed methods research. Indeed, triangulation, complementarity and developmental designs are not new. They have long operated outside any formal mixed methods nomenclature.

Expansion studies use different methods in the project that are most intuitively aligned with each component of the problem. Expansion designs focus on the same concept or problem (cf. triangulation) but deploy different methods in order to address different facets of the same phenomenon (cf. complementarity). Such designs usually determine the best methods practically, in isolation, and without considering how the different methods can, or should, be integrated. Mixed methods expansion studies create challenges regarding connecting different data and results from the different methods. Although these issues are usually commented on, seldom are satisfactory responses to questions of ontology, epistemology and methodology given.

Lastly, contrary to triangulation, complementarity, development and expansion mixed methods which seek consistency and convergence of results as a hallmark of accuracy and quality (or perhaps more accurately reliability and validity), *initiation* studies seek disagreement and incongruity. Inconsistency, then, is the basis for interesting and valuable exploration in initiation studies. Therefore, inconsistency designs are not only likely to be improved by using methods that are significantly different from each other but also it is the only design with an in-built appreciation of paradigmatic differences as its starting point.

Developing mixed methods research projects

As indicated above, we are not interested in determining what counts as mixed methods research. Like Bergman (2011, p. 272), we believe "it may be just as interesting and complex to integrate a qualitative thematic analysis with a narrative analysis, or a random controlled trial experiment with a questionnaire" as seeking to mix quantitative and qualitative methods, precisely because "not all research methods are compatible with all paradigmatic assumptions and all methodologies" (Sparkes, 2015, p. 50). We are, therefore, advocating that the goals of the research, the role of the researcher, values, theory, and corresponding judgement criteria are considered carefully in all research practice. Indeed, ethnography, for example, is universally accepted as a qualitative research approach, yet is conducted according to numerous ontological, epistemological and methodological assumptions, and combinations thereof. As such, most methods can cross paradigms relatively untroubled. However, methodologies pose more significant challenges, which are amplified in mixed methods. For example, researchers seeking to use methodologies requiring detached and disinterested observers alongside methodologies valuing non-neutrality and explicitly politicised positions will face significant challenges regarding both how the research is to be conducted and its successful evaluation. Indeed, comparing and/or combining 'objective' data from rigorous experimental methods with data from semi-structured interviews without appreciation and discussion of how different philosophical paradigms can be integrated, leads inevitably to discussions about 'truth' of the situation, experience, event, process or happening versus experience (Gibson, 2012). Such work can be both important and illuminating. It can be, and more often is, un-informing, of poor quality, and a pale imitation of the possibilities because of the absence of thoughtful consideration of paradigmatic issues.

Beyond research practice and quality, failing to engage in conscious and meaningful ways with methodological assumptions has created a troubling post-positivist "methodological orthodoxy" (Hesse-Biber, 2010) that reinforces disciplinary hierarchies and privileges particular ways of knowing, which is antithetical to many values and practices of interpretivist researcher. Mixed methods research accentuates the importance of identifying assumptions underpinning our research endeavours. As such, for both purists and philosophical pragmatists, epistemological and ontological assumptions matter, while mixed methods pragmatists believe epistemological and ontological assumptions do not. However, that does not mean that adopting a mixed methods pragmatism means anything goes. We reiterate not offering a position does not mean you do not have one. Therefore, (mixed methods) researchers in outdoor studies will benefit from, firstly, articulating their assumptions and how they prejudice what we see as valuable, desirable and achievable in research as well as explaining legitimacy and quality criteria for their research. Second, carefully and considerately identifying where and how crossing boundaries might enable the creation of new knowledge and acknowledging where differences might be incommensurable. Third, embracing intellectual curiosity to gain understanding of the broadest possible range of processes, theories and perspectives used in research in part by engaging with research and researchers outside our own disciplinary and paradigmatic areas. Finally, extending the respect and value shown to methodological multiplicity in mixed methods research to recognising and appreciating ontological, epistemological and theoretical diversity.

Doing mixed methods research in outdoors studies

Given the careful considerations needed for mixed methods research as discussed above, it is still intuitively appealing and a highly useful approach to research. One theoretical and practical engagement with mixed methods research is underpinned by Dewey's (1910, pp. 72–78) five distinct steps in reflection. Morgan (2014, p. 30) defines this as "Dewey's Five step Model of Inquiry". Firstly, encountering a situation, the problem or research question, outside of our current experience, with no appropriate course of action. Secondly, reflecting on the nature of the problem, and using existing beliefs to think about why the situation is problematic. Thirdly, recognising what possible actions (research questions and research tools) would address the problem; the suggested solution. Fourthly, using existing beliefs to think about likely outcomes of the action, reflecting on the effects of the solution. Finally, following through on the suggested solution(s) to address the problem, the action taken. Dewey (1910, p. 79) titled this "The Double Movement of Reflection". For outdoor studies practitioner-researchers this process may well be familiar and useful, especially considering the foundational work of Dewey's pragmatic philosopy in outdoor studies (see, for example, Quay, 2013, and Quay and Seaman, 2013. However, as Morgan (2014) highlights, it is important here to differentiate pragmatism as a philosophical system from simple ideas about what is 'pragmatic' – and while there is an overlap in meaning, from a philosophically pragmatic point of view, there is no way that any human action can ever be separated from past experiences and the beliefs that have arisen from them. As such a broad definition of pragmatism is where the meaning of actions and beliefs is found in their consequences, and more specifically; actions cannot be separated from the situations and contexts in which they occur; actions are linked to consequences in ways that are open to change; actions depend on worldviews that are socially shared sets of beliefs (Morgan, 2014).

Recent examples of mixed methods research in outdoor studies

There is a wealth of research papers that can be found with 'mixed methods research', 'outdoor' and 'outdoors' as keywords in library database searches. We selected from this a number that may be of interest to the reader of this chapter both for content, and to show how mixed methods research is used and discussed across the range of research areas and contexts for outdoor studies, coming from five different academic journals. This range highlights the popularity of this approach to research. The full references are below and include: Atencio and Tan (2016) on PE teachers in Singapore; Collins, Carson, Amos and Collins (2018) on mountain leaders in the UK; Gull, Goldenstein and Rosengarten (2018) on the benefits and risks of tree climbing on child development and resiliency; Jirásek, Veselský and Poslt (2017) on the spiritual aspects of environmental education in winter outdoor trekking; and Savery et al. (2017) on how Forest School engagements influence the perceptions of risk.

When considering the methods sections of these papers, it appears to us that the term mixed methods research is only used to outline the research tools used; a survey of some description followed by interviews – often semi-structured. There is no acknowledgement of the paradigmatic tensions of mixed methods research, as discussed above, and none of these articles reference the engagement with mixed methods specific literature as cited in this chapter. They are all peer-reviewed outdoor studies academic papers and it is apparent that the paradigmatic complexities of mixed methods research are not considered. Engagement with other supporting academic research texts is clear for the data analysis sections. It seems that mixed methods research is adopted as an intuitively appealing way to develop robust answers in a most pragmatic manner. However, the pragmatist philosophy applied to authors' own beliefs and values as to the ways of truth and knowing is missing; ontological, epistemological and methodological considerations are not discussed – perhaps because they just want to get on with their research? For us, this underscores the need and purpose of this chapter: to highlight the need for researchers to be more mindful and ontologically considered when opting for a mixed methods approach.

Conclusion

Commonly, mixed methods research purposes relate to improving accuracy, generating a fuller understanding through combining sources of data, mitigating biases or shortcomings in individual methods, developing findings through exploring contrasting data, and/or refining inclusion and exclusion criteria. However, hopefully we have shown that this is not a simplistic or straightforward process. Further, we hope we have demonstrated that mixing methods is potentially valuable and worthwhile and while paradigmatic differences matter it is important that "to argue that it is paradigms that are in contention is probably less useful than to probe where and how paradigms exhibit confluence and where and how paradigms exhibit differences, controversies, and contradictions" (Lincoln et al., 2011, p. 97). Similarly, we again underscore that adherence to particularised, exclusive or limiting definitions of what counts as mixed methods research and/or obligations to theoretical or methodological positions is not important compared to the need to engage with them. In other words, while critical difference exist both within and between research paradigms this should not deter researchers from exploring the potential of combining and transgressing ontological, epistemological and methodological lines.

References

Atencio, M. & Tan, Y.S.M. (2016). Teacher deliberation within the context of Singaporean curricular change: Pre-and in-service PE teachers' perceptions of outdoor education. *The Curriculum Journal, 27*(3), 368–386.

Bergman, M.M. (2011). The good, the bad, and the ugly in mixed methods research and design. *Journal of Mixed Methods Research, 5*(4), 271–275.

Collins, L., Carson, H.J., Amos, P. & Collins, D. (2018). Examining the perceived value of professional judgement and decision-making in mountain leaders in the UK: a mixed-methods investigation. *Journal of Adventure Education and Outdoor Learning, 18*(2), 132–147.

Creswell, J.W. (2011). Controversies in mixed methods research. In N.K. Denzin & Y.S. Lincoln (Eds.). *The SAGE handbook of qualitative research*. London: SAGE

Creswell, J.W. & Clark, V.L.P. (2017). *Designing and conducting mixed methods research* (3rd ed.). London: SAGE.

Creswell, J.W. & Creswell, J.D. (2017). *Research design: Qualitative, quantitative, and mixed methods approaches*. London: SAGE.

Denzin, N.K. (2012). Triangulation 2.0. *Journal of Mixed Methods Research, 6*(2), 80–88.

Dewey, J. (1910). *How we think*. Boston, MA: D.C. Heath Publishers.

Gibson, K. (2012). Two (or more) feet are better than one: Mixed method research in sport and physical culture. In K. Young and M. Atkinson (Eds.), *Qualitative research on sport and physical cultures* (pp. 213–232). London: Elsevier.

Gibson, K. (2016). Mixed methods research: Integrating qualitative research. In B. Smith & A. Sparkes (Eds.) *International handbook of qualitative methods in sport and exercise* (pp. 382–396). London: Routledge.

Greene, J., Caracelli, V. & Graham, W. (1989). Toward a conceptual framework for mixed-method evaluation designs. *Educational Evaluation and Policy Analysis, 11*(3), 255–274.

Gull, C., Goldenstein, S.L. & Rosengarten, T. (2018). Benefits and risks of tree climbing on child development and resiliency. *International Journal of Early Childhood Environmental Education, 5*(2), 10–29.

Hesse-Biber, S.N. (2010). *Mixed methods research: Merging theory with practice*. London: The Guilford Press.

Hesse-Biber, S.N. & Johnson, R.B. (Eds.). (2015). *The Oxford handbook of multimethod and mixed methods research inquiry*. Oxford: Oxford University Press.

Jirásek, I., Veselský, P. & Poslt, J. (2017). Winter outdoor trekking: spiritual aspects of environmental education. *Environmental Education Research, 23*(1), 1–22.

Lincoln, Y., Lynham, S. & Guba, E. (2011). Paradigmatic controversies, contradictions, and emerging confluences, revisited. In N.K. Denzin & Y.S. Lincoln (Eds.). *The SAGE handbook of qualitative research* (pp. 97–128). London: SAGE.

Mertens, D. (2010). Philosophy in mixed methods teaching: The transformative paradigm as illustration. International Journal of Multiple Research Approaches, 4(1), 9–18.

Mertens, D. & Hesse-Biber, S. (2012). Triangulation and mixed methods research: Provocative positions. *Journal of Mixed Methods Research, 6*(2), 75–79.

Morgan, D.L. (2014). *Integrating qualitative and quantitative methods: A pragmatic approach*. London: SAGE.

Quay, J. (2013). *Education, experience and existence: Engaging Dewey, Peirce and Heidegger*. Abingdon: Routledge.

Quay, J. & Seaman, J. (2013). *John Dewey and education outdoors: Making sense of the "educational situation" through more than a century of progressive reforms*. Rotterdam: Sense Publishers.

Sale, J.E., Lohfeld, L.H. & Brazil, K. (2002). Revisiting the quantitative-qualitative debate: Implications for mixed-methods research. *Quality & Quantity, 36*(1), 43–53.

Savery, A., Cain, T., Garner, J., Jones, T., Kynaston, E., Mould, K., Nicholson, L., Proctor, S., Pugh, R., Rickard, E. & Wilson, D. (2017). Does engagement in Forest School influence perceptions of risk, held by children, their parents, and their school staff? *Education 3–13, 45*(5), 519–531.

Smith, B. & McGannon. K.R. (2017). Developing rigor in qualitative research. *International Review of Sport and Exercise Psychology*. Advance online publication. doi:10.1080/1750984X.2017.1317357

Smith, B., Sparkes, A., Phoenix, C. & Kirby, J. (2012). Qualitative research in physical therapy: A critical discussion on mixed-method research. *Physical Therapy Reviews, 17*(6), 374–381.

Sparkes, A.C. (2012). The paradigms debate: An extended review and a celebration of difference. *Research in physical education and sport. Exploring alternative visions* (pp. 9–60). Abingdon: Routledge.

Sparkes, A. (2015). Developing mixed methods research in sport and exercise psychology: Critical reflections on five points of controversy. *Psychology of Sport and Exercise, 16*(3), 4–59.

Sparkes, A.C. & Smith, B. (2009). Judging the quality of qualitative inquiry: Criteriology and relativism in action. *Psychology of Sport and Exercise, 10*, 491–497.

Sparkes, A. & Smith, B. (2014). *Qualitative research in sport, exercise & health sciences. From process to product*. London: Routledge.

Tashakkori, A. & Creswell, J.W. (2007). The new era of mixed methods. *Journal of Mixed Methods Research, 1*(1), 3–7.

Teddlie, C. & Tashakkori, A. (2010). Major issues and controversies in the use of mixed methods in the social and behavioral sciences. In Tashakkori, A. & Teddlie, C. (Eds.) Handbook of mixed methods in social and behavioral research (2nd ed., pp. 3-50). Thousand Oaks, CA: SAGE.

25

MIXED METHODS RESEARCH IN OUTDOOR STUDIES

Practical applications

Suzanne Peacock and Eric Brymer

This chapter outlines the practical implications of mixed methods research (MMR) in outdoor studies. A distinction between research methodology and research method is made (Creswell & Plano Clark, 2011; Wahhyuni, 2012). First, we briefly discuss the importance of research paradigms in MMR. Second, we overview and provide a rationale for MMR. A case study exploring the practicalities and challenges of MMR illustrates the process.

Research paradigms

A research paradigm is a philosophical belief system (or worldview) characterised by epistemological (philosophy of knowledge), ontological (philosophy of reality), axiological (role of values) and methodological (practices used to gain knowledge) assumptions that guide and direct researchers (Denzin & Lincoln, 2017). A major concern for MMR is the researcher's paradigmatic foundation (Shannon-Baker, 2016; Sparkes, 2015). A "researcher's ontological, epistemological and methodological commitments will constrain which methods can be used" (Sparkes, 2015, p. 50). Tashakkori and Teddlie (1998) identified four major research paradigms: (a) positivism; (b) post-positivism; (c) interpretivism; and (d) pragmatism.

Briefly, *positivism* is associated with empiricism, where observation and measurement are paramount. Knowledge and verifiable claims are only credible through quantifiable measures and systematic analysis (Krauss, 2005; Patton, 2002; Tashakkori & Teddlie, 1998). Positivists claim one objective reality and reject speculations and abstractions (Krauss 2005). The approach seeks regularities and causal-relationships, using deductive reasoning to test hypothesis or theory and reduce phenomena to the simplest form (Krauss, 2005). Findings reflect 'reality' and universal time and context-free generalisations are made (Wahyuni, 2012).

Post-positivism, emerging from criticisms of positivism (Guba & Lincoln, 1994) does not simply collate data to arrive at 'truth' and make law-like generalisations. Post-positivism aims to generate meaning and understanding and is associated with discovery, explanation and interpretation (McGregor & Murnane, 2010). Knowledge is relative and socially constructed, measurement is fallible and research is influenced by the values and experiences of the researcher (Patton, 2002). Truth can only be understood imperfectly, with varying degrees of certainty (Mertens & Wilson, 2012).

Post-positivists draw upon quantitative and qualitative methods to develop empirical evidence in real world settings (Patton, 2002). The approach provides a comprehensive and holistic insight to phenomena. However, critics suggest that inquiry is value-laden. Post-positivists aspire to achieve research objectivity, often by limiting their contact or 'involvement' with participants (Mertens & Wilson, 2012).

Interpretivism is associated with social science and practical reason, favouring qualitative research approaches to develop rich descriptions to understand subjectively meaningful experiences. Interpretivists recognise that human behaviour and experience are complex phenomena. With an emphasis on meaning, this holistic approach draws on the lived experience of individuals and generates connections between psychological, social, cultural and historical elements of life while also considering the context in which these occur (Ritchie & Lewis, 2003). Inquiry is value-bound, with the researcher being part of what is being researched (Frels & Onwuegbuzie, 2013). Interpretivists recognise the inseparability of the researcher and researched and the influence each has on data collection and analysis (Wahyuni, 2012).

The underlying assumptions associated with positivism and interpretivism are fundamentally different. The debate between the two stances has been termed the "paradigm war" (Reichardt & Rallis, 1994) which lead to the "incompatibility thesis" (Tashakkori & Teddlie, 1998), where 'purists' argue for the total incompatibility of these two forms of enquiry (Salkind, 2010; Frels & Onwuegbuzie, 2013).

Although perspectives (sometimes referred to as paradigms) such as transformative-emancipation, critical realism and dialectics warrant attention in MMR, *pragmatism* is most often associated with MMR. Advocates of pragmatism argue for a continuum approach between positivism and interpretivism. They reject the incompatibility thesis, believing that quantitative and qualitative research approaches should be viewed as complementary and suitable for use within a researcher's 'toolkit' (Ritchie & Lewis, 2003). Pragmatists initially focus on the research question before considering ontological and epistemological assumptions (Wahyuni, 2012). Through the combination of inductive and deductive thinking, their approach focuses on a methodological appropriateness approach to address specific research problems. In this chapter we focus on the application of the pragmatic approach.

Mixed methods research

Mixed methods have emerged as popular for researchers interested in gaining a rich understanding of social phenomena that could not be obtained by using one approach alone (e.g. Creswell & Plano Clark, 2011; Johnson, Anthony & Turner,

2007). This relatively new research approach provides a framework to guide the researcher interested in combining qualitative and quantitative research techniques to explore a phenomenon (Tashakkori & Teddlie, 2010; Teddlie & Tashakkori, 2009). Although the definition of mixed methods research is contested Johnson et al's. (2007) analysis of 19 existing definitions suggest that MMR is:

> ...an intellectual and practical synthesis based on qualitative and quantitative research; it is the third methodological or research paradigm (along with qualitative and quantitative research). It recognizes the importance of traditional quantitative and qualitative research but also offers a powerful third paradigm choice that often will provide the most informative, complete, balanced and useful research results.
>
> p. 129

Combining qualitative and quantitative methods is thought, by some, to generate the most complete analysis of a phenomenon, and provide a greater comprehensive insight and understanding of a topic (see Greene et al., 1989, and Bryman, 2006, summarised in Table 25.1).

Although MMR has strengths and benefits, there are a number of challenges. Mixing methods could be viewed as limited by methodological purists (Sparkes, 2015). MMR is often more time consuming, labour intensive and (sometimes) expensive (Teye, 2012). Teye (2012) found that the anticipated time and energy for quantitative and qualitative strands may alter once data collection begins. Consequently, although equal priority might have initially been allocated to each strand, researchers may find the weighting changes once research is initiated.

Effective MMR requires effective integration of findings (Bryman, 2006; Teye, 2012). Each strand must be developed fully, in terms of data collection, analysis and inferences, before integrating findings at the final stage (Tashakkori & Teddlie, 2003). Researchers are also challenged when handling conflicting data (Teddlie & Tashakkori, 2009; Teye, 2012). However, discrepancies between the quantitative and qualitative findings do not have to be viewed as a methodological limitation (Meetoo & Temple, 2003). When equal weighting is allocated to the quantitative and qualitative strand, conflicts can be highlighted, discussed and viewed as an opportunity for further exploration. Effective MMR design is important.

Types of research designs

There are a number of approaches to MMR design (e.g. Creswell & Plano Clark, 2011; Hesse-Biber, 2010). The most appropriate designs for research in the outdoor context are: (i) convergent parallel design (collection and analysis of quantitative and qualitative data remains independent before combining for an overall interpretation); (ii) explanatory sequential design (collection and analysis of quantitative data is followed by collection and analysis of qualitative data to help explain quantitative data); (iii) exploratory sequential design (collection and analysis of qualitative data is

TABLE 25.1 Reasons for using mixed method research

Greene et al., 1989 seeks:	
Triangulation	Convergence, corroboration, correspondence of results from different methods.
Complementarity	Elaboration, enhancement, illustration, clarification of the results from one method with the results from another.
Development	To use the results from one method to help develop or inform the other method, where development is broadly construed to include sampling and implementation, as well as measurement decisions.
Initiation	The discovery of paradox and contradiction, new perspectives of frameworks, the recasting of questions or results from one method with questions or results from the other method.
Expansion	To extend the breadth and range of enquiry by using different methods for different inquiry components.
Bryman 2006 refers to:	
Triangulation	Quantitative and qualitative research might be combined to triangulate findings in order that they may be mutually corroborated.
Offset	The research methods associated with both quantitative and qualitative research have their own strengths and weaknesses so that combining them allows the researcher to offset their weaknesses to draw on the strengths of both.
Completeness	The researcher can bring together a more comprehensive account of the area of enquiry in which he or she is interested if both quantitative and qualitative research are employed.
Process	Quantitative research provides an account of structures in social life, but qualitative research provides sense of process.
Different research questions	Quantitative and qualitative research can each answer different research.
Explanation	When one is used to help explain findings generated by the other.
Unexpected results	Quantitative and qualitative research can be fruitfully combined when one generates surprising results that can be understood by employing the other.
Instrument development	Contexts in which qualitative research is employed to develop questionnaire and scale items.
Sampling	Situations in which one approach is used to facilitate the sampling of respondents or cases.
Credibility	Suggestions that employing both approaches enhances the integrity of findings.
Context	Cases in which the combination is rationalised in terms of qualitative research providing contextual understanding coupled with either generalisable, externally valid findings or broad relationships among variables uncovered through a survey.

TABLE 25.1 (Cont.)

Bryman 2006 refers to:	
Illustration	Use of qualitative data to illustrate quantitative findings, often referred to as putting 'meat on the bones' of 'dry' quantitative findings.
Utility findings	Combining the two approaches will be more useful to practitioners and others.
Confirm and discover	Using qualitative data to generate hypotheses and using quantitative research to test them within a single project.
Diversity of views	Includes two slightly different rationales – namely, combining researchers' and participants' perspectives through quantitative and qualitative research respectively, and uncovering relationships between variables through quantitative research while also revealing meanings among research participants through qualitative research.
Enhancement	Entails a reference to making more of or augmenting either quantitative or qualitative findings by gathering data using a qualitative or quantitative research approach.

followed by collection and analysis of quantitative data to test the relationship found in quantitative data); and (iv) embedded design (addition of quantitative data to a largely qualitative study, or vice versa) (see Figure 25.1).

Mixed methods research in action: A case study

In 2011 The Royal British Legion (TRBL) initiated a five-day adapted sport and adventurous training (AS & AAT) coaching recovery programme for wounded (battle casualties), injured (non-battle causalities) and sick (mental/physical illness) (WIS) in-Service personnel through the Battle Back Centre (BBC). The research aimed to identify the psychological wellbeing impact of the intervention and explicate participant experiences during the programme. It also aimed to provide insight regarding the underlying processes or mechanisms, which generated programme outcomes.

Research design

Pragmatism was deemed appropriate because it facilitated quantitative *and* qualitative approaches, which offered the possibility of generating a more complete and comprehensive understanding of participant experiences. A pragmatic worldview also meant that the research question guided the choice of appropriate methods, techniques and procedures. Quantitative and qualitative approaches were combined to capitalise their strengths and facilitate the development of knowledge.

The pragmatist stance taken recognised that the research would be value-bound, because the researcher and participant are linked. Values and beliefs of the researcher

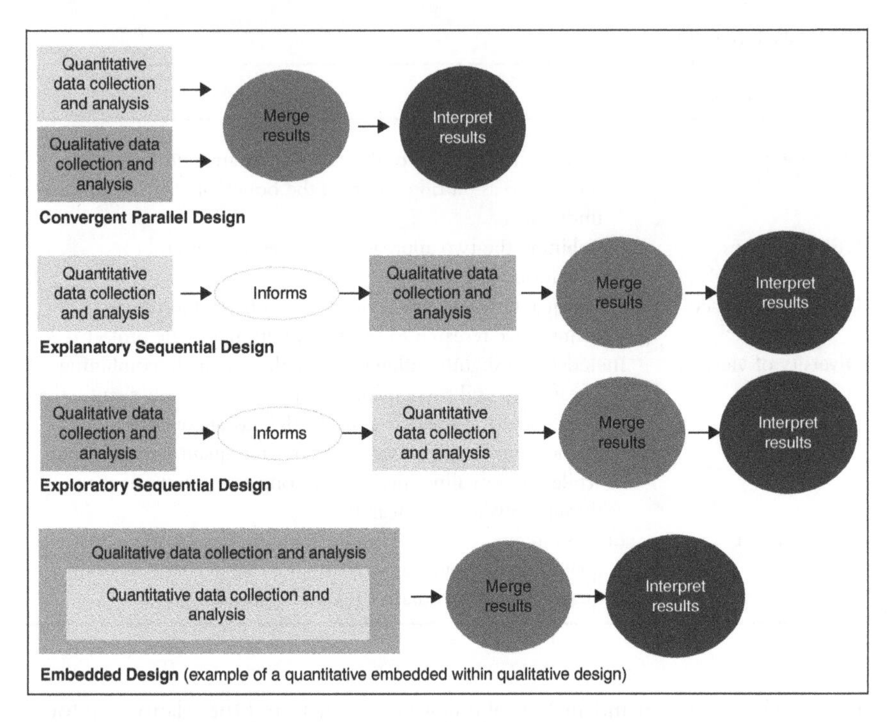

FIGURE 25.1 Mixed method research design frameworks
Source: Developed from Creswell & Plano Clark, 2011

would influence the researched and knowledge of existing frameworks and theories would impact observations and interpretation.

Rationale for a mixed methods research design

The contractual requirements for the project required quantifiable programme benefits. A pre-post quantitative design quickly highlighted the impact of a programme across a large sample and findings could be generalised and compared to other research using the same measure. This approach also generated graphical summative representations of data, which were necessary when presenting findings to various audiences and stakeholders. The findings also established the merit of the programme and had implications for the ongoing financial investment.

However, the quantitative approach was ineffectual for gathering participant experiences and perceptions during the programme. Quantitative methods were also limited in their capacity to explore how or why the programme benefited/hindered development and address the role of the programme within the context of an individual's life trajectory. For this reason, a qualitative strand was also included in the research.

Referring to Bryman's (2006) detailed framework (Table 25.1), this research used a mixed method design for the purpose of triangulation, completeness and explanation. Combining quantitative and qualitative research strands provided a comprehensive account of the phenomenon and enabled findings to be synthesised, compared and explained. This approach generated context and illustration, with the two strands complementing each other and providing breadth and depth to the phenomenon. The mixed method approach also capitalised on the strengths of each method, adding credibility to the research. Finally, the approach increased the utility of the findings and generated unexpected results.

Developing a mixed methods research design

Initially, the lead author adopted the role of participant observer. Participant observation (PO) is a process of immersing oneself into a setting to achieve a level of understanding and insight about the people and environment that will be shared with others (Schensul, Schensul & LeCompte, 2013; Watts, 2011; Wolcott, 2005) (see Chapter 10). For 18 months Suzanne was an active member of the setting and was involved in the day-to-day life of each programme. She engaged in all the briefing sessions and participated in the AAT and AS activities alongside participants. She shared her meals with participants and joined the group for social events.

Adopting this role meant Suzanne was the primary instrument for data collection. She focused on listening to participant conversations during the week and recorded these accounts and conversations in a fieldwork diary (Figure 25.2). In particular, she focused on comments which shed light on participant's experiences and perceptions.

While the PO process created an overall impression of participant responses and perceptions, semi-structured interviews facilitated deeper explorations. A conversational strategy was employed with an interview guide approach (Patton, 2002). The interview guide provided provisional structure while still supporting freedom and flexibility to explore and develop conversation naturally and spontaneously. The interview guide was framed to explicate: 1) participant background and their experience of injury and/or illness; 2) expectations of the BBC; and 3) BBC experience.

Initial findings from the PO suggested that the programme was consistent with mechanisms of change associated with Self-Determination Theory (Ryan & Deci, 2000). This initiated a change in the research strategy with the addition of a pre-post assessment using validated questionnaires. Psychological needs satisfaction was measured through the 21-item Basic Need Satisfaction in General Scale (BNSG-S) (Gagné, 2003) and positive mental health was measured via the Warwick-Edinburgh Mental Well-Being Scale (WEMWBS) (Tennant et al., 2007).

Quantitative and qualitative strands continued in parallel (see Figure 25.3). Ultimately, the pre-test post-test added breadth by identifying immediate impact on the recovery of WIS personnel attending the Centre. The 18 months of fieldwork provided depth by generating insight into participant perceptions of, and responses to, the programme.

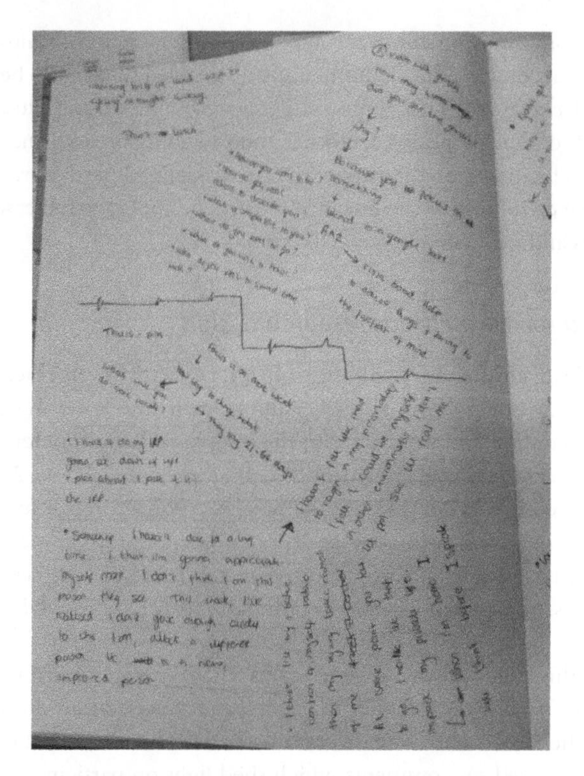

FIGURE 25.2 Example field notes taken from fieldwork diary

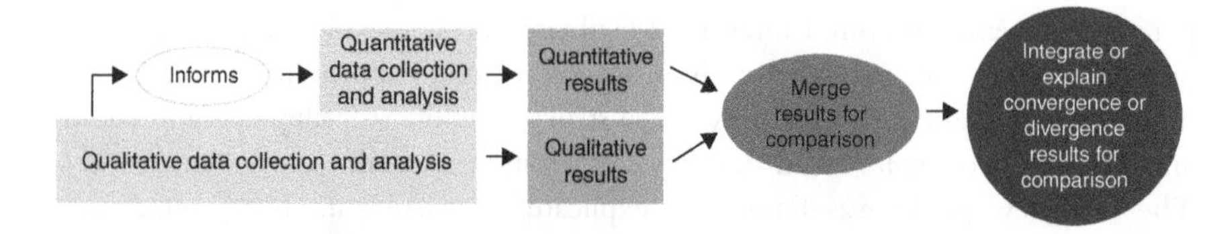

FIGURE 25.3 Procedural diagram for the mixed methods research design

Data analysis and integration

Quantitative and qualitative strands were analysed separately, and integrated later. Quantitative findings provided an initial snap-shot of impact, qualitative findings provided depth, context and meaning. The synthesising process followed a seven-stage process (Onwuegbuzie & Teddlie, 2003):

1. *Data reduction*: reduces the dimensionality of the data.
2. *Data display*: provides pictorial representations of the data.
3. *Data transformation*: qualitative data are converted to numerical data, or quantitative data are qualitised.

4. *Data correlation*: quantitative data are correlated with qualitised data or qualitative data are correlated with quantitised data.
5. *Data consolidation*: quantitative and qualitative data are combined to create new or consolidated variables or data sets.
6. *Data comparison*: data from the qualitative and quantitative data sources are compared.
7. *Data integration*: quantitative and qualitative data are integrated into a coherent whole.

Following stages 1 and 2, quantitative data was statistically analysed, with findings displayed in tables and figures and qualitative data was thematised, summarised and presented within a thematic map. Stages 3 and 4 are optional and were omitted. Since this research was underpinned by pragmatism, placing equal weight on quantitative and qualitative data, transforming the data would compromise this balance. Also, the aim of triangulation, completeness, explanation and illustration (Bryman, 2006), would be compromised. Finally, quantitative and qualitative findings were consolidated, compared and integrated (stage 5 to 7).

Quantitative findings

Paired sample t-tests were conducted on each of the dependent variables (Positive Mental Health [PMH], Autonomy, Competence and Relatedness) to identify statistically significant changes between arrival and course completion. Effect size was presented to establish the magnitude of the intervention effect (Thalheimer & Cook, 2002) (see Table 25.2 and Figures 25.4 and 25.5 for overview). Measures of PMH and psychological need satisfaction showed statistically significant increases from baseline to course completion. Effect size was large, the largest increase was in PMH, rising by 16%. For psychological needs, Competence increased most, rising by 14%, followed by Autonomy and Relatedness.

Qualitative findings

A thematic analysis was conducted on qualitative data following recommendations made by Braun and Clarke (2006). The thematic analysis identified five broad themes (see Figure 25.6): 1) "It's not what I expected"; 2) "I don't know where I'm going; but I like what I'm seeing now"; 3) "The 'belief' factor has been resurrected"; 4) "The attitudes towards people being ill or sick are terrible"; and 5) "It's all about the people and the interaction".

Integration of the quantitative and qualitative strands

The findings were then integrated. The research found that suffering and growth can go hand in hand. The quantitative and qualitative strands collectively indicate personal progress. Although 'progress' is a personally meaningful and individualised outcome it was reflected in a renewed or restored hopeful sense of self (Figure 25.7).

TABLE 25.2 Descriptive statistics, reliability coefficients and change in scores of Positive Mental Health and Basic Psychological Need Satisfaction (LOCF)

Measure & Variable	*n*	*Time 1*		*Time 2*		*Change in Score*				
		Mean (SD)	α	*Mean (SD)*	α	*Mean (SD)*	*%*	*95% CI*	*p- value*	*eta (effect size)*
PMH	971	45.05 (11.37)	.95	52.24 (10.3)	.95	7.19 (9.61)	15.9	-7.8 to -6.59	.000	.36 (large)
Autonomy	957	4.41 (.98)	.72	4.63 (.93)	.72	.27 (.84)	6.1	-.32 to -.22	.000	.10 (large)
Competence	957	4.26 (.97)	.65	4.73 (.96)	.68	.46 (.90)	10.87	-.53 to -.41	.000	.21 (large)
Relatedness	957	4.93 (1.03)	.67	5.19 (.97)	.66	.26 (.86)	5.27	-.21 to -.20	.000	.08 (moderate)

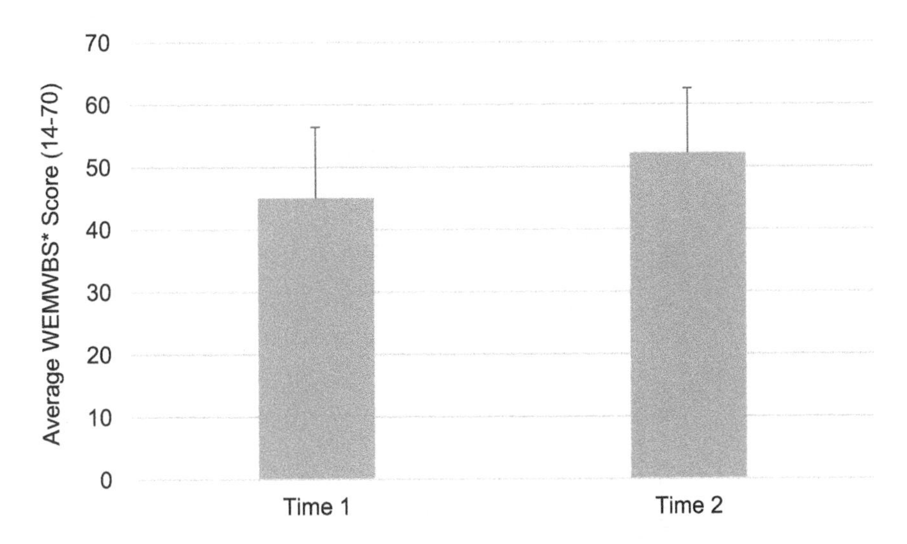

FIGURE 25.4 Changes to positive mental health (*Warwick-Edinburgh Mental Well-Being Scale)

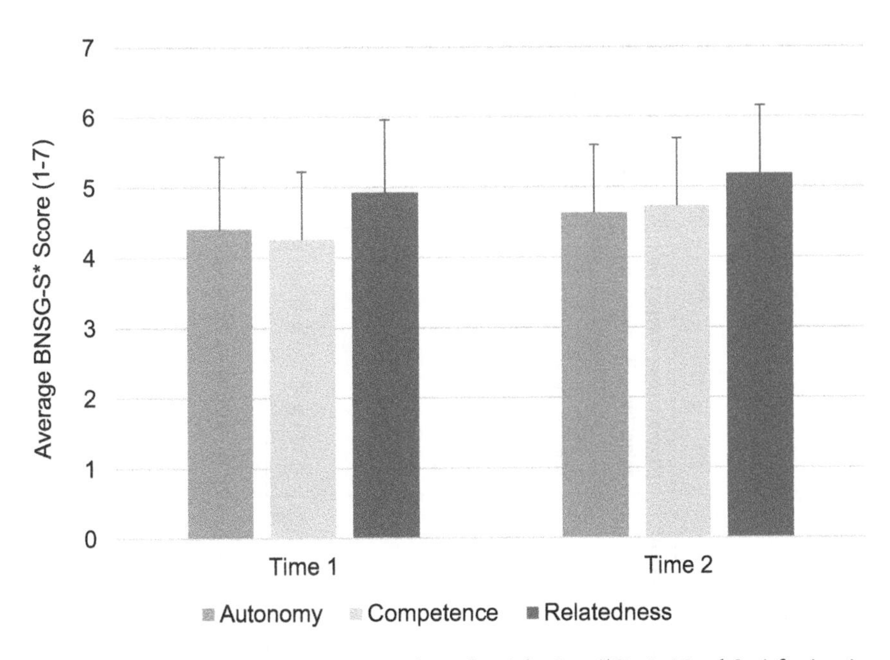

FIGURE 25.5 Changes to psychological need satisfaction (*Basic Need Satisfaction in General Scale)

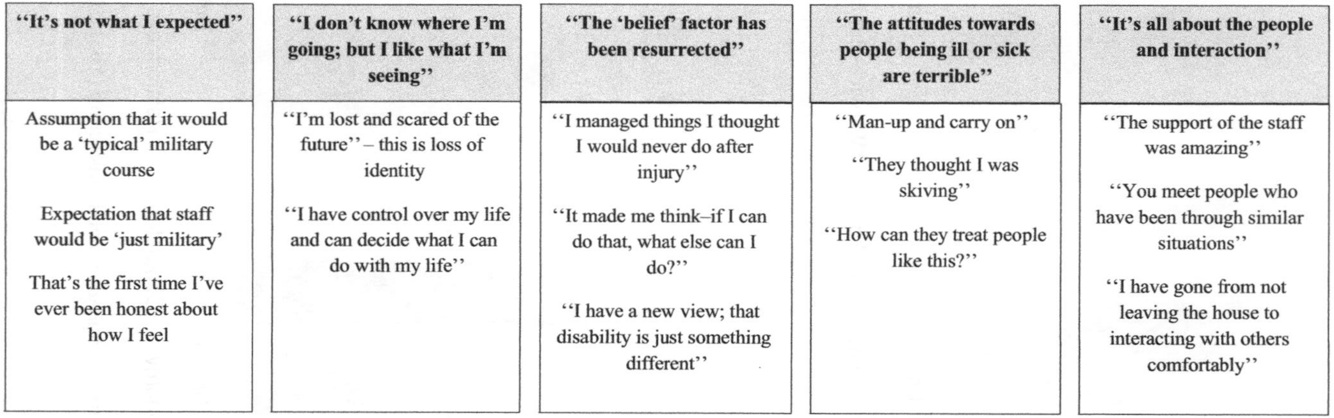

"It's not what I expected"

Assumption that it would be a 'typical' military course

Expectation that staff would be 'just military'

That's the first time I've ever been honest about how I feel

"I don't know where I'm going; but I like what I'm seeing"

"I'm lost and scared of the future" – this is loss of identity

"I have control over my life and can decide what I can do with my life"

"The 'belief' factor has been resurrected"

"I managed things I thought I would never do after injury"

"It made me think–if I can do that, what else can I do?"

"I have a new view; that disability is just something different"

"The attitudes towards people being ill or sick are terrible"

"Man-up and carry on"

"They thought I was skiving"

"How can they treat people like this?"

"It's all about the people and interaction"

"The support of the staff was amazing"

"You meet people who have been through similar situations"

"I have gone from not leaving the house to interacting with others comfortably"

FIGURE 25.6 Thematic map of qualitative findings

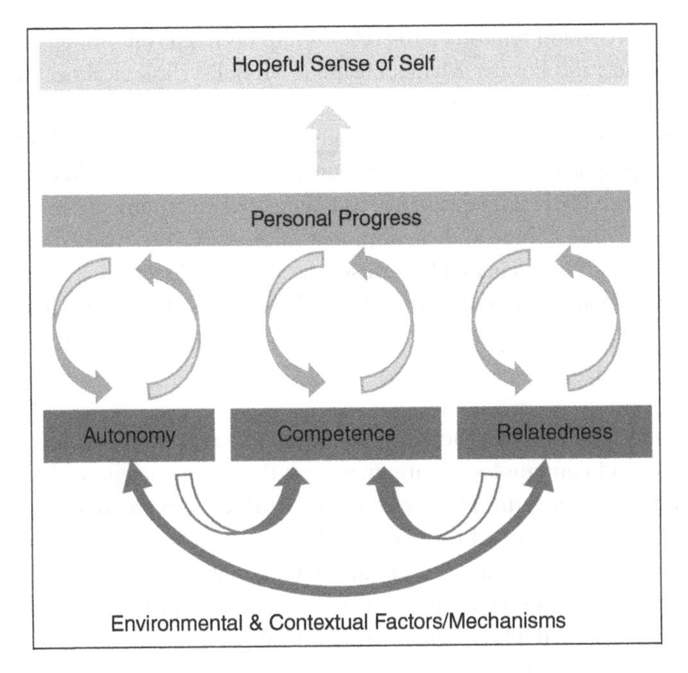

FIGURE 25.7 Graphical representation of the impact of the Battle Back Centre

Personal progress

Findings from the quantitative and qualitative strand suggest personal progress was marked by i) an altered perspective on life; ii) positive changes to self-perception; iii) feelings of optimism about the future; iv) the development of new goals and a reintegrated sense of purpose; v) an enhanced sense of connectedness and feelings of acceptance; vi) a restored or enhanced sense of belief and capability; vii) a willingness to be open and emotionally expressive; and viii) the expression of joyful emotion. Outcomes often considered minor or trivial for 'healthy' individuals, were frequently perceived as significant recovery milestone for WIS participants.

Hopeful sense of self

Findings suggested that personal progress linked to the restoration of a hopeful sense of self lost following injury. A positive sense of self underpins personal and social skills essential for wellbeing (Muenchberger, Kendall & Neal, 2008).

Autonomy, competence and relatedness

In line with literature, which indicates that psychological need satisfaction is necessary when striving for optimal psychological health and wellbeing (Ryan & Deci, 2000), the quantitative strand reinforced the relationship between Autonomy, Competence and Relatedness and Positive Mental Health (PMH). Psychological need satisfaction accounted for 41% of the variance in changes to PMH. Autonomy was the largest predictor of change, followed by Relatedness and Competence. However, there was a large proportion of variance left unexplained suggesting other psychosocial factors influence PMH. The qualitative data suggested environmental and contextual factors, possibly the culture and environment of the BBC, was markedly different to previous military experiences. The course was also outside a formal structure, non-militarised, informal and grounded in the philosophy "challenge-by-choice".

Summary

This chapter has outlined the thoughts and actions required when considering MMR. Ontological and epistemological arguments on the appropriateness of MMR were discussed. Recognising the purist critique of MMR we presented the established approach of pragmatism and then provided a real world example of how research following this approach could be conducted. Undertaking MMR effectively in the outdoor sector is in its infancy and as a consequence the principles guiding MMR are less well established. This is sure to change. Well thought out and designed MMR has significant benefits for research in outdoor studies, providing researchers with a valuable research tool. Poorly structured MMR is fraught with danger.

References

Braun, V. & Clarke, V. (2006). Using thematic analysis in psychology. *Qualitative Research in Psychology, 3*, 77–101

Bryman, A. (2006). Integrating quantitative and qualitative research: How is it done? *Qualitative Research, 6*(1), 97–113.

Creswell, J.W. & Plano Clark, V. (2011). *Designing and conducting mixed methods research.* Thousand Oaks, CA: SAGE.

Denzin, N. & Lincoln, Y. (2017). *The SAGE handbook of qualitative research* (5th ed.). Thousand Oaks, CA: SAGE.

Frels, R.K. & Onwuegbuzie, A.J. (2013). Administering quantitative instruments with qualitative interviews: A mixed research approach. *Journal of Counselling and Development, 91*(2), 184–194.

Gagné, M. (2003). The role of autonomy support and autonomy orientation in pro-social behavior engagement. *Motivation and Emotion, 27*(3), 199–223.

Greene, J., Valerie, C. & Graham, W. (1989). Toward a conceptual framework for mixed-method evaluation designs. *Educational Evaluation and Policy Analysis, 11*(3), 255–274.

Guba, E. & Lincoln, Y. (1994). Competing paradigms in qualitative research. In N.K. Denzin & S. Lincoln (Eds.) *The SAGE handbook of qualitative research.* Thousand Oaks, CA: SAGE.

Hesse-Biber, S. (2010). *Mixed methods research: Merging theory with practice.* London: The Guilford Press.

Johnson, B., Anthony, O. & Turner, L. (2007). Toward a definition of mixed methods research. *Journal of Mixed Methods Research, 1*(2), 112–133.

Krauss, S. (2005). Research paradigms and meaning making: A primer. *The Qualitative Report, 10*(4), 758–770.

McGregor, S. & Murnane, J. (2010). Paradigm, methodology and method: intellectual integrity in consumer scholarship. *International Journal of Consumer Studies*, *34*(4), 419–427.

Meetoo, D. & Temple, B. (2003). Issues in multi-method research: Constructing self-care. *International Journal of Qualitative Methods*, *2*(3), 1–21.

Mertens, D. & Wilson, A. (2012). *Programme evaluation theory and practice: A comprehensive guide*. New York, NY: The Guilford Press.

Muenchberger, H., Kendall, E. & Neal, R. (2008). Identity transition following traumatic brain injury: A dynamic process of contraction, expansion and tentative balance. *Brain Injury*, *22*, 979–992.

Onwuegbuzie, A. & Teddlie, C. (2003). A framework for analysing data in MM research. In A. Tashakkori & C. Teddlie (Eds.) *Handbook of mixed methods in social and behavioral research* (pp. 351–384). London: SAGE.

Patton, M.Q. (2002). *Qualitative research & evaluation methods*. London: SAGE.

Reichardt, C. & Rallis, S. (1994). The relationship between the qualitative and quantitative research traditions. *New Directions for Program Evaluation, 61*, 5.

Ritchie, J. & Lewis, J. (2003). *Qualitative research practice: A guide for social science students and researchers*. London: SAGE.

Ryan, R. & Deci, E. (2000). Self-determination theory and the facilitation of intrinsic motivation, social development, and well-being. *American Psychologist*, *55*(1), 68–78.

Salkind, N. (2010). *Encyclopedia of research design*: Los Angeles, CA: SAGE.

Schensul, S., Schensul, J. & LeCompte, M. (2013). *Initiating ethnographic research: A mixed methods approach*. Lanham, MD: Altamira Press.

Shannon-Baker, P. (2016). Making paradigms meaningful in mixed methods research. *Journal of Mixed Methods Research*, *10*(4), 319–334.

Sparkes, A. (2015). Developing mixed methods research in sport and exercise psychology: Critical reflections on five points of controversy. *Psychology of Sport and Exercise*, *16*(3), 49–59.

Tashakkori, A. & Teddlie, C. (1998). *Mixed methodology: Combining qualitative and quantitative approaches*. London: SAGE.

Tashakkori, A. & Teddlie, C. (2010). *Handbook of mixed methods in social and behavioral research* (2nd ed.). London: SAGE.

Teddlie, C. & Tashakkori, A. (2009). *Foundations of mixed methods research: Integrating quantitative and qualitative approaches in the social and behavioral sciences*. London: SAGE.

Tennant, R., Hiller, L., Fishwick, R., Platt, S., Stephen, J., Weich, S., Stewart-Brown, S. (2007). The Warwick- Edinburgh Mental Well-being Scale (WEMWBS): Development and UK validation. *Health and Quality of Life Outcomes*, *16*(9), 606–613.

Teye, J. (2012). Benefits, challenges, and dynamism of positionalities associated with mixed methods research in developing countries: Evidence from Ghana. *Journal of Mixed Methods Research, 6*(4), 379–391.

Thalheimer, W. & Cook, C. (2002). *How to calculate effect sizes from published research: A simplified methodology*. Work-Learning Research. Retrieved from: www.bwgriffin.com/gsu/courses/edur9131/content/Effect_Sizes_pdf5.pdf

Wahyuni, D. (2012). The research design maze: Understanding paradigms, cases, methods and methodologies. *Journal of Applied Management Accounting Research*, *10*(1), 69–80.

Watts, J.H. (2011). Ethical and practical challenges of participant observation in sensitive health research. *International Journal of Social Research Methodology: Theory & Practice*, *14*(4), 301–312.

Wolcott, H.F. (2005). *The art of fieldwork* (2nd ed.). Walnut Creek, CA: Altamira Press.

26

QUANTITATIVE ANALYSES OF SMALL SAMPLES WITH COMPLEX DATA-STRUCTURES

Ulrich Dettweiler

Evidence, facts and values in educational research

> Contemporary educational research has been experiencing an explosion of new methodologies and approaches to inquiry. Many of these approaches have drawn from philosophical or theoretical positions that underlie their determinations of research methods, aims, and criteria of validity. Yet the substance of these philosophical or theoretical assumptions is not always made clear to the reader, and so it is difficult [...] to judge those assumptions.
>
> Phillips & Burbules, 2000, p.vii

This uneasiness about empirical methodologies in education applies both to the stereotypical distinction of so-called 'quantitative' and so-called 'qualitative' approaches, as well as to the different methodologies within each paradigm. A distinction between 'fact' and 'value' still lingers in (educational) empirical research. However, any concept of the rationality of science is in fact deeply irrational, if it considers methods to be purely formal, distinct and free from value-judgements. Every researcher needs to specify his or her prior beliefs, and there is in fact a rich tradition, also in outdoor studies, to do so in so-called qualitative research approaches. Yet, in so-called quantitative designs, many researchers seem to believe that the encoding of observations in numbers and their statistical processing produces 'neutral facts' that warrant the language of 'evidence'. Admittedly, this naïve realist concept of 'evidence' is rather problematic, and especially in outdoor studies, many researchers doubt the significance of 'evidence' as an epistemic concept (Scrutton & Beames, 2015). This is clearly an obstacle to link outdoor studies and outdoor educational research within the wider discourse of education.

Within the outlined context, I will critically examine some core concepts of quantitative statistical analysis and demonstrate their usefulness and limitations.

I will then extend those basic concepts to more complex hierarchical models, which in most cases represent the data in a better way. With respect to the language of 'evidence', I will finally refer to Bayesian inference and demonstrate how statistical analyses can be conducted such that 'evidence' can be understood in constructivist rather than naïve realist terms.

Some very basic statistical methods: How and when to use them

The standard statistics course for graduate students typically will briefly cover research designs, and then go straight on to introducing different statistical methods and how to run the analyses in standard software packages. Often, statistics courses are disastrous, teaching neither concepts nor skills adequately (Gelman & Nolan, 2017), and thus, students (and very often their supervisors, too) think little about the statistical concepts and focus only on clicking the necessary options in the graphical user interface (GUI) of their respective statistical software package. Within this chapter, I will invite you to think in models, rather than in methods, and conceptualise research problems with their epistemic claims.

When we use statistical methods on data, we assume that we can express something of interest that is going on the world, mathematically. For this we use a model that enables us to reduce complexity, whilst keeping some core features of interest. Moreover, mathematical models are very precise and can quantify potential differences between groups or data-levels. Models are all around us; an architect, for example, builds a model of a house or a bridge for the sake of presenting, or testing the planned structure. The mathematical reduction here is the down-scaling of size. In such models, we define assumptions together with implications drawn from them by mathematical reasoning (Rodgers, 2010). A very basic statistical model for research data is the arithmetic mean value $\bar{x}$ of a variable x for $x_1, x_2, x_3, \ldots x_n$ observations. The mean is defined as the sum of the sampled values divided by the number of items in the example:

$$\bar{x} = \frac{1}{n}\left(\sum_{i=1}^{n} x_i\right) = \frac{x_1 + x_2 + x_3 + \ldots x_n}{n}.$$

The mean of a variable gives us an average value of an observation of interest, and it is very often the starting point of all statistical analyses. When looking at the arithmetic mean of a variable, we also want to know something about the range of the values that define the arithmetic mean, i.e. we need a measure (another model) that summarises the range or dispersion of the sampled values. Typically, we use the 'standard deviation'. A low standard deviation indicates that the data points tend to be close to the mean (also called the expected value) of the set, while a high standard deviation indicates that the data points are spread out over a wider range of values. The mathematical definition of the standard deviation, denoted as SD, s

or σ is a little bit more complicated, since it is the square root of its variance, i.e. the expectation of the squared deviation of the variable from its mean.

With the mean x, often also noted as μ, and the standard deviation σ, we have the basic concepts in order to test the relationship of two variables of interest. Here, two elementary operations are important, covariance and correlation. Both describe the degree to which two variables or sets of variables tend to deviate from their expected values in similar ways. For the two variables X and Y, with means (expected values) μ_X and μ_Y and standard deviations σ_X and σ_Y, respectively, their covariance and correlation are as follows:

$$\text{Covariance}: \quad \sigma_{XY} = E[(X - \mu_X) \times (Y - \mu_Y)]$$

$$\text{Correlation}: \quad \rho_{XY} = E\left[\frac{(X - \mu_X) \times (Y - \mu_Y)}{\sigma_X \times \sigma_Y}\right],$$

$$\text{so that} \quad \sigma_{XY} = \rho_{XY} \times \sigma_X \times \sigma_Y,$$

with $E[X]$ as the expected value operator.

Let us go a step further and look at some real examples. Typically, in outdoor studies research, we will have some sort of intervention that is going on outdoors for which we define outcome variables that are of interest with respect to some 'normal' indoor setting.

Most often, such studies are conceived as pre-post-(post) measurements, with the first time point of measurement before ('pre-') the intervention, also called 'baseline', the next time point of measurement shortly after ('post-') the intervention, and a follow-up measurement some sensible time after the intervention ('post-post'). The first thing for the data analysis is to decide which empirical model would best capture the research question. Most probably, we would be interested if the values directly after the intervention ('post-') are higher than before ('pre-'), and if an assumed higher level sustains over time ('post-post'). Maybe, we want to know about gender-differences or other categorical grouping factors (age group, ethnicity, previous experience …) or continuously scaled moderators (e.g. physical exercise levels, perceived stress, etc.).

Such an empirical model would translate into a statistical model, called 'regression' and we use it to predict a continuous outcome on the basis of one or more continuously scaled predictor variables. In contrast, an analysis of variance (ANOVA) is the statistical model that we use to predict a continuous outcome on the basis of one or more categorical predictor variables. A special case of ANOVA is if we have only a single categorical variable which only has two levels; then most people would describe the method/approach as a two-sample t-test. ANOVA-designs with both categorical and continuous predictor variables are called analyses of covariance (ANCOVA).

The simplest regression or ANOVA model for a continuous outcome variable Y for each observation i is

$$Y_i = X_i\beta + \varepsilon_i,$$

with normally distributed independent errors

$$\varepsilon_i \sim N(\mu, \sigma^2)$$

The term X_i defines the vector, i.e. the ith row in the matrix X, which is comprised of k predictor variables, such that

$$X_i\beta = \beta_1 X_{i1} + \beta_2 X_{i2} + \ldots \beta_k X_{ik}.$$

The β's are called 'slope'-parameters. They indicate the strength of the respective effects for each of the X's. The N in the error-term denotes the form of distribution of the errors or residuals, with N for normal.

In order to set up such models, we first have to specify a likelihood function for our data, which will give us the probability of observing the data given the parameters. Classically, this is defined as normal distribution for the model with a uniform distribution on the infinite range $(-\infty, \infty)$ for each of the components of β, and a uniform distribution $(0, \infty)$ for σ_i^2 as well. The likelihood function for the regression or ANOVA for the outcome variable Y_i can then be specified as

$$Y_i \sim N(\alpha_i + \beta x_i, \sigma^2),$$

where α_i is the intercept or arithmetic mean when βx_i equals zero, and βx_i is the (slope) parameter β for each observation i for the observed variable x. As stated above, this model can be expanded freely with as many ("k") predicting variables – categorically, ordinally or continuously scaled – that make empirical sense. Since all those models can be expressed in the same mathematical notation (Gelman, 2005), and share more or less the same assumptions, we can simply call them generalised linear models (GLM).

With the *decision* to model the data linearly, according to a normal function within an infinite range of possible values, there are already a number of value-driven presuppositions in the model before we even have started entering the data. The rationale behind the uniform probability functions used in standard GLM is that they contain as little information as possible, in order to make it a 'neutral', 'value-free' procedure. However, we could well conceive of many other distribution functions, with more specified parameters, that might even better fit the data, i.e. the encoded empirical world. And in fact, we should!

Let us go back to our example: We will furthermore have to decide – and this is a rather philosophical question – if we assume to measure *the same thing* at the three time points or not. In fact, the classical approach is somewhat counter-intuitive, coding those observations in different variables, assuming thus that they are not the same thing, and storing the data in wide format. A different approach would be to store the data in long format. Hereby, we treat each measure at the different timepoints as being *the same,* i.e. subsume it under one variable, but code this measure with a time-variable. This time-variable can be continuous (the factual date of measurement) or dichotomous (pre-, post-, post-post). Coding the 'real time' as continuous would give us the flexibility to test (and correct) for distances in time if this is appropriate for our research question. For example, let us say we are measuring anxiety level in an outdoor intervention during weeks 1, 2, 4 (pre-), 8 (post-), and 16 (post-post). Using the long format, we have the choice to treat those five time points as either five discrete categories or as true numbers, which accounts for the different spacing of the weeks.

In classical analyses of (co-)variances, this is not possible. Here, group means in a sample are compared, and with multiple measurements, those are only corrected with respect to the significance level. Hereby, the observed variance in a variable is partitioned into components attributable to different sources of variation. In its simplest form, ANOVA provides a statistical test of whether the population means of several groups are equal or not. So, if there are missing data, the default approach in ANOVA is listwise deletion, which drops any observation with any missing data on any variable involved in the analysis. If the percentage missing is small and the missing data are a random sample of the data set, this is a reasonable approach. But with small samples, each missing data cell is crucial. In this multivariate approach, if a child is missing one time point, then the whole data-row would be dropped from the entire analysis.

Let me give you an example from my own research. In a recent pilot-study on physical activity and mental health, we fitted an analysis of covariance (ANCOVA) in order to evaluate the influence of the eight-week intervention. Data were taken at baseline, ten days into the intervention, and directly after the intervention. The measurement at ten days was used as the response, treatment ('intervention group'/'control group', 'between-subject design') was the design factor, and the baseline score was a covariate. This would answer the question if people with the same entrance level of the mental health score in the respective treatment groups are expected to have the same score ten days into the programme. The baseline-covariate secures the individual entrance level for each participant. The ANCOVA was appropriate, since it was a randomised sample without missing data. Moreover, we wanted to compare our results to important reference studies which used this ANCOVA design, a statistical practice that is still very common in medical research.

For the explorative analysis of the trajectories of the outcome variable over the whole period of the intervention, however, we fitted multi-level linear models. Hereby, the number of training units at the actual number of days in the programme after baseline were used as covariates. The GLM was better fitted to answer the exploratory research question of whether there was an interaction effect between the actual number of training units and the number of days. And since a log-transformed outcome variable showed better data-fit, an exponential decay-model can be expected. If more time points are available, we can even include a time function in the model and explore which (linear, quadratic, exponential) fits the data best.

Clearly, the GLM is much more flexible and allows much richer analyses. Only within very limited circumstances, as with those above, would I use the ANCOVA design, and I would strongly recommend performing such analyses in the long format, i.e. specifying an appropriate (continuous or dichotomous) time-variable encoding the timepoint of measurement, for multiple measurements.

Hierarchical structures in data

With respect to the two above-mentioned examples, we want to have the data in the long format using generalised linear models, not only because it is less sensitive to missing data, but for a number of other reasons, too. First of all, in GLMs, the outcome variable does not need to be continuous, and those models work for other types of dependent variables too: categorical, ordinal, discrete counts, etc. So, if we have one of these outcomes, ANOVA is not an option. There is no Repeated Measures ANOVA equivalent for count or logistic regression models. Moreover, in many designs, as in our example above, we have multiple measures over time for each subject, but the subjects can also be clustered in some other grouping, for example students within classes, and classes in schools, schools in school districts, and so forth. Alternatively, as we have done in the second example above, we can allow individual values for each participant and cluster those in the respective treatment group. Such a structure is called hierarchical and, in GLMs, we can account for this hierarchical structure, basically by assigning an additional probability function for the intercept capturing this dependency structure. This is called a mixed model, for we 'free' the intercept while keeping the predictors 'fixed'. Those 'freed' parameters are also called 'random effects', as opposed to 'fixed effects'. Mathematically, this simple varying-intercept model with one predictor on the individual level i and one at the group level j can be expressed as:

$$Y_i \sim N(\alpha_{j[i]} + \beta x_i, \sigma_y^2)$$

and the auxiliary function for the intercept:

$$\alpha_j \sim N\left(\mu_\alpha, \sigma_\alpha^2\right).$$

The normal distribution for the α_j's can be thought of as a prior distribution for these intercepts. The parameters of this prior distribution, μ_α and σ_α^2, are called hyperparameters and are themselves estimated from the data. This auxiliary function, informed by the hyperparameters, in fact smooths the differences between the respective groups by 'borrowing strength', i.e. statistical power, between them. Statistical power is defined as the probability that the test correctly rejects the null hypothesis, H_0 (that there is no difference between groups) when a specific alternative hypothesis, H_1, is true (correct rejection at least 80% of the time is usually an accepted level of accuracy).

In order to illustrate why one should know and use the full hierarchical structure in the data, let me give another example from my research. We were interested in the motivational behaviour of children during a residential research week in contrast to their experiences in 'normal' science class (within-subjects design). We had n=281 pupils from ten different schools attending the courses, over three summer seasons. See the problem? One cannot possibly control such important factors as weather, course dynamics, the current political or societal situation. One research week is somehow different from the other, while we assume that the programme itself is similar. So, what to do? We could try to describe all those differences and define them as variables in the model. This would obviously not make much sense and would require thousands of observations to achieve adequate statistical power for such complex models.

So, what we can do is to use a statistical model that allows us to define specific intercept values for each research week. As explained above, the intercept is that value of the outcome variable when all explanatory parameter values are zero. By 'nesting' individuals in groups (research weeks), we can account for all kinds of differences, observed and unobserved, in that specific situation with respect to the dependent variable, because the intercepts are not fixed but defined for each individual in this group.

A repeated measures ANOVA cannot incorporate this extra clustering of subjects/groups in some other clustering, but hierarchical models can, and this kind of clustering can indeed get quite complicated, especially when you do not only allow for individual intercepts but also for individual slopes, in the so-called varying-intercept, varying-slope model, i.e. specify yet another probability function for the respective slope-parameter β.

You can see that those kinds of statistical models give us exactly that flexibility we need for complex empirical research settings.

Null-hypothesis-significance testing (NHST), p-values, and the Bayesian alternative

The statistical operations behind ANOVA and (multiple-) regression designs are based on null-hypothesis testing, i.e. the hypothesis that there is no significant difference between specified populations, and any observed difference being due to sampling or experimental error. Very often, the significance of this difference is expressed by means of a p-value. This practice has been criticised for a number of reasons and is subject to an ongoing debate in science, going so far that some journals do not accept p-values from NHST any longer (Trafimow & Marks, 2015) or have produced guidelines for proper use (Wasserstein & Lazar, 2016). Let me sum up the most important points of critique:

Since the p-value is basically a function of sample size, all null hypotheses can be rejected with a large enough sample. In a very useful simulation study, testing the p-value on different sample sizes, Krueger and Heck (2017) state that p-values perform "quite well as a heuristic cue in inductive inference, although there are identifiable limits to its usefulness" (p. 908). They conclude that despite its general usefulness, the p-value cannot bear the full burden of inductive inference. Rejecting the null does not provide logical or strong support for the alternative, since NHST is backwards: it evaluates the probability of the data by means of the hypothesis. Rodgers (2010) states that exactly the opposite should be done in science, we should always test the probability of the hypothesis by means of the data.

A first step towards this goal is to set up GLMs with all available variables of interest with their interaction terms, define their hierarchical structure, and determine how well the model fits the data. By stepwise reducing complexity in the data, and by comparing the respective models with each other – this can be done by means of a simple ANOVA between the models – we can determine which model fits the data best. In addition to the p-value, the confidence interval (CI; i.e. that range of values that might contain the true value of an unknown population parameter/mean given a confidence level), mostly 95%, should be determined and reported. The true value of the population mean is something obscure because it cannot, in practice, ever be determined. But all NHST procedures aim to exactly define that – what would be the 'true' value if we tossed the coin infinitely often. That is why such techniques are also called 'frequentist', and the analyses in this frame depend strongly on the sampling plan.

You can clearly see that such frequentist statistical approaches are quite orientated towards finding a 'real' or 'true' value, which reminds us of the above-mentioned language of 'evidence'.

In a recent paper, Wagenmakers et al. (2017) have summarised nicely the disadvantages of such NHST techniques, and proposed an alternative, i.e. Bayesian inference. For our purpose here, where we deal with complex situations with small samples, some of their arguments will be highlighted.

Firstly, the Bayesian alternative does not depend on a sample plan. We can simply add more data when they are available, or collect data from so-called 'convenience samples', e.g. school classes that take part in an intervention, which cannot be randomised. The second advantage is that Bayesian inferences yield robust results even with relatively small samples. This is because, in Bayesian statistics, the parameters in a given model are determined from the data, and then a simulation is run on randomly generated values within the range of the observed parameters with a huge ($n > 25.000$) sample. This simulation is run several times and the results are checked to see whether or not they converge agreeably. Then, the simulated results are compared to the observed ones, and when there is no significant difference, i.e. the Bayesian p-value is about 0.5, we have justification to trust our results from the small sample. Moreover, the evidence that the data provide for H_0 vs. H_1 can be straightforwardly quantified. And, perhaps the most important advantage of Bayesian inferences is that we can, and in fact (even better), we must, include expert knowledge, results from literature, or complementary data from mixed methods observations into our statistical models. Since we have to specify a prior probability function for each parameter in a Bayesian GLM, as we have done for the so-called 'random effects' above, we are free to also specify the distribution parameters. Very often, we have specific information at hand that limits the range of empirically sensible data. In such cases, it would seem irrational to allow the whole infinite range of values in our distribution function and not to use this extra information. If it makes sense to centre one parameter on a certain value, we can and should do so. And if we know about a sensible range of those parameter values, we can/should specify the standard deviation accordingly. Alternatively, if no information is available, we can define hyperparameters, i.e. inform each parameter from the data. For many researchers, this form of Bayesian model specification is too 'subjective'. In fact, the logic behind this form of scientific reasoning is deeply rooted in pragmatism, and endorses a concept of "subjective probability" (De Finetti, 1974; Jeffrey, 2004) that allows for a notion of 'evidence' that is generated through many subjective decisions in the parameter definitions.

However, this process is very clear and transparent on every parameter function used in the model and performed with utmost methodological rigour. As a side note: there is also a branch in Bayesian statistics that defines so-called 'default' or 'objective' prior probability functions, basically as an 'optimised' mix over different distribution forms (Jeffreys, 1961; Rouder & Morey, 2012). These methods best suit situations where researchers want to use the benefits of Bayesian inference but do not trust their subjective expert knowledge enough to follow the fully subjectivist pathway.

So, my advice is to begin each analysis with a brief correlation analysis of those variables that do empirically make sense to be compared. Then run quick ANOVA or regression analysis in order to check for statistically 'significant' co-dependency of several variables and/or interactions. In some cases, you may stop your analyses right here and go on and publish your 'significant' ($p < 0.05$) results. However, in other cases you may want to go one step further to account for the hierarchical structure in your data by setting up mixed-effects models which you can then compare and identify the one that best fits the data. Once you have done this, the extra step to do the full Bayesian analysis is quite simple, with new software packages on the market. There are very good free software options available, with great community support, such as JASP (JASP Team, 2016) if you want a GUI, or R (R Development Core Team, 2008) if you are happy with writing code and want more flexibility. Also, the classical IBM-SPSS package (IBM Corp., 2017) has included standard Bayesian models. Happy coding!

References

Beames, S., Higgins, P. & Nicol, R. (2012). *Learning outside the classroom: Theory and guidelines for practice*. London: Routledge.

Becker, C., Lauterbach, G., Spengler, S., Dettweiler, U. & Mess, F. (2017). Effects of regular classes in outdoor education settings: A systematic review on students' learning, social and health dimensions. *International Journal of Environmental Research and Public Health, 14*(5), 485.

De Finetti, B. (1974). *Theory of probability. A critical introductory treatment* (Vol. 1). New York, NY: John Wiley & Sons.

Gelman, A. (2005). Analysis of variance? Why it is more important than ever. *The Annals of Statistics, 33*(1), 1–53.
Gelman, A. & Nolan, D.A. (2017). *Teaching statistics: A bag of tricks* (2nd ed.). Oxford: Oxford University Press.
IBM Corp. (2017). *BM SPSS Statistics for Windows, Version 25.0* [Computer software]. Armonk, NY: IBM Corp.
JASP Team. (2016). *JASP (Version 0.8.0.1)* [Computer software].
Jeffrey, R.C. (2004). *Subjective probability: The real thing.* Cambridge, New York, NY: Cambridge University Press.
Jeffreys, H. (1961). *Theory of probability* (3rd ed.). Oxford: Oxford University Press.
Krueger, J.I. & Heck, P.R. (2017). The heuristic value of p in inductive statistical inference. *Frontiers in Psychology, 8*(908).
Phillips, D.C. & Burbules, N.C. (2000). *Postpositivism and educational research.* Lanham, MD, Oxford: Rowman & Littlefield Publishers.
R Development Core Team. (2008). *R: A language and environment for statistical computing* [Computer software]. Vienna: R Foundation for Statistical Computing.
Rodgers, J.L. (2010). The epistemology of mathematical and statistical modeling: A quiet methodological revolution. *American Psychologist, 65*(1), 1–12.
Rouder, J.N. & Morey, R.D. (2012). Default Bayes factors for model selection in regression. *Multivariate Behavioral Research, 47*(6), 877–903.
Scrutton, R. & Beames, S. (2015). Measuring the unmeasurable: Upholding rigor in quantitative studies of personal and social development in outdoor adventure education. *Journal of Experiential Education, 38*(1), 8–25.
Trafimow, D. & Marks, M. (2015). Editorial. *Basic and Applied Social Psychology, 37*(1), 1–2.
Wagenmakers, E.J., Marsman, M., Jamil, T., Ly, A., Verhagen, J., Love, J., … Morey, R.D. (2017). Bayesian inference for psychology. Part I: Theoretical advantages and practical ramifications. *Psychonomic Bulletin and Review, 25*(1), 35–57.
Waite, S. (2011a). *Children learning outside the classroom: From birth to eleven.* London: SAGE.
Waite, S. (2011b). Teaching and learning outside the classroom: Personal values, alternative pedagogies and standards. *Education 3–13, 39*(1), 65–82.
Wasserstein, R.L. & Lazar, N.A. (2016). The ASA's statement on p-values: Context, process, and purpose. *The American Statistician, 70*(2), 129–133.

PART V

Disseminating, communicating and sharing research

27

PUBLISHING AND DISSEMINATING OUTDOOR STUDIES RESEARCH

Linda Allin, Heather Prince and Barbara Humberstone

Introduction

The inclusion of a chapter on publishing in outdoor studies is testament to the growth of outdoor studies as an academic discipline and its visibility over the past decade. There is now a wide range of publishing outlets for outdoor studies research and an increasing number of quality academic journals. Publishers are becoming more sophisticated and finding new ways to access a wide readership (see Prince, Christie, Humberstone & Gurholt, 2018; Taylor & Francis, 2018). These include having online and Open Access options as well as print presence (e.g. tandfonline.com) and use of social media such as twitter (e.g. @tandfsport). Alongside the increase in published material, outdoor research has also become more rigorous and ambitious, with the editorial board of outdoor journals such as the *Journal of Adventure Education and Outdoor Learning* encouraging "papers engaging with critical, theoretical and methodological perspectives" (Taylor & Francis, 2018). With current societal issues such as the decline in children's outdoor play, a rise in mental health problems, concerns over risk, equity and social justice, sustainability and global environmental challenges, contemporary research in outdoor studies can have great benefit and influence. Publishing for research impact and the metrics used for judging outdoor research quality needs to be understood. It is also important to recognise that publishing is a vital feature of an increasingly competitive research culture in universities (Yokoyama, 2006), where evidence of high-quality publications and the "effect on, change or benefit to the economy, society, culture, public policy or services, health, the environment or quality of life, beyond academia" are both markers of academic reputation, and inform funding decisions (Research Excellence Framework, 2021). As such, the need to publish in outdoor studies and show its value has arguably become greater than ever.

Why publish outdoor research?

There are many reasons why people choose to publish research. At a fundamental level, one may ask why undertake research at all if the knowledge is not to be shared or disseminated? For those employed in a university, publishing is what is expected in the role of a lecturer or researcher and could be considered a defining characteristic of working in academia. Research publications also underpin research-rich teaching expectations in universities, which are increasing across academia in many Western nations. Eley, Wellington, Pitts and Biggs (2012) highlight a number of additional extrinsic motivations for publishing associated with a university job role: improving one's CV, gaining promotion and/or gaining respect in the field. An external driver for many researchers in universities is typically to be able to contribute to an external assessment body such as the Research Excellence Framework (REF) in the UK, the results of which are the main criteria by which both they and their institution will be judged and funding allocated.

It is also likely that as an outdoor researcher there are other more internal motives. For example, sharing ideas among the widest possible outdoor community, reaching other academic disciplines and influencing policy; the satisfaction of seeing outdoor ideas in print; hoping findings reach outdoor practitioners and lead to changes in the field; or trying to form a bridge between academics and professional organisations. Is it most important to be making a difference or building networks or framing a sense of community with other outdoor researchers?

There may be different combinations of motivations for different pieces of writing or a single main driving force. The important aspect is to know the reasons for publishing and be able to balance internal motivations with any external drivers to help in targeting the right publication outlet for a particular piece of writing. This can also help with any longer-term research strategy. In the next section we will identify some of the main publication outlets and discuss some of the implications for publishing in each type.

Outdoor publication outlets

Outdoor magazines

For the outdoor enthusiast or practitioner who wants to disseminate his or her thoughts or experiences to the wider outdoor community, then 'non-academic' magazines are a good starting point. These are usually published quarterly and may be related to a particular outdoor governing body – for example, *Climber* or *The Paddler* – or may be published independently. These typically include articles about trips, gear, expeditions, different disciplines within the outdoor activity, coaching or environmental issues, in an easy to read format, designed for a 'lay' audience.

Alongside the 'non-academic' magazine is the practitioner magazine, whose intended audience is those working in the outdoor profession. In the UK, the journal is *Horizons*, published by the professional body, the Institute for Outdoor

Learning (IOL). IOL advertise for "contributors to share good practice, expertise or experience of their work in the outdoors" (Institute for Outdoor Learning, 2018) and the focus here is for publishing articles that emphasise vocational and peer relevance over theoretical underpinning (although they may have both). Writing for practitioner journals needs to be informative and in a style that can be understood by an audience of professionals.

Outdoor studies books or book chapters

There are an increasing number of academic textbooks, handbooks and other books related to outdoor studies that have been published in recent years. This reflects the growing demand for such texts with the number of university courses offering outdoor degree programmes. Examples of recent handbooks include Humberstone, Prince and Henderson (2016) and Gray and Mitten (2018). Book chapters may be invited or authors may work alone or together with academic colleagues in order to present an idea for an entire book to a well-known publishing house, such as Routledge, SAGE or Elsevier. These publishing houses will have guidelines for authors or editors in relation to book proposals on their websites. Indeed, Routledge has recently added a book series *Advances in Outdoor Studies* to encourage such proposals. The first step in a book proposal is to put forward an idea or respond to a publisher's request, usually providing a proposed title, synopsis, contents list and proposed abstracts or example chapters. The book proposer will also have to identify the intended audience and market competitors, and suggest reviewers. The publisher will then distribute the book proposal through its internal approvals processes and networks to establish whether there is a demand for the publication topic, and will seek reviews of the proposal. If the feedback and reviews are favourable it may result in a contract and being asked to develop the idea into full book form, with a proposed timetable towards estimated publication. It can sometimes take two years to go from book proposal to published book, and the process can go back and forth between editors and the authors of the different book chapters, so it is a considerable undertaking and there will be a time lag in the resultant published material. However, books and book chapters can be rewarding writing. Often the main academic publishers can distribute work to libraries and online in order to maximise sales. They are also able to reach audiences and libraries, which may not be able to access or subscribe to particular academic journals, and/or promote it as a core text book for students on outdoor studies courses. While books and book chapters are not always valued in the same way as journal articles in universities, due to their relatively low status in terms of research rankings, if popular they can be effective in helping to gain visibility, wider impact and recognition as an author in the field.

Outdoor conference proceedings

Conference proceedings are typically a book of summaries (abstracts), extended abstracts or full texts of articles presented at an academic conference. They can be published at the time and/or afterwards. For example, the International Outdoor Research Conference (2018) website (www.usc.edu.au) publishes both a list of accepted abstracts for the specific conference and links to the proceedings for some of the previous conferences, and the European Institute of Outdoor Adventure Education and Experiential Learning (www.eoe-network.eu/home/) publishes conference abstracts and proceedings. Conference proceedings are a good starting point for early career researchers to get work in progress published and an initial publication record as well as reaching a conference audience. Some conference proceedings are advertised as 'peer-reviewed', with some form of review process prior to acceptance, whilst other conference proceedings may simply publish the abstract (or article) submitted for the conference itself or may have a less stringent single peer-review process post-conference (see Brown & Boyes, 2013). There is usually a short time frame after the end of the conference to submit the final article for publication, often a matter of weeks rather than months. As such, conference proceedings can be a relatively quick way of getting research ideas out in print.

Outdoor peer-reviewed academic journals

For academic researchers in outdoor studies, the apex of publication is usually in peer-reviewed academic journals. These journals have a rigorous editorial and review process whereby any submitted article is usually scrutinised by at least two experts in the field, with feedback on the quality of the paper and a recommendation in relation to publication provided to the author(s). As such, some perceive that gaining publication in an academic journal means that it has greater 'quality'. Authors may have to respond to reviewer feedback comments on one or more occasions before eventual acceptance and at any stage, acceptance for publication is not guaranteed. Thus there is usually a relatively lengthy period of time between submission of the journal article and possible publication, which would need to be factored in to a publication strategy, although online publication usually precedes incorporation into a volume.

Historically, journals specifically in the outdoor field have been few, but the reputation of these has grown in the last decade, as the outdoor field has become more established. Perhaps the most well-known international outdoor journals are the *Journal of Adventure Education and Outdoor Learning* published in the UK, the *Journal of Outdoor and Environmental Education* (previously the *Australian Journal of Outdoor Education*) published in Australia and the *Journal of Experiential Education* in the US. A good start is to offer to review books for a journal or submit to a special issue on a specific theme. The latter is often published quickly and can attract more readers than general submissions. Further details of the purpose, history and content of each of these journals can be found in Thomas, Potter and Allison (2009). In addition to outdoor specific journals, outdoor studies research can also be published in journals focusing on the broader disciplines on which they are based; for example, education, sociology or psychology. This is useful as it gets outdoor studies recognised by a wider audience. The English Outdoor Council (www.englishoutdoorcouncil.org) also lists a number of refereed journals that publish outdoor learning based research.

Publishing, access and research metrics

Which journals or outputs to target will depend again on the motivations to publish, the intended readership and consideration of the quality or status of the journal. This means paying attention to what measures are used to indicate research impact, productivity or status in the field. Some of these are identified and discussed below:

Open Access options

The concept of Open Access is based on the principle that "A commitment to the value and quality of research carries with it a responsibility to extend the circulation of such work as far as possible and ideally to all who are interested in it and all who might profit by it" (Willinsky, 2006, p. xii). Making publications Open Access has gained momentum in recent years, due to what Harnard et al. (2008) called the "journal affordability" and "access/impact" problems (p. 36). That is, whilst the number of publications may be around two and a half million per year (Harnard et al., 2008), university libraries cannot afford the costs of journal access to them all, hence the potential readership and impact of the research in such journals is often lost. At the same time, there is a demand by funders for knowledge gained through publically funded research to be freely accessible as soon as possible (Terry & Kiley, 2006; UK Research and Innovation, 2018).

Some publication outlets publicise themselves as being fully Open Access, meaning there are no restrictions on being able to read or download the material they publish, which is usually published online relatively quickly after acceptance. Other journals advertise different levels of access known as 'Green' or 'Gold' access; the former allowing self-archiving on a personal or university repository and the latter usually asking for a fee to enable the text to be made fully available. Costs, therefore, might influence a decision about where to publish. It may be useful to note that the REF (2021) has as a condition of entry that authors must have their accepted manuscripts deposited in an institutional or subject repository.

The impact factor

Another key consideration in publishing choices, particularly for those in higher education institutions, is the 'impact factor' (Thomson, 2018) of a particular journal. This is the most commonly used metric for research impact and historically has been based on citations – the number of times an article, or articles in a particular journal, are cited. It is one of the key metrics used in judging the rank and status of journals and hence important for researchers involved in such assessments as the REF. Unfortunately, due to outdoor studies being a relatively young and small area of research and the comparatively recent development of its academic journals (Prince et al., 2018), the 'impact factor' of publishing in any of the main international outdoor journals has tended to be considerably less than others with a broader discipline base. Brookes and Stewart (2016) also reviewed the citation patterns of outdoor education journals 2000–2013 to suggest "with the exception of a few articles, any impact of Outdoor Education (OE) research and scholarship outside of the OE journals, theses, or OE conferences, is highly diffuse" (p. 12). It is also the case that some of the psychological or physical science discipline-based journals publishing perceived greater 'scientific' research using a quantitative approach tend to have a greater 'impact factor' than those in education or sociology due to a historical bias towards this approach. A similar problem for publishing in a relatively small, young and applied field is identified in relation to publishing sports coaching research (Trudel, Culver & Gilbert, 2014). As such, in the competitive academic environment, university researchers will often be encouraged to place outdoor articles with a strong theoretical base in a more highly rated psychology, sociology, science or education-based journal.

As outdoor journals and the quality of the articles within them have increased over time, their impact factors and citation figures have begun to show progress. For example, the *Journal of Adventure Education and Outdoor Learning* (*JAEOL*) had an impact factor of just 0.29 in 2003 but showed a linear increase to 0.98 in 2014 based on ResearchGate citations (see paragraph below on broader impact measures). However, whilst 'impact factors' in terms of journal ratings are typically important for academic institutions, they are not the only criterion or metric to consider. It is worth also considering who you want your work to be read by and the likelihood of readers citing or using the research findings you present. The notion of the 'impact factor' has also been criticised recently as a greater number of potential other metrics become available. There is some debate as to whether Open Access leads to greater citations of the article, with some authors suggesting that it does and others highlighting weaknesses in the methodologies used to make these claims (Craig, Plume, McVeigh, Pringle & Amin, 2007). Vanclay (2013) indicates that whilst strategically considering where to publish in environmental science, the evidence for Open Access remains inconclusive.

Citation databases and journal rankings

When choosing a journal it may also be important to know whether it subscribes to a citation database such as *Scopus* or *Web of Science*. These are abstract and citation databases of peer-reviewed literature which serve as a basis for journal rankings. For example, ranking based on *Scopus* is available through the *SCImago* Journal and Country rank web portal. *Scopus* covers scientific journals, books and conference proceedings, and calculates its score based on average citations known as the Source-Normalised Impact per Paper (SNIP). The advantage of this score is that it considers discipline differences in its calculations. For example, *JAEOL* is ranked both in Education and also in the health professions under Physical Therapy, Sports Therapy and Rehabilitation. In such journal citation rankings, the top 25% of journals are placed in the uppermost quartile (Q1), with the next 25% as Q2, the next 25% as Q3 and the lowest 25% of journals in the bottom quartile (Q4). *JAEOL* currently scores in the second quartile (Q2) of each area. The Citescore is also relevant – the number of citations received by a journal in one year to documents published in the three previous years, divided by the number of documents indexed in Scopus published in those same years. *JAEOL* has risen from 0.48 (2014) to 1.51 (2017).

H-index

The H-index is calculated by considering the number of articles an author has, and how many of them have been cited the same number of times: for example, an author with an H-index of five, will have at least five articles that have been cited at least five times each. An author with ten publications each cited at least ten times will have an H-index of ten, and so on. The H-index only increases when the next milestone is reached. That is, an author who has ten publications, but only five are cited at least five times, will remain with an H-index of five until the sixth publication reaches six citations. H-indices can also be calculated for journals in the same way and can contribute to an overall assessment of the journal impact.

Broader measures of impact

There are a number of other data points that can give some indication of reach, impact or influence beyond journal citations. These include the amount of media or social media coverage a publication or piece of work may attract, the number of reads or downloads from publication websites, blogs or personal research sites such as ResearchGate or Academia. These 'altmetrics' although becoming more recognised and enabling a broader range of publication outlets to be considered, have yet to take over from the more traditional forms of research impact. This may change as the ways in which people access research findings continues to diversify.

Considerations in research design

A final point is associated with the concept of 'engagement,' which has particular relevance to research in outdoor studies, especially in those areas examining the impact of outdoor environments or outdoor programmes on learning. 'Engagement' refers to the involvement of stakeholders, usually the end users, in the research being published. Research councils and REF particularly value research that includes, or works in partnership with, stakeholders or beneficiaries. For outdoor studies research this may include, for example, children and young people, environmental bodies, outdoor organisations, teachers or members of communities. Involving the potential research users in different ways in your research can also extend its reach and increase its possible impact. It is also useful to remember that whilst journal rankings, impact factors and citations are all important when considering which journal to place completed research, some deep thought around the potential 'significance, originality and rigour' of work in the design stage can pay dividends later.

How to be successful in getting outdoor studies research published

In this final section, we explain in more detail the process of publication in peer-reviewed journal articles and provide some suggestions on how outdoor researchers can enhance their chances of success. The suggestions here are based on the authors' personal experiences of co-editorship of the *Journal of Adventure Education and Outdoor Learning* for a number of years and of publishing outdoor-related research.

Most academic journals will have guidelines for authors on their web pages. The first step is to understand the aims and scope of the journal itself. It may sound surprising, but many articles submitted to journals are rejected at the first step simply because they are not aligned with the aims of the journal or the type of article the journal publishes. Hence reading the journal aims and having a scan through the types of articles already published in the journal to see how an article would fit are useful reconnaissance activities which can save much time later. It can also be worth noting the editorial board members and reviewers to get an idea of the types of areas and approaches (qualitative or quantitative) in that journal. Look through past issues for the kinds of themes or topics that seem to recur.

Additional guidelines for authors usually include such aspects as expected font, referencing format and word length. Although these might seem like minor points, from an editor's point of view, having a paper that is considerably over-length raises questions as to whether the author has properly considered the journal. Receiving an article with many incorrectly formatted or missing references provides much additional work and is unlikely to set a good first impression.

Having identified your target journal, there are a few other key areas that reviewers will consider when deciding whether to accept your article. These include ensuring the paper is well written, has a clear focus, theory and argument, and provides a contribution to literature and knowledge or practice. Make the contribution clear in the conclusion, but ensure not to overstate the claims of findings – no research is perfect! Finally, if possible get someone to read through the article before submission; it is easy to miss minor typographical errors or issues.

Handling the review process

It is important to take any comments from reviewers as a learning experience. Rejection from anything can be a disheartening experience. However, rarely do articles pass through the reviewer process without any changes at all, and reviewer comments are there to help improve your paper and ultimately the quality of your work and the published article. Unless the paper receives an outright 'reject' (which can happen for a number of reasons, including the article not being an appropriate 'fit' for the journal), there remains the potential for work to be published if authors carefully consider and respond to reviewer comments. The best approach is typically to make a written response against each comment in turn from each reviewer, explaining clearly what changes have been made to the text or to rebuff the comment and provide justification. Where both reviewers' comments are in agreement, this is straightforward. In cases where reviewers differ, a judgement may be needed as to which reviewer comments take precedence and how to make changes that will be acceptable to both. Authors may receive guidance from the editor on this point. Ideally, both reviewers will have identified similar points, giving a clear steer on the kind of improvements needed to make to get your work accepted. Reviewers give their time freely and reviewing can be an edifying process.

Conclusion

This chapter has provided a brief overview of the different publication outlets and considerations when making choices about where to publish outdoor studies research. It also describes what is meant by research 'impact' and some of the common ways in which it is calculated. The need for some outdoor researchers to be able to demonstrate research impact means that having a publication strategy that includes considerations of some of these metrics is now vital in a competitive research environment. However, whilst broader discipline-based journals may have a greater 'impact factor' it is also important to consider the standing of journals in the outdoor field and the kind of audience the work is intended for. More recent broader ways of reaching audiences include the use of blogs, social media and/or online profiles where ideas and articles can be read, downloaded and used by readers. The chapter concluded with a reminder about the value of 'engagement' when designing outdoor research and some hints and thoughts about how to be successful in getting ideas and findings published in a peer-reviewed journal.

References

Brookes, A. & Stewart, A. (2016). What do citation patterns reveal about the outdoor education field? A snapshot 2000–2013. *Journal of Outdoor and Environmental Education, 19*(2), 12–24.

Brown, M. & Boyes, M. (2013, November). Future faces: Outdoor education research innovations and visions. In *Proceedings and extended abstracts of the Sixth International Outdoor Education Research Conference 2013*. Dunedin, New Zealand.

Craig, I.D., Plume, A.M., McVeigh, M.E., Pringle, J. & Amin, M. (2007). Do open access articles have greater citation impact? A critical review of the literature. *Journal of Infometrics, 2*(3), 239–248.

Eley, A., Wellington, J., Pitts, S. & Biggs, C. (2012). *Becoming a successful early career researcher.* London: Routledge.

Gray, T. & Mitten, D. (Eds.). (2018) *The Palgrave handbook of international women in outdoor learning.* London: Palgrave Macmillan.

Harnad, S., Brody, T., Vallieres. F., Carr, L., Hitchcock, S., Gingras, Y, Oppenheim, C., Hajjem, C. & Hilf, E.R. (2008). The access/impact problem and the green and gold roads to open access: An update. *Serials Review, 31*(1), 36–40.

Humberstone, B. Prince, H. & Henderson, K.A. (Eds.) (2016) *International handbook of outdoor studies.* Oxford, New York, NY: Routledge.

Institute for Outdoor Learning (2018). *Horizons.* Retrieved from: www.outdoor-learning.org/Good-Practice/Research-Publications/Horizons

Prince, H., Christie, B., Humberstone, B. & Gurholt, K.P. (2018). Adventure education and outdoor learning: examining journal trends since 2000. In P. Becker, B. Humberstone, C. Loynes & J. Schirp (Eds.) *The changing world of outdoor learning in Europe* (pp. 144–160). London: Routledge.

Research Excellence Framework (REF) (2021). Retrieved from: www.ref.ac.uk

Taylor & Francis. (2018). *Journal of Adventure Education and Outdoor Learning.* Retrieved from: www.tandfonline.com/action/journalInformation?show=aimsScope&journalCode=raol20

Thomas, G., Potter, T. & Allison, P. (2009). A tale of three journals: A study of papers published in AJOE, JAEOL and JEE between 1998 and 2007. *Australian Journal of Outdoor Education, 13*(1), 16–29.

Terry, R. & Kiley, R. (2006). Open access to the research literature: a funder's perspective. In N. Jacobs (Ed.) *Open Access: key strategic, technical and economic aspects* (pp. 101–109). Abington: Woodhead Publishing Ltd.

Thomson (2018). Journal impact factor. Retrieved from http://ipscience-help.thomsonreuters.com/inCites2Live/indicatorsGroup/aboutHandbook/usingCitationIndicatorsWisely/jif.html

Trudel, P., Culver, D. & Gilbert, W. (2014). Publishing coaching research. In L. Nelson, R. Groom & P. Potrac (Eds.) *Research methods in sports coaching* (pp. 251–260). London: Routledge.

UK Research and Innovation (2018). Open access. Retrieved from: www.ukri.org/

Vanclay, J.K. (2013). Factors affecting citation rates in environmental science. *Journal of Informetrics,* 7(2), 265–271.

Willinsky, J. (2006). *The access principle. The case for open access to research and scholarship.* London: MIT Press.

Yokoyama, K. (2006). The effect of the research assessment exercise on organisational culture in English universities: Collegiality versus managerialism. *Tertiary Education and Management, 12*(4), 311–322.

28

RESEARCH HUBS

The theory-practice nexus

Carrie Hedges, Chris Loynes and Sue Waite

Introduction

Many professions, teaching and youth work amongst them, are keen to support the raising of standards through professional development by bringing the evidence of research closer to practitioners in the field. A number of strategies have proved effective, including the development of strong social networks between researchers and practitioners, local action research hubs, participative enquiry and the development of practitioner-researchers.

Two recent studies in the field of Outdoor Learning (OL) have been successful at using an action research approach to support professional development leading to increased take up and raised standards. These are Natural Connections (Waite, Passy, Gilchrist, Hunt & Blackwell, 2016), advocating for learning outside the classroom in natural environments (LiNE) in primary schools in South West England, and Learning Away (Kendall & Rodger, 2015), encouraging 'brilliant residentials' in schools throughout the UK. Both projects brought teachers and schools together in local hubs supported by advisors, evaluators and researchers in order to develop and disseminate best practices.

The success of, and the lessons from, these two projects has led to the piloting of the regional research hub concept in the UK for outdoor researchers and practitioners. The aim is to support local research that informs practice and enhances the quality of provision. In addition, the project intends to aggregate and analyse the data from local small-scale studies in order to create a larger evidence base to inform and influence strategic developments in outdoor learning nationally.

This chapter explores the approaches and impact of Natural Connections, Learning Away and the regional outdoor learning research hubs and outlines lessons learned for future practice.

Natural Connections

The White Paper, *The Natural Choice: Securing the value of nature* (HM Government, 2011) affirmed the UK government's commitment to "remove barriers to learning outdoors and increase schools' abilities to teach outdoors when they wish to do so". In response, the Department for the Environment, Food and Rural Affairs (DEFRA), Natural England and Historic England commissioned the Natural Connections Demonstration Project, an ambitious outdoor learning project delivered by the University of Plymouth between 2012 and 2016 (Waite et al., 2016). The project engaged over 125 schools across South West England in developing outdoor learning through stimulating school demand for LiNE, providing support to incorporate it into planning and practices, and brokering outdoor learning services.

The project included delivery support and evaluation to facilitate future wider development of curriculum-based outdoor learning. A contributory factor in the award of the contract to the university was its successful track record of practice/research interaction through the Outdoor and Experiential Learning Research Network.[1] The network had been established in 2006 to facilitate and enhance mutual understanding of research needs and the evidence base in the field through regular workshops, seminars and conference with a regular email digest of relevant information to over 200 practitioners and researchers. Research reports were made freely available on the university website and there was ongoing collaboration between academic and practice communities in funded projects and in writing articles and books together.

The project plans were informed by scoping research (Rickinson, Hunt, Rogers & Dillon 2012) and reviews of outdoor learning and educational innovation literature, resulting in a distributed model of leadership, ownership and support.

Key stakeholders, including the funders and the Council for Learning Outside the Classroom, monitored and guided progress of the project. Participants, through regular local cluster group and hub leader meetings, informed its development. The structure for information flowing between these links meant that the project's direction and methods could respond rapidly to changing needs (see Figure 28.1). Educational attainment prioritisation, cost and risk (Waite, 2010) combine within school contexts with low staff confidence and experience (Dillon, 2010) to create barriers to outdoor learning. The model with regional brokerage and peer support, whereby recruitment was gradual, and schools developed preferred ways of using their school grounds and local community spaces for outdoor learning with tailored support from hub leaders and external sources, was well suited to meeting their specific issues, encouraging gradual development and securing sustainability by enabling teachers to own the process of change (Gilchrist & Passy, 2018). For example, some schools wanted to raise funds to redesign their outdoor learning environments to maximise available time for teaching outside, so fundraising courses were run in several hubs. In view of the coastal proximity of some schools, whose pupils might have never visited the seaside, Teach on the Beach professional development sessions were held to inspire staff.

The original plan was to ground evaluation within school-level action research and aggregate data using the decision theoretic technique, which balances importance of

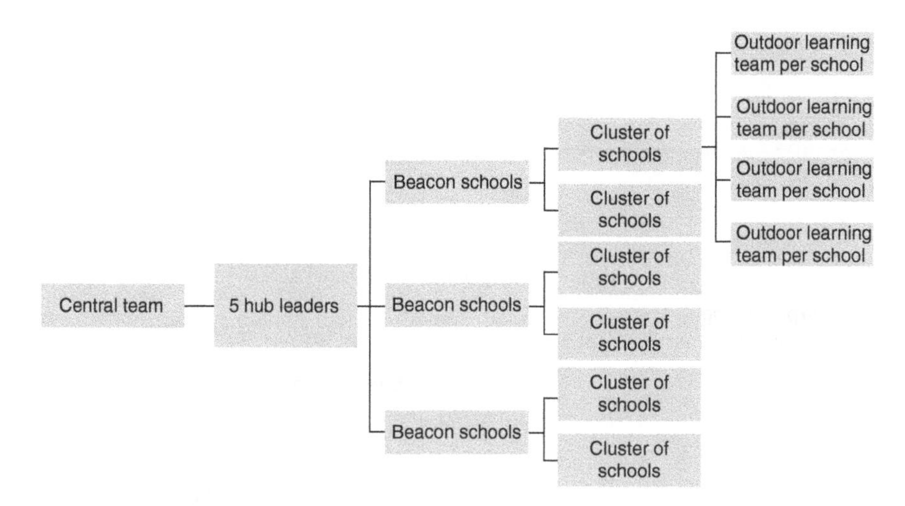

FIGURE 28.1 Natural Connections model

Source: Waite et al., 2016

outcomes with the likelihood of them being achieved within a programme and, by attributing a numerical value to this combined value, enables relative achievement of differing goals to be aggregated across schools (Waite, Bromfield & McShane, 2005). The opportunity to gather data across common outcome measures in a big sample of schools was unusual and highly valuable. The evaluation design was therefore guided by Natural England's comprehensive evidence requirements with 100 key evaluation questions to gain traction for roll-out nationally (Gilchrist et al, 2017) balanced by sensitivity to the burden of extensive data collection for schools (Waite, Passy & Gilchrist, 2014). Quantitative electronic surveys at staged intervals monitored activity and impacts throughout the project's lifetime. They provided feedback to hubs and schools about use of different spaces for OL, increased investment in OL environments within school grounds, and perceptions of impact of outdoor experiential learning for children. We found that staff concerns about providing OL generally lessened as their confidence and practice in using outdoor spaces grew. As an indication of commitment, school grounds were often modified to meet learning needs, using grants or school budgets (Gilchrist, Passy, Waite & Cook, 2016). The following positive benefits for children were reported by 85–95% of respondent schools: enjoyment of lessons; engagement with and understanding of nature; social skills; engagement with learning; health and wellbeing; and behaviour. No school reported OL had had a negative impact on attainment; many said it was difficult to attribute attainment to a single cause.

The data enabled us to develop a model (Figure 28.2) of how curriculum OL might lead to raised attainment, building on existing research about links between enjoyment and engagement and the development of non-cognitive foundational skills (Gutman & Schoon, 2013).

Case studies of 24 schools (19 primary, 2 secondary and 3 special schools) provided more detailed staff, volunteer, parent and pupil perspectives of the main

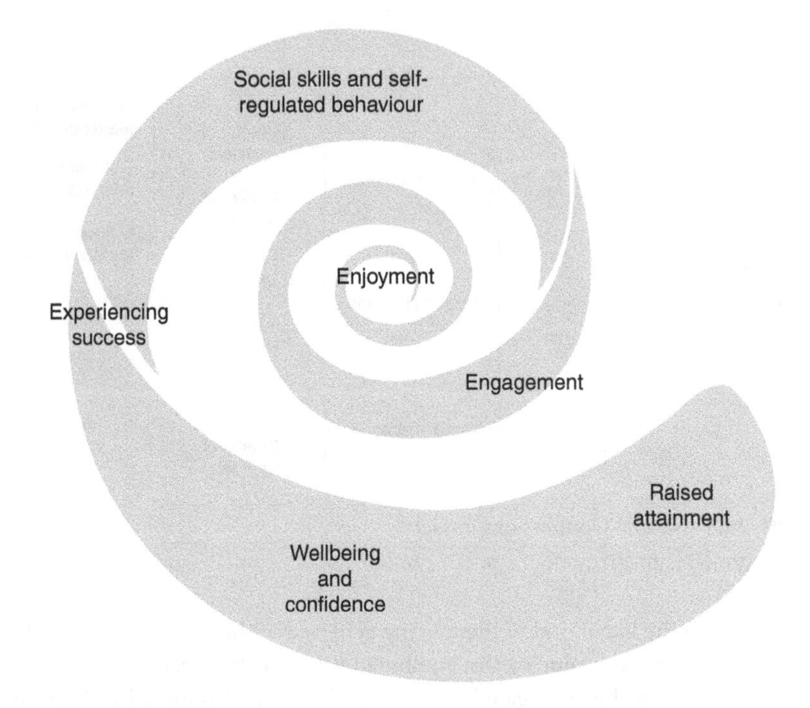

FIGURE 28.2 The pathway to raised attainment through outdoor learning
Source: Waite et al., 2016, p. 10

benefits and challenges of introducing OL. We used a schedule for consistency in semi-structured interviews and focus groups about pupil benefits, whether and how OL supported teaching and learning, and challenges faced integrating OL within school practices, gathering 119 staff views. Some case study visits included focus groups with children and parental questionnaires. Salient points from the data generated were transcribed into a standard template, yielding detailed summaries.[2]

Edwards-Jones, Waite and Passy (2018) discuss some of the emergent challenges and responses to embedding LiNE. Significantly, while most initial barriers such as lack of teacher confidence and uncertainty about how to link outdoor learning to curriculum objectives were overcome, time remained a challenge as the will to include more OL grew with experience of its benefits.

Learning Away: Collaborative action research impacts on students, teachers, schools and policy

The Learning Away Initiative worked with over 60 primary, secondary and special schools in 13 cluster partnerships across the UK, aiming to enhance young people's learning, achievement and wellbeing by developing, piloting and evaluating the impact of residential experiences as an integral part of the curriculum. An action research approach was developed so that schools could deploy a continuous development model using an evidence-based approach to their ideas. In addition, the aggregated data provided a data set for generic analysis that could be fed back to the schools and policy makers to underpin a campaign for residentials as a practice. The research design created a virtuous circle of learning influencing practice and the quality of teaching and learning for the students involved.

Prior to this initiative, evidence for residential experiences and their impacts was largely anecdotal, though these had stood the test of time. Additionally, conventional models of staffed residential centres, often in places remote from schools, had become time consuming (especially at secondary school level), administratively complex and expensive. The aim was therefore to use a diverse set of schools to demonstrate the value of the residential approach whilst finding new ways to make them work.

In order to draw on the wealth of largely unresearched experience, the initiative invited a number of leading practitioners to imagine the best provision they could. Embedded in their many proposals were a set of generic criteria: That residential experiences should be progressive throughout primary and secondary school, inclusive of all students and integrated with the curriculum. These criteria framed the call for schools to partner with the five-year initiative. Other key requirements were the full support and engagement of the school's senior leadership and partnership between schools across a local area to provide support and critical mass.

After two years, case studies of each cluster indicated the diversity of partnerships, approaches and intended impacts of the plans that schools had begun to implement. The initiative had been successful at provoking engaged partners and set out to determine what the impacts actually were. At the annual gathering each cluster was invited to make explicit its theory of change. This led to the development of nine hypotheses for the impacts of residential experiences on students and teachers and three further hypotheses related to whole school change (Table 28.1).

Nine online surveys, pre-, post- and long-term post, were developed to test each hypothesis across as many clusters as wished to participate in each theme. This survey was given to students (n = 5,821 pre-residential, 4,652 post-residential and 988 long-term follow-up), and every teacher (n = 285 pre-residential and 254 post-residential) taking part in a residential over three years. Approximately 20% of students and staff were from secondary schools. The remainder were primary plus three special needs schools. The results of these surveys were supplemented by focus groups of students (63, involving 398 students from 27 schools) and staff (40, involving 192 staff across 37 schools). Over 100 case studies of individuals, classes, trips and schools were completed as were a number of observations of residentials and classes. Schools received feedback on their residentials on an annual basis. Data were also aggregated and analysed by cluster, year, hypothesis and key stage.

Some clusters added additional practitioner research projects to test their own specific questions. These smaller studies covered topics such as attainment in Years 10 and 11 (14–16 years old), transition from primary to secondary school (10–11 years old), behaviour change, attendance, exclusion and bullying. These new data were sometimes added to existing data such as progress scores that were already

TABLE 28.1 The Learning Away hypotheses (Adapted from Carne, Loynes & Williams, 2015)

Learner Achievement and Engagement
- Progress and attainment
- Knowledge and understanding
- Skills development
- Learner engagement

Learning Experience
- Relationships with others
- Transition
- Leadership, co-design and facilitation
- Resilience, self-confidence and wellbeing

Transforming Schools
- School improvement
- Pedagogical skills
- Cohesion, interpersonal skills and belonging

collected but not always previously evaluated by the schools concerned. Findings included significant jumps in progress and attainment post-residential in maths (Year 10) and literacy (Years 5 and 6, 9–11 years old); significant steps in progress for underachieving students (Years 6 and 10); positive social behaviours including elective mutes becoming voluble post-residential; improved attendance rates alongside decreased truancy rates and incidents of anti-social behaviour in class and in the playground. One cluster reported exclusions dropping to zero on the introduction of their programme.

In addition, all data were combined, which enabled an overall theory of change describing some impacts of residential experiences that occur no matter what is done, how often, where and at what cost (Table 28.2) and some indicators of approaches that heightened these generic impacts. These were the presence of the students' class teachers on the residential with their students, low cost approaches especially camping, student leadership within peer groups and for younger pupils, and the co-construction of residentials between students, teachers and specialists.

Comparisons between clusters or years was not possible. Also, despite the long-term nature of the study, it was still difficult to evaluate the impact of residential experiences on attainment in examinations.

The approach was a success. Residentials became embedded and sustained in most schools. An evidence base for their impacts was developed and staff skills were cemented. Innovative approaches overcame resourcing issues, raised confidence in the impact of the experiences and addressed staff concerns with new teaching approaches, safety and class control. The main threat to sustained provision was a change of senior leadership in a school.

The evidence from the initiative went on to underpin a campaign to involve more schools and influence national educational policy. The annual cycle of

TABLE 28.2 "Why brilliant residentials?" The Learning Away Theory of Change (Adapted from Carne, Loynes & Williams, 2015)

Residentials provide the opportunity and experience of living with others. This transforms relationships and develops a strong sense of community and belonging between the staff and students involved. The findings of the evaluation confirm that this sense of community supports a wide range of positive social and learning outcomes long after the return to school.	Residentials bring...	which, in the short and medium term, lead to...	which, in the longer term, lead to...
	→	→	—
	The overnight stay and an intensity of experience	Enhanced relationships	Improved achievement, progress and attainment
	A new context for relationships	Improved engagement and confidence in learning	Improved knowledge, skills and understanding
	Different and varied opportunities to experience success	New and developing skills and understanding	Improved relationships
	New ways of learning		Improved engagement
			Improved behaviour and attendance
			More successful transition experiences at all key stages
			Raised aspirations
			Greater cohesion and a sense of belonging
			Enhanced trajectories to work, sixth form, further and higher studies

evidence-based reflection and sharing in clusters created an effective action research environment from which academics, practitioners and policy makers benefitted.

The Outdoor Learning Research Hub project

Inspired by the success of the cluster model adopted by Natural Connections and the Learning Away initiatives, the Institute for Outdoor Learning (IOL), the Council for Learning Outside the Classroom and Natural England, together with a national network of researchers – the LiNE Strategic Research Group (LiNE SRG) – proposed the idea of regional Research Hubs to bring researchers and practitioners together. Andy Robinson, Chief Executive of IOL, which funded the pilot year, encapsulated the idea:

> If Outdoor Learning is to be valued more highly by UK society it needs to be better understood and more consistently delivered to high standards. I think the work of the Research Hubs will support these aims by providing better dialogue between different research institutes and between researchers and practitioners.
>
> Robinson, A., 2018

Launched in 2017, in partnership with the University of Cumbria, the Research Hub project aimed to facilitate and co-ordinate researcher-practitioner engagement to drive the nationwide development of an evidence base to support the delivery of high-quality, frequent and progressive outdoor learning experiences for children and young people. It aspires to raise the standards of professional delivery in outdoor learning research and practice.

The regional research hubs. A network of regional researcher-practitioner hub groups, research 'hubs', have been established that will enable academics and practitioners from across the sector to discuss needs and priorities. These research hubs are tasked with identifying local priorities, supporting evidence gathering, and progressing local action research and evaluation (Hedges, 2018). Each regional research hub arose spontaneously and is developing autonomously in order to:

1. Build links between local research-practice communities with an interest in OL – universities, providers, professionals, researchers, postgraduate students, etc. to stimulate action research capacity and activity that meets local needs;
2. Capture the scope of research in their area and feed any publications in the public domain to the central research coordinator;
3. Create plans to support evidence gathering in areas of national priority;
4. Disseminate information about evidence and good practice within their hub area, among the network of pilot hubs and to the LiNE SRG;
5. Provide a bridge between local and national needs for research findings.

Hubs are meeting several times during the year and their discussions and developments so far have been both encouraging and insightful. Already, they have identified previously unknown research, debated the most important questions to ask – and how best to ask them, begun to develop action research workshops to support new projects and build new practitioner-researcher partnerships. Feedback indicates that these meetings promote confidence amongst practitioners in undertaking research, raise awareness of existing research evidence and stimulate discussion about the focus for further research activity. Practitioners also report feeling more articulate about what they do and what it achieves.

The Central OL Research Coordinator. A Central Research Coordinator was appointed to bring together local evidence via the hub network and to summarise those findings to inform and stimulate local hub-driven research, research and evaluation priorities, and national policy and practice. The coordinator will also develop an online toolkit that provides the various forms of OL practice with underpinning evidence for impact and case studies that model good practice. This can feed into national policy development and provides insight into UK-wide research needs and priorities. Simultaneously, the regional hub is already helping answer key questions posed by national policy makers.

It is hoped that action research and an evidence-based approach will support the quality and reach of OL for all in the UK.

The benefits and barriers of a collaborative research model

Just as Natural Connections and Learning Away identified barriers to the take up of their respective interventions that they were able, in significant ways, to overcome for their participating schools, the OL Research Hubs project found similar barriers to practitioner engagement with research. To summarise the findings of the Hub coordinator's first report, practitioners:

- were unsure of what counted as 'research';
- lacked confidence in their findings because they were not sure if they were rigorous;
- were not sure if their work was ethically arranged or had sufficiently large sample sizes;
- were not sure what questions to ask or how best to ask them;
- felt that quantitative approaches were more valid than qualitative ones;
- found research hard to read and were unsure of the value of repeat studies.

These concerns applied to previous research not shared externally or, sometimes, internally, and to the initiation of new projects. Practitioners also expressed concern that they might find that their implicit theories of change, once made explicit and tested, might not meet with the approval of colleagues or that the findings of research would prove them to be of limited educational value. Lack of resources, especially time, was also mentioned.

It also became clear that schools and other organisations, routinely collect data that, if applied to specific projects or groups, could provide qualitative and quantitative evidence of great value in answering practical questions about effective teaching strategies. In this case, practitioners were unsure of the theoretical frameworks that might help them to understand the data or approaches with which to undertake an analysis.

Learning Away drew on the experience of Natural Connections and took an evidence-based and explicit approach to change using an iterative, action research model. The partnership between experienced evaluators and researchers provided practitioners in both projects with the skills, knowledge, additional resource and a constructive approach to working with the findings. The experience in schools was of an evidence-based approach to the transformation of pedagogy, the development of students as learners and staff as teachers in a framework of a supportive organisation. In addition, the Learning Away project reported significant positive and sustainable impacts across the culture of whole schools. In some cases, the Learning

Away project reported that teachers moved beyond collaborative approaches and initiated research projects of their own as practitioner-researchers. Staff became articulate advocates of the pedagogical changes they were making, influencing other staff, other schools and policy makers.

The Hubs project has taken these findings and applied them broadly to the professional development and strategic planning of OL more widely. The success of a collaborative action research approach has been applied to the local context of an institution such as a school or outdoor centre with the intention of building researcher-practitioner partnerships that will provide the skills, knowledge, resources and confidence to implement small-scale research that can make a difference to practice. In addition, the role of the central coordinator means that, like the Learning Away and Natural Connections projects, the small-scale findings can be aggregated in ways that allow for a larger picture to emerge of the difference outdoor experiences can make to students, teachers and organisations.

Conclusion

In our view, collaborative action research has the following advantages:

- Bringing researchers and practitioners together enhances the quality of practice and provides a deeper understanding of complex educational approaches;
- It encourages ongoing reflection amongst practitioners and gives them confidence in their articulating their approach and making claims for impacts;
- By integrating small-scale studies, a larger picture can emerge underpinned by a greater evidence base.

Outdoor education has a long history with many different forms and enactments. We have mainly focused in this chapter on outdoor learning, Learning in Natural Environments and residential experiences. The visibility of OL has increased with a growing international evidence base and the Natural Connections project and Learning Away projects in the UK contributed to this. Their collaborative approaches have helped to bring together research and practice, although they have also highlighted tensions between the demands of different audiences, such as policy developers and teachers, which need to be negotiated carefully. The formation of the Natural England Strategic Research Group has enhanced evidence-based strategic influence and supported the OL Research Hubs initiative in partnership with the Institute for Outdoor Learning and the University of Cumbria. In combination, these factors have given lobbyists and government the confidence to include OL and LiNE as strategies within the 25-year Environment Plan (HM Government, 2018) with the ambition to provide progressive outdoor experiences for all young people in the UK. The OL Research Hubs project, and initiatives like it, with a collaborative, evidence-based, action research approach, can continue to support the exploration of how OL can provide young people with a progression of relevant experiences. From the practice perspective, it has helped to make evidence more relevant and applicable to specific contexts within outdoor studies. Research can inform the quality of practice, the narratives to advocate for these practices and the direction and expansion of provision. It can also offer robust evaluative feedback about innovative practices, giving stakeholders at a local level the confidence and knowledge to build effective provision for all into the future. Both the Natural Connections and the Learning Away projects highlighted the potential for practitioners to become involved in small-scale research and evaluation in collaboration with researchers and also as research practitioners in their own right. A key element in this has been that schools have gained the confidence to ask questions that matter to them, to trust the results and to value the way in which research can be a powerful tool in enhancing practice. This capacity is a key aspect of the OL Research Hubs' ambitions to encourage further action research. However, the time needed to engage with research in practice cannot be underestimated and can conflict with other priorities.

Consideration of how methodologies can have relevance and utility at multiple levels is worthwhile. In this way, aggregation and synthesis of the findings can continue to inform policy makers and strategic planning at local, regional, national and international levels. As such, the collaborative action research model has benefits to the whole eco-system of education, informing and supporting change from the student learner to the national and international policy maker. The evidence also suggests that the outcome of implementing initiatives in this way is one of embedded organisational change that raises standards and is sustained beyond the life of the formal intervention (Loynes, 2017). Collaborative action research becomes rooted and integrated with practice in professional and organisational reflective cycles, a capacity that has the potential to influence change beyond the aspirations of outdoor learning as schools and other organisations apply these approaches to other subject areas and pedagogies as well as to whole school transformations.

Notes

1 https://www.plymouth.ac.uk/research/peninsula-research-in-outdoor-learning
2 See https://learningoutsidetheclassroomblog.org/category/case-studies/

References

Carne, P., Loynes, C. & Williams S. (2015) *Learning Away: Impact.* Paul Hamlyn Foundation. Retrieved from: http://learningaway.org.uk/impact/

Dillon, J. (2010). *Beyond barriers to learning outside the classroom in natural environments.* Retrieved from: http://publications.naturalengland.org.uk/publication/4524600415223808

Edwards-Jones, A., Waite, S. & Passy, R. (2018). Falling into LiNE: school strategies for overcoming challenges associated with learning in natural environments (LiNE). *Education 3–13, 46*(1), 49–63.

Gilchrist, M. & Passy, R. (2018). *Natural Connections Demonstration Project: Sustainability report.* Retrieved from: www.plymouth.ac.uk/uploads/production/document/path/10/10809/NCDP_sustainability_report_2018.pdf

Gilchrist, M., Passy, R., Waite, S. & Cook, R. (2016). Exploring schools' use of outdoor spaces. In C. Freeman, & P. Tranter & T. Skelton (Eds.) *Risk, protection, provision and policy. Geographies of Children and Young People Vol. 12.* Singapore: Springer.

Gilchrist, M., Passy, R., Waite, S., Blackwell, I., Edwards-Jones, A., Lewis, J. & Hunt, A. (2017). *Natural Connections Demonstration Project, 2012–2016: Analysis of the key evaluation questions.* London: Natural England Commissioned Reports, Number 215 Annex 1.

Gutman, L.M. & Schoon, I. (2013). *The impact of non-cognitive skills on outcomes for young people: A literature review.* London: Educational Endowment Foundation.

Hedges, C. (2018). *Outdoor Learning Hubs project developments.* Institute for Outdoor Learning Report, April 2018. Retrieved from: www.outdoor-learning-research.org/Portals/0/Research%20Documents/Research%20Hubs/Hubs%20Project%20Report-%20April%2018.pdf?ver=2018-05-23-124429-900

HM Government (2011). *The Natural Choice: Securing the value of nature.* London: Natural Environment White Paper, HM Government CM8082.

HM Government (2018). *The 25 Year Environment Plan.* London: Department for Environment, Food and Rural Affairs.

Kendall, S. & Rodger, J. (2015). *Paul Hamlyn Foundation evaluation of Learning Away: Final report.* London: Paul Hamlyn Foundation.

Loynes, C. (2017). The renaissance of residential experiences: Their contribution to outdoor learning: In S. Waite (Ed.) *Children learning outside the classroom: From birth to eleven* (2nd ed., pp. 209–221). London: SAGE.

Rickinson, M., Hunt, A., Rogers, J. & Dillon, J. (2012). *School leader and teacher insights into Learning Outside the Classroom in Natural Environments.* London: Natural England Commissioned Reports, Number 097.

Robinson, A. (2018, 19 March). Personal communication.

Waite, S. (2010). Losing our way?: Declining outdoor opportunities for learning for children aged between 2 and 11. *Journal of Adventure Education and Outdoor Learning, 10*(2), 111–126.

Waite, S., Bromfield, C. & McShane, S. (2005). Successful for whom?: A methodology to evaluate and inform inclusive activity in schools. *European Journal of Special Needs Education, 20*(1), 71–88.

Waite, S., Passy, R. & Gilchrist, M. (2014). Getting it off PAT: Researching the use of urban nature in schools. In E. Backman, B. Humberstone & C. Loynes (Eds.) *Urban nature: Inclusive learning through youth work and school work* (pp. 35–49). European Outdoor Education Network, Stockholm: Recito Forlag.

Waite, S., Passy, R., Gilchrist, M., Hunt, A. & Blackwell, I. (2016). *Natural Connections Demonstration Project 2012–2016: Final report.* Natural England Commissioned Report, Number 215. Available at: http://publications.naturalengland.org.uk/publication/6636651036540928

29

KNOCKING ON DOORS IN THE POLICY CORRIDOR – CAN RESEARCH IN OUTDOOR STUDIES CONTRIBUTE TO POLICY CHANGE?

A professional narrative on shaping educational policy and practice in Scotland

Peter Higgins

Introduction

Ask any professional outdoor educator about the effectiveness of outdoor learning – and you will likely be treated to a glowing endorsement of the impact on personal and social skills, academic achievement, health and wellbeing, attitudes to sustainability and the environment, community engagement. They may even resort to an argument like McDonald's (1997, p. 377) – "We don't need independent research to prove the value of outdoor education; we believe in it". Ask any researcher in the field and they are likely to tell you there is some evidence of some aspects of such claims. Ask policy makers and they are likely to say that outdoor learning is a wonderful thing, nice to have, but expensive and not really 'proper' education.

Much of what follows is a personal reflection on over 25 years of effort to bridge the gap between existential belief, evidence (or lack of), public perception, policy development and implementation in outdoor learning (OL) and the related area of 'learning for sustainability' (LfS) (which in Scotland has a very close relationship to OL). The context is personal in that I live and work in Scotland, and I draw heavily on that experience, but also have policy-related experience in the UK, European Union and through a UNESCO organisation. It is important to state that whilst I attempt to maintain a broad perspective on what is meant by 'outdoor learning' and the many international nuances of the field, I am primarily focusing on provision for young people of school-age (3–18 years), who have an entitlement to state-funded education. The context for any outdoor provision for this community is therefore one where there are competing demands for such funds, where a wide range of educational and developmental arguments may be pertinent, and policy can be subject to rapid political change.

I begin with a thought piece; a letter I wrote about 10 years ago …

A LETTER TO MY DAUGHTER'S HEAD-TEACHER

Thank you for your letter inviting me to allow my daughter to attend your annual Primary 7 outdoor learning residential.

I am happy to pay the £300 and allow her to go but I note that this trip takes place in term-time, and before I agree I would like you to please explain the following:

1. What is the educational purpose of the trip?
2. Assuming there is an educational purpose, in what way are the activities appropriate, and what evidence is there for this?
3. How are the staff who work with the pupils qualified to deliver these educational aims? (Please don't tell me about their outdoor activities qualifications – I am sure they are competent and she will be safe in their hands).
4. What aspects of the curriculum and her general education will she miss whilst she is away from school for this week?
5. Why is this trip deemed to be more valuable than a week in school?
6. If it is 'educational' then why am I paying £300 for it?
7. If you are unable to answer any of the above, perhaps you will explain to me how this trip differs from a holiday?
8. If it *is* a holiday I would prefer to take Ella out of school for the week and go away with her myself. I would be grateful if you would agree to me doing so as I would rather have your permission than a letter from the school complaining that this is an 'unauthorised absence'.

I didn't send it – I never intended to. It is a good school which has the best of intentions in organising such trips, but I have used the 'letter' to considerable effect in challenging myself and other colleagues to recognise that belief and tradition drive practice more than evidence. So, in the example of the residential centre visit above, the purposes (e.g. educational, personal/social development etc.) were largely unstated, and so continuance of the practice requires the school, the centre staff and the parents all to 'collude' to simply believe (and not question) its worth.

Whilst this is a specific example of residential outdoor learning, similar concerns may well be levelled at school-based (academic or other) experiences, 'wilderness'/therapeutic, place-based, culturally focused and other programmes. More or less universally, the choice of place, duration, and conduct of activities undertaken is, understandably, left to the specialist staff, who may or may not make decisions based on specific evidence on the merits of one choice over another. In my experience there is also a high likelihood that there will be little or no evaluation of a given programme's effectiveness, or even a shared understanding between funder, provider, and participant/'client' of what outcomes are intended.

Outdoor learning research

Research support for the value of many aspects of outdoor learning has grown in the ten years since I wrote that letter, and the philosophical rationales are now more robust, but so has the demand for supporting evidence before funding and time are committed to staff, resources and facilities. Provision of various aspects of OL has grown and others faded in most countries where it exists, and it is plain that there is no consistent international approach or philosophy in which to locate our work. This is at odds with formal education where there is far more coherence and strong evidence bases in support.

There are sound reasons for this – meaningful outdoor learning research is difficult, the field is complex and multidimensional, empirical research is expensive and poorly funded. Further, university degree programmes with a strong research orientation are scarce and located in few countries, as are senior academic positions. Consequently, we see the growth in research outputs, as evidenced by peer-reviewed publications in specialist outdoor journals (themselves a relatively recent phenomenon), the emergent interest in related disciplinary journals (education, psychology etc.), and Doctoral studies. Findings from the good-quality research that does exist are rarely communicated through research briefings, professional publications or popular media (but see Chapter 27 on increasing visibility). Little wonder that colleagues 'doing the day job' lack the evidence base they need to make professional judgements, and argue why their budgets, facilities and posts should not be cut. Similarly, how can a policy maker or budget holder be expected to make informed decisions concerning policy development or funding on the strength of limited evidence of impact, which even if it does exist is highly likely to be of short-term benefit?

The variable quality of informal outdoor learning evaluations

The need to sustain support or funding can lead to a perfectly understandable tendency amongst providers to make claims for a programme, intervention or approach that are not supported by evidence – either because it is not gathered (through an evaluation) or because the claims are disproportionate to the quality of the evidence, or perhaps even due to 'belief' as McDonald (1997) argues. So for example, a residential provider may state something like – "Through a range of challenging adventurous activities participants increase self-confidence, self-esteem and develop leadership skills". Whilst it may well be justifiable to say this is the *aim* of a programme, this is of course a *claim* that is difficult to support with evidence of even a temporary, let alone lasting effect.

Many providers and teachers are increasingly tending to evaluate the effectiveness of their work to satisfy funders or their own curiosity. Such studies can be very valuable if conducted properly and any claims made are appropriate to the evidence. In order to assist in planning such evaluations, Nicol and Higgins (2002) promoted an approach to encourage providers to think through the process, based on planning to achieve any developmental outcome they claim to facilitate.

	Aims	Assumptions	Content	Methods	Evaluation	Claims
Insert developmental or learning outcome						

Whilst this approach was designed originally as a daily planning tool for any session, with potentially multiple rows relating to specific developmental or learning outcomes, it is equally valuable in planning a whole evaluation strategy that encourages realistic claims to be made.

Policy relating to outdoor learning

For those who are 'believers' it is difficult to understand why policy makers are not immediately and consistently supportive. However, particularly in formal education, there are political pressures, and an increasing array of competing demands on resources. New subject disciplines have emerged, there is an expectation amongst pupils and parents of increasing choice, and curricular content continues to expand. In a school context, it is difficult to judge which non-statutory activities might be most beneficial (in what way, and for which pupils?) – how do teachers and management teams decide whether to offer drama, music or outdoor learning experiences?[1] Add to this the complexity of timetabling necessary to build such activities (especially outdoor experiences) into school programmes, staff-cover, costs, perceptions of risk, and it seems remarkable that policy makers and implementers are ever able to overcome the inertia to support outdoor learning in schools, residentials etc.

Whilst all of the above are no doubt significant and feature in the decisions of policy makers and managers, they may well be trivial in relation to the curricular requirements of a given national or regional education authority. By far the strongest rationale for including a subject, material or an approach is that it is an existing policy requirement or a cultural norm. In the case of outdoor learning both are rarities globally, though in the case of Scotland, it is the focus of a specific national policy document[2] (Curriculum for Excellence through Outdoor Learning) and another (summarised by Higgins & Christie, 2018) on Learning for Sustainability (LfS), which state:

> The journey through education for any child in Scotland must include a series of planned, quality outdoor learning experiences.
>
> *Learning and Teaching Scotland, 2010, p. 6*

> The Scottish Government established a model of LfS that integrated three equally important facets – Sustainable Development, Global Citizenship Education and Outdoor Learning with an overarching aim to develop: "a whole school approach that enables the school and its wider community to build the values, attitudes, knowledge, skills and confidence needed to develop practices and take decisions which are compatible with a sustainable and more equitable future".
>
> *Higgins & Christie, 2018, p.555*

Whilst it would seem self-evident that outdoor learning would be thriving in all Scottish schools, this is not the case. The policy-related reasons for this are discussed further below.

In the Nordic countries, time in the outdoors, *friluftsliv*, is a long-established cultural norm, much envied in other nations. Here, making arguments for outdoor learning experiences would seem straightforward, and these have indeed been an established feature of education (particularly in the early years and primary) for many years. However, the mere 'normality' of recreational time spent in nature makes it vulnerable to the argument that it does not need to be a feature of school provision.

Negotiating the policy 'corridor'

The process of policy development requires a number of factors to come together. There are 'doors' along the 'corridor' that have to be opened and passed through, though not necessarily sequentially, for the policy to be formulated, agreed and then implemented. I illustrate this process with reference to specific events in Scotland with emphasis on the role of research and dissemination. The policy context has been the subject of a recent Doctoral study by Baker (2015), and the contemporary situation is outlined by Higgins and Nicol (2018) and Higgins and Christie (2018).

A good idea Outdoor learning (and learning for sustainability) are plainly good ideas. They have a long tradition in Scotland and are well understood as valuable aspects of education provision. The arguments in favour are generally accepted by policy makers, who may well have positive memories themselves or have children or family members who have recently had such experiences.

A simple message Whist those of us who work in the field have no problem recognising the commonalities within the wide diversity of provision and activities, explaining this to a politician policymaker who needs a simple short explanation is not easy. The range of terms and the 'subject or approach' debate does not help. Until around 2005 in Scotland we used the term 'outdoor education' – and it carried a range of connotations, particularly relating to personal and social development primarily through residential provision. As curricular change took place through Curriculum for Excellence, the term 'outdoor learning' gained popularity, and conveyed its potential in formal education and in schools, using the grounds and local area as well as further afield, residentials and expeditions. This was conveyed using a simple "concentric circles" model (Higgins & Nicol, 2002). These tactics (the new term and image) significantly reduced the need to explain the concept in words. What also helped considerably was the growing research evidence base nationally and internationally which whilst at this stage was not substantial, was indicative of the potential benefits. Concurrently, the number and rigour of academic outdoor learning and outdoor studies conferences increased, facilitating the argument that the benefits of local outdoor learning experiences were well established elsewhere (e.g. in the Nordic countries and Germany).

A sympathetic politician – a champion Following a period of almost 300 years when Scotland was governed entirely from the UK Parliament in London, devolution led to Scottish Government elections and a parliament in Edinburgh 1999. Soon after, the new Minister for Education (Peter Peacock) met with me to seek informal advice on how to "give outdoor education a boost". This was followed by a range of modest commitments (with very little funding), but his advocacy and that of a range of subsequent education and environment Ministers has been consistent and significant. Of particular importance in the following ten years was independent research and that commissioned by the government's education and environment advisory bodies (summarised by Nicol, Higgins, Ross & Mannion, 2007), as it facilitated a debate on the role of outdoor learning in the revised national curriculum.

Seizing the moment Following a national debate on education, a new Scottish Curriculum for schools was published in 2004. Over the following five years the policy and practice architecture of Curriculum for Excellence (CfE) was developed (Education Scotland, 2017). This policy made explicit that whilst content mattered, the development of pupil "capacities" (helping children and young people to become: "successful learners, confident individuals, responsible citizens and effective contributors") was fundamental. This gave outdoor educators something to tag their approach to, and following considerable work between that community and the government's education advisory agency a specific policy was written and published. Curriculum for Excellence through Outdoor Learning (CfEtOL), asserts that:

> The core values of Curriculum for Excellence resonate with long-standing key concepts of outdoor learning. Challenge, enjoyment, relevance, depth, development of the whole person and an adventurous approach to learning are at the core of outdoor pedagogy. The outdoor environment encourages staff and students to see each other in a different light, building positive relationships and improving self-awareness and understanding of others.
>
> *Learning & Teaching Scotland, 2010, p.* 7

The mere existence of this document would have been unthinkable before 2000, and it represents both a rationale for outdoor learning and specific guidance to link such work to the curriculum. More particularly CfEtOL led to further policy development architecture, considerable expansion of resources and advice on the government's website, and an extensive programme of government-funded professional development for teachers.

Working with other educational priorities Whist much of the research and policy development outlined above applied specifically to outdoor learning, there have been and continue to be broader implications for related educational priorities such as 'learning for sustainability'. Due primarily to extensive lobbying, in 2011 the Scottish National Party committed to explore sustainability education if they won the election – which they did. A Ministerial Advisory Committee worked for a year and argued successfully that 'learning for sustainability' should be an entitlement of all pupils, and a responsibility of all teachers and all schools (Scottish Government, 2012). Remarkably, this report's unique set of 31 recommendations was accepted by Scottish Ministers the following March (Scottish Government, 2013). Early in this process the advisory group sought external funding for two commissioned reports on outdoor learning – one on the "impact of outdoor learning on attitudes to sustainability", and the other on the "impact of outdoor learning experiences on attainment and behaviour in schools" (Christie & Higgins, 2012a, 2012b). Both of these were fundamental to our decision to develop an integrated approach which includes "education for sustainable development", "global citizenship" and "outdoor learning" (Higgins & Christie, 2018). This makes learning for sustainability a globally unique approach, where outdoor learning is included as a sensitising element, to build awareness of the significance of the natural world.

One key facilitating factor was the establishment of a United Nations Regional Centre of Expertise in Education for Sustainable Development for Scotland (known locally as Learning for Sustainability Scotland,[3] located at Edinburgh University. This too was the result of collaboration by a range of stakeholders. It has played an important role in many of the subsequent developments outlined below.

One key recommendation of the 2012 report was that there should be a "strategic national approach" to learning for sustainability. This became the work of the Learning for Sustainability Ministerial Implementation Group for the following four years, and the Scottish Government commitment to the United Nations Sustainable Development Goals (SDGs) (United Nations, 2015), provided the ideal link. Our final report, "Vision 2030+" (Scottish Government, 2016), provided a route-map for the coming years, and this has resulted in further government-commissioned research, specifically on attainment.

Throughout these processes there has been a consistent commitment to engage with all relevant stakeholders. Consequently, the membership of the Ministerial committees included all key agencies, and was supported by an additional group of representatives of the third sector and other bodies involved in outdoor environmental and sustainability education with which joint meetings were held periodically. This collaborative approach led to learning for sustainability and its principles being embedded in the General Teaching Council for Scotland's Professional Standards,[4] making it the only teacher accreditation agency in the world to do so.

The Scottish Government has made a significant commitment to early childhood education provision for all children aged 3 to 5 years, at a time when the health and educational benefits of time in the outdoors have become evident and well-publicised (e.g. through published research from medical and social fields, and from the outdoor studies field a JAEOL (Journal of Adventure Education and Outdoor Learning) special issue on early childhood), and an effective campaign to expand nature kindergartens has been very successful. Here too research evidence has been crucial[5] and the result is a substantial policy and funding commitment (£860,000), supported by Ministers and the Chief Medical Officer[6]. This invites the prospect of a range of outdoor learning opportunities becoming available for Scottish children and young people throughout their school life.

The process of developing and implementing learning for sustainability described above has now spawned specific research on the impacts, a special issue of the Scottish Educational Review, a Master's degree and Doctoral studies. One specific study by Education Scotland, Learning for Sustainability Scotland and the University of Edinburgh established further evidence of the significance of the approach to pupil attainment, school culture, staff satisfaction, and reputation in the community (Education Scotland, 2015a). This in turn has become part of a major 18-nations study led by UNESCO (Laurie, Nonoyama-Tarumi, McKeown & Hopkins, 2016) which provides international support for the positive impacts of education for sustainable development. The significant implication of this 'virtuous circle' of research informing policy which then leads to more research examining impacts, is that it provides direction and reassurance for those throughout the policy-provision-research community, and the funders of professional development and further research.

Professional development Professional development opportunities provide both support for teachers and signal the significance of policy developments. Early in the roll-out of CfEtOL the Government funded some outdoor learning professional development opportunities for teachers, but since then there has been little further central funding. However, the British Council has provided substantial support for in-service training for Scottish teachers through awards for 2015–2018 and 2018–2021. This has led to over 500 teachers completing blended learning or online courses, many gaining GTCS Professional Recognition for their commitment.

The GTCS and Learning for Sustainability Scotland sought charitable funding to develop a 'self-evaluation tool' for Teacher Education Institutes, and though this has considerable value, re-orienting these is a slow process (see Nicol, Rae, Murray & Higgins, In press). Similarly, there is slow progress in gaining acceptance that changes to the Scottish Qualifications Authority school curricula and assessments are necessary.

Blowing the trumpet Ensuring that the research that led to, and emanated from, policy development is disseminated is crucially important in both demonstrating transparency and supporting wider communities involved in research and policy development in other policy or geographical areas. To support the policy developments above we (myself, colleagues in the University of Edinburgh and Learning for Sustainability Scotland) published a series of Research into Action Briefings[7] and

Election Briefings[8] to provide policy makers and professionals with easy access to recent research evidence. This has been valued by the Scottish Government which has commissioned updates and new briefings on other topics (e.g. relating to early years education etc.).

Credibility is of course further enhanced if peer-reviewed articles reflect positively on the policy developments, but only if these are available or summarised for policy makers. In the present case there are a number of published studies (and several ongoing) and some have been brought to the attention of policy makers through UNESCO briefings (e.g. Martin, Dillon, Higgins, Peters & Scott, 2013a), which have a certain cache, followed by peer-reviewed publications (e.g. Martin, Dillon, Higgins, Peters & Scott, 2013b).

Academics are often invited to present policy developments and related research findings at meetings and conferences. Whilst this is attractive to those involved, encouraging policy makers and politicians to do so (maybe also providing appropriate materials) at events, national and international conferences adds further to a sense of 'ownership' of any policy that they have committed to.

Maintaining momentum It would seem self-evident from the above that outdoor learning and learning for sustainability would be thriving in all Scottish schools, but this is far from the truth. Provision is very patchy, and whilst there are some examples of great commitment and excellent practice, the majority of teachers do not use the outdoors at all, and many pupils have at best only a single outdoor residential course in their school life. Even though policy may exist, teachers and other providers have many pressures on their time, and resources may only be prioritised if it is someone's job and there are checks through inspection processes. Embedding examination of practice in formal processes is a clear signal that it is important and 'required'. The recently published guidance on self-inspection in Scottish schools, How Good is Our School 4' (Education Scotland, 2015b), emphasises the importance of and expectations concerning learning for sustainability and outdoor learning. This is all helpful, but whilst much rests on the school culture, staff confidence etc., in the end it will be the decision of the staff as to what they prioritise. In that context, awareness of and easy access to summaries of research on the value of outdoor learning may be a key factor.

Delivering demonstrable benefits to wider communities of interest Beyond outdoor learning and learning for sustainability, there are a number of related areas in the broad field of outdoor studies where the benefits of published research are evident. These include outdoor recreation, health and wellbeing, environmental awareness, therapeutic use of the outdoors, nature and adventure tourism etc. Wherever possible we have tried to make evident these links, but without seeming to argue that the outdoors is a panacea. However, the argument that Scotland's landscape affords considerable national benefits, and that education and recreation contribute to this is attractive to policy makers as it has potential for integrated and effective policy development, and Figure 29.1 has been valuable in communicating this concept. One key factor has been that customary traditions of access to the countryside

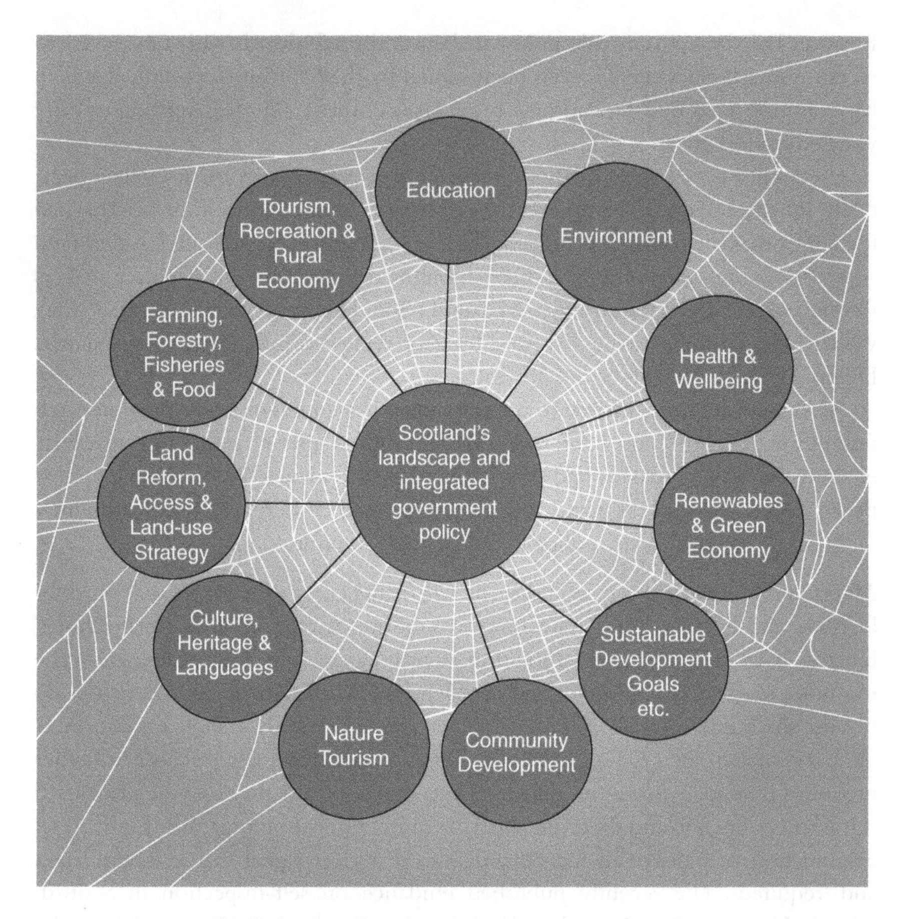

FIGURE 29.1 Scotland's landscape and integrated government policy

have been enshrined in law through the Land Reform (2003) Scotland Act and an associated Access Code.[9] This has been significant in developing outdoor recreation, education and tourism as well as the potential to (re)connect people with their history and culture (Higgins, 2017). Recently, the Scottish Government has consulted on and devised a new National Performance Framework,[10] with which all their policies intended to align. It is built around the UN SDGs (United Nations, 2015) framework and hence the outdoor and sustainability policies outlined above have considerable contemporary relevance.

Avoiding collateral damage One hazard of policy development and funding commitments in a given educational field is the potential loss from another. In the Scottish context there have been fears that local outdoor learning and learning for sustainability developments would have a negative impact on other forms of provision; however, bookings remain high in residential centres and for independent providers.

Concluding comments

As the Russian proverb goes – "The past is unpredictable". The reflections above should be set in this context, and so for an excellent historical overview from the early origins of education outdoors to the policy developments above, see Baker (2015).

So, what is the view from the far end (for now) of the 'policy corridor'? First, looking back (and forward) from this perspective I feel a certain empathy with policy makers. We ask a lot of them in terms of commitment, costs, risks, disruption, and without the benefit of decent and convincing research, all the more so. To paraphrase President John F. Kennedy, I have become increasingly sympathetic to the view that we should "Ask not what policy can do for us, but what we can do for policy". That is not to say that we should not attempt to bring about worthwhile change, but also that there are usually mainstream policies, systems and structures we might usefully work with rather than against.

However, I worry that in the process of 'policy chasing', outdoor learning may weaken what I have always valued as a major strength, that it can be a radical alternative to conventional education. Perhaps it is possible to be both, to work to change the system whilst working with it. If this can be done along with meeting the challenge 'Why outdoors?' with 'Why Indoors?' and stand a chance of bringing about appropriate coherent policy change in mainstream education, there needs to be more research – good-quality qualitative and quantitative studies across a wide range of outdoor studies. To develop such research further means working with academics and professionals in a range of disciplines across the social sciences. As with the policy makers we seek to influence, the problem is not a lack of enthusiasm, but time, funding and resources. Whilst academics may be short of all three, supporting the development of research-informed policy is imperative. Doing so may well help local head-teachers (and of course others) to respond positively to challenges from awkward parents like me; to support, facilitate, and celebrate learning and recreation outdoors which, as seems indicated by recent research, leads to a wide range of significant benefits for pupils, staff and the whole school community.

Notes

1 There are forms of guidance on a range of educational interventions and approaches such as the Education Endowment Foundation Toolkit (https://educationendowmentfoundation.org.uk/evidence-summaries/teaching-learning-toolkit) which considers metrics such as cost, effectiveness and duration. Outdoor Adventure Learning is considered to have "moderate impact, for moderate cost based on moderate evidence" and comparatively lasting effects. However, there are a limited range of interventions and approaches to compare this with.
2 Whilst Scotland is part of the United Kingdom, it has 'devolved powers' to make decisions regarding education and some other matters.
3 http://learningforsustainabilityscotland.org

4 www.gtcs.org.uk/professional-standards/learning-for-sustainability.aspx
5 See, for example, supporting evidence in the National Position Statement www.inspiringscotland.org.uk/wp-content/uploads/2018/10/Coalition-Position-Statement.pdf
6 www.gov.scot/news/learning-outdoors/
7 http://learningforsustainabilityscotland.org/?s=Research+into+action+briefings
8 https://blogs.glowscotland.org.uk/glowblogs/lfsblog/files/2018/01/Holyrood-Briefing-Learning-for-Sustainability-29-03-16.pdf
9 www.outdooraccess-scotland.scot/
10 https://nationalperformance.gov.scot/

References

Baker, M. (2015). *Policy development of outdoor learning in Scotland*. (Doctoral thesis). University of Edinburgh, UK.

Christie, E. & Higgins, P. (2012a). *The impact of outdoor learning on attitudes to sustainability*. Commissioned report for the Field Studies Council.

Christie, E. & Higgins, P. (2012b). *The impact of outdoor learning experiences on attainment and behaviour in schools*. Commissioned report for the Forestry Commission Scotland.

Education Scotland. (2015a) *Conversations about Learning for Sustainability*. Retrieved from: https://education.gov.scot/improvement/Pages/lfs3-conversations-about-learning-for-sustainability.aspx

Education Scotland. (2015b) *How Good is our School*? (4th edition). Retrieved from: https://education.gov.scot/improvement/documents/frameworks_selfevaluation/frwk2_nihedithgios/frwk2_hgios4.pdf

Education Scotland. (2017) Curriculum for Excellence. Retrieved from: https://education.gov.scot/scottish-education-system/policy-for-scottish-education/policy-drivers/cfe-(building-from-the-statement-appendix-incl-btc1-5)/What%20is%20Curriculum%20for%20Excellence?

Higgins, P. (2017, August). Outdoor education and the Scottish countryside. *The Geographer: Fifty years of conservation: the legacy of the Countryside (Scotland) Act 1967*. 21.

Higgins, P. & Christie, E. (2018). Learning for sustainability. In T. Bryce, W. Humes, D. Gillies & A. Kennedy (Eds.) *Scottish Education* (5th ed., pp. 554–564). Edinburgh: Edinburgh University Press.

Higgins, P. & Nicol, R. (2018). Outdoor learning. In T. Bryce, W. Humes, D. Gillies & A. Kennedy (Eds.) *Scottish Education* (5th ed., pp. 538–544). Edinburgh: Edinburgh University Press.

Laurie, R., Nonoyama-Tarumi, Y., McKeown, R. & Hopkins, C. (2016). Contributions of ESD to quality education: A synthesis of research. *Journal of Education for Sustainable Development, 10*(2), pp. 1–17.

Learning & Teaching Scotland. (2010). *Curriculum for excellence through outdoor learning*. Glasgow: Learning & Teaching Scotland.

McDonald, P. (1997). Climbing lessons: Inside outdoor education. New Zealand: Pete McDonald (private publication).

Martin. S., Dillon, J., Higgins, P., Peters C. & Scott, W. (2013a). Education for Sustainable Development Policy in the United Kingdom: Current Status, Best Practice, and Opportunities for the Future. UK National Commission for UNESCO. Policy Brief

No. 9. London: UNESCO. Retrieved from: (www.unesco.org.uk/publication/policy-brief-9-education-for-sustainable-development-esd-in-the-uk-current-status-best-practice-and-opportunities-for-the-future/)

Martin, S., Dillon, J., Higgins, P., Peters C. & Scott, W. (2013b). Divergent evolution in education for sustainable development policy in the United Kingdom: Current status, best practice, and opportunities for the future. Sustainability, *5*(4), pp. 1522–1544.

Nicol, R. & Higgins, P. (2002). A framework for the evaluation outdoor education programmes. In P. Higgins & R. Nicol (Eds.) *Outdoor education: Authentic learning in the context of landscapes, Vol. 2* (pp. 29–36). Sweden: Kunskapscentrum.

Nicol, R., Higgins, P., Ross, H. & Mannion, G. (2007). *Outdoor education in Scotland: A summary of recent research.* Perth: Scottish Natural Heritage.

Nicol, R., Rae, A., Murray, R. and Higgins, P. (In press). Moving beyond the printed word of policy to practice: an exploration into the conditions in which learning for sustainability might flourish in initial teacher education. *Scottish Educational Review.*

Scottish Government. (2012). Learning for Sustainability – Report of the One Planet Schools Ministerial Advisory Group. Retrieved from: https://education.gov.scot/improvement/Documents/One-planet-schools-report-learning-for-sustainability.pdf

Scottish Government. (2013). Ministerial response to the One Planet Schools Report (Learning for Sustainability). Retrieved from: https://education.gov.scot/improvement/learning-resources/A%20summary%20of%20learning%20for%20sustainability%20resources

Scottish Government. (2016). Vision 2030+: The concluding report of the Learning for Sustainability Implementation Group. Retrieved from: https://education.gov.scot/improvement/documents/res1-vision-2030.pdf

United Nations. (2015). Sustainable Development Goals (SDGs) (2015). www.un.org/sustainabledevelopment/sustainable-development-goals/

INDEX

Note: Page numbers in **bold** refer to tables and in *italics* to figures.

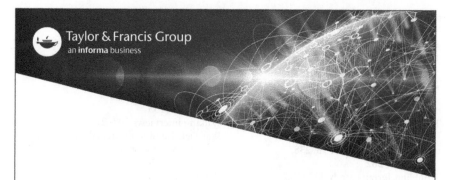